LONDON MATHEMATICAL SOCIETY LECTURE NOTE SERIES

Managing Editor: Professor J.W.S. Cassels, Department of Pure Mathematics and Mathematical Statistics, University of Cambridge, 16 Mill Lane, Cambridge CB2 1SB, England

The books in the series listed below are available from booksellers, or, in case of difficulty, from Cambridge University Press.

34 Representation theory of Lie groups, M.F. ATIYAH *et al*
36 Homological group theory, C.T.C. WALL (ed)
39 Affine sets and affine groups, D.G. NORTHCOTT
46 p-adic analysis: a short course on recent work, N. KOBLITZ
49 Finite geometries and designs, P. CAMERON, J.W.P. HIRSCHFELD & D.R. HUGHES (eds)
50 Commutator calculus and groups of homotopy classes, H.J. BAUES
57 Techniques of geometric topology, R.A. FENN
59 Applicable differential geometry, M. CRAMPIN & F.A.E. PIRANI
66 Several complex variables and complex manifolds II, M.J. FIELD
69 Representation theory, I.M. GELFAND *et al*
74 Symmetric designs: an algebraic approach, E.S. LANDER
76 Spectral theory of linear differential operators and comparison algebras, H.O. CORDES
77 Isolated singular points on complete intersections, E.J.N. LOOIJENGA
79 Probability, statistics and analysis, J.F.C. KINGMAN & G.E.H. REUTER (eds)
80 Introduction to the representation theory of compact and locally compact groups, A. ROBERT
81 Skew fields, P.K. DRAXL
82 Surveys in combinatorics, E.K. LLOYD (ed)
83 Homogeneous structures on Riemannian manifolds, F. TRICERRI & L. VANHECKE
86 Topological topics, I.M. JAMES (ed)
87 Surveys in set theory, A.R.D. MATHIAS (ed)
88 FPF ring theory, C. FAITH & S. PAGE
89 An F-space sampler, N.J. KALTON, N.T. PECK & J.W. ROBERTS
90 Polytopes and symmetry, S.A. ROBERTSON
91 Classgroups of group rings, M.J. TAYLOR
92 Representation of rings over skew fields, A.H. SCHOFIELD
93 Aspects of topology, I.M. JAMES & E.H. KRONHEIMER (eds)
94 Representations of general linear groups, G.D. JAMES
95 Low-dimensional topology 1982, R.A. FENN (ed)
96 Diophantine equations over function fields, R.C. MASON
97 Varieties of constructive mathematics, D.S. BRIDGES & F. RICHMAN
98 Localization in Noetherian rings, A.V. JATEGAONKAR
99 Methods of differential geometry in algebraic topology, M. KAROUBI & C. LERUSTE
100 Stopping time techniques for analysts and probabilists, L. EGGHE
101 Groups and geometry, ROGER C. LYNDÓN
103 Surveys in combinatorics 1985, I. ANDERSON (ed)
104 Elliptic structures on 3-manifolds, C.B. THOMAS
105 A local spectral theory for closed operators, I. ERDELYI & WANG SHENGWANG
106 Syzygies, E.G. EVANS & P. GRIFFITH
107 Compactification of Siegel moduli schemes, C-L. CHAI
108 Some topics in graph theory, H.P. YAP
109 Diophantine analysis, J. LOXTON & A. VAN DER POORTEN (eds)
110 An introduction to surreal numbers, H. GONSHOR
111 Analytical and geometric aspects of hyperbolic space, D.B.A. EPSTEIN (ed)
113 Lectures on the asymptotic theory of ideals, D. REES
114 Lectures on Bochner-Riesz means, K.M. DAVIS & Y-C. CHANG
115 An introduction to independence for analysts, H.G. DALES & W.H. WOODIN
116 Representations of algebras, P.J. WEBB (ed)
117 Homotopy theory, E. REES & J.D.S. JONES (eds)
118 Skew linear groups, M. SHIRVANI & B. WEHRFRITZ
119 Triangulated categories in the representation theory of finite-dimensional algebras, D. HAPPEL
121 Proceedings of *Groups - St Andrews 1985*, E. ROBERTSON & C. CAMPBELL (eds)
122 Non-classical continuum mechanics, R.J. KNOPS & A.A. LACEY (eds)
124 Lie groupoids and Lie algebroids in differential geometry, K. MACKENZIE
125 Commutator theory for congruence modular varieties, R. FREESE & R. MCKENZIE
126 Van der Corput's method of exponential sums, S.W. GRAHAM & G. KOLESNIK
127 New directions in dynamical systems, T.J. BEDFORD & J.W. SWIFT (eds)
128 Descriptive set theory and the structure of sets of uniqueness, A.S. KECHRIS & A. LOUVEAU
129 The subgroup structure of the finite classical groups, P.B. KLEIDMAN & M.W.LIEBECK
130 Model theory and modules, M. PREST

131 Algebraic, extremal & metric combinatorics, M-M. DEZA, P. FRANKL & I.G. ROSENBERG (eds)
132 Whitehead groups of finite groups, ROBERT OLIVER
133 Linear algebraic monoids, MOHAN S. PUTCHA
134 Number theory and dynamical systems, M. DODSON & J. VICKERS (eds)
135 Operator algebras and applications, 1, D. EVANS & M. TAKESAKI (eds)
136 Operator algebras and applications, 2, D. EVANS & M. TAKESAKI (eds)
137 Analysis at Urbana, I, E. BERKSON, T. PECK, & J. UHL (eds)
138 Analysis at Urbana, II, E. BERKSON, T. PECK, & J. UHL (eds)
139 Advances in homotopy theory, S. SALAMON, B. STEER & W. SUTHERLAND (eds)
140 Geometric aspects of Banach spaces, E.M. PEINADOR and A. RODES (eds)
141 Surveys in combinatorics 1989, J. SIEMONS (ed)
142 The geometry of jet bundles, D.J. SAUNDERS
143 The ergodic theory of discrete groups, PETER J. NICHOLLS
144 Introduction to uniform spaces, I.M. JAMES
145 Homological questions in local algebra, JAN R. STROOKER
146 Cohen-Macaulay modules over Cohen-Macaulay rings, Y. YOSHINO
147 Continuous and discrete modules, S.H. MOHAMED & B.J. MÜLLER
148 Helices and vector bundles, A.N. RUDAKOV *et al*
149 Solitons nonlinear evolution equations & inverse scattering, M. ABLOWITZ & P. CLARKSON
150 Geometry of low-dimensional manifolds 1, S. DONALDSON & C.B. THOMAS (eds)
151 Geometry of low-dimensional manifolds 2, S. DONALDSON & C.B. THOMAS (eds)
152 Oligomorphic permutation groups, P. CAMERON
153 L-functions and arithmetic, J. COATES & M.J. TAYLOR (eds)
154 Number theory and cryptography, J. LOXTON (ed)
155 Classification theories of polarized varieties, TAKAO FUJITA
156 Twistors in mathematics and physics, T.N. BAILEY & R.J. BASTON (eds)
157 Analytic pro-*p* groups, J.D. DIXON, M.P.F. DU SAUTOY, A. MANN & D. SEGAL
158 Geometry of Banach spaces, P.F.X. MÜLLER & W. SCHACHERMAYER (eds)
159 Groups St Andrews 1989 volume 1, C.M. CAMPBELL & E.F. ROBERTSON (eds)
160 Groups St Andrews 1989 volume 2, C.M. CAMPBELL & E.F. ROBERTSON (eds)
161 Lectures on block theory, BURKHARD KÜLSHAMMER
162 Harmonic analysis and representation theory for groups acting on homogeneous trees, A. FIGA-TALAMANCA & C. NEBBIA
163 Topics in varieties of group representations, S.M. VOVSI
164 Quasi-symmetric designs, M.S. SHRIKANDE & S.S. SANE
165 Groups, combinatorics & geometry, M.W. LIEBECK & J. SAXL (eds)
166 Surveys in combinatorics, 1991, A.D. KEEDWELL (ed)
167 Stochastic analysis, M.T. BARLOW & N.H. BINGHAM (eds)
168 Representations of algebras, H. TACHIKAWA & S. BRENNER (eds)
169 Boolean function complexity, M.S. PATERSON (ed)
170 Manifolds with singularities and the Adams-Novikov spectral sequence, B. BOTVINNIK
171 Squares, A.R. RAJWADE
172 Algebraic varieties, GEORGE R. KEMPF
173 Discrete groups and geometry, W.J. HARVEY & C. MACLACHLAN (eds)
174 Lectures on mechanics, J.E. MARSDEN
175 Adams memorial symposium on algebraic topology 1, N. RAY & G. WALKER (eds)
176 Adams memorial symposium on algebraic topology 2, N. RAY & G. WALKER (eds)
177 Applications of categories in computer science, M.P. FOURMAN, P.T. JOHNSTONE, & A.M. PITTS (eds)
178 Lower K- and L-theory, A. RANICKI
179 Complex projective geometry, G. ELLINGSRUD, C. PESKINE, G. SACCHIERO & S.A. STRØMME (eds)
180 Lectures on ergodic theory and Pesin theory on compact manifolds, M. POLLICOTT
181 Geometric group theory I, G.A. NIBLO & M.A. ROLLER (eds)
182 Geometric group theory II, G.A. NIBLO & M.A. ROLLER (eds)
183 Shintani zeta functions, A. YUKIE
184 Arithmetical functions, W. SCHWARZ & J. SPILKER
185 Representations of solvable groups, O. MANZ & T.R. WOLF
186 Complexity: knots, colourings and counting, D.J.A. WELSH
187 Surveys in combinatorics, 1993, K. WALKER (ed)
189 Locally presentable and accessible categories, J. ADAMEK & J. ROSICKY
190 Polynomial invariants of finite groups, D.J. BENSON
191 Finite geometry and combinatorics, F. DE CLERCK *et al*
192 Symplectic geometry, D. SALAMON (ed)
197 Two-dimensional homotopy and combinatorial group theory, C. HOG-ANGELONI W. METZLER & A.J. SIERADSKI (eds)
198 The algebraic characterization of geometric 4-manifolds, J.A. HILLMAN

London Mathematical Society Lecture Note Series. 189

Locally Presentable and Accessible Categories

Jiří Adámek
Czech Technical University, Prague

Jiří Rosicky
Masaryk University, Brno

Published by the Press Syndicate of the University of Cambridge
The Pitt Building, Trumpington Street, Cambridge CB2 1RP
40 West 20th Street, New York, NY 10011-4211, USA
10 Stamford Road, Oakleigh, Melbourne 3166, Australia

First published 1994

Library of Congress cataloguing in publication data available

British Library cataloguing in publication data available

ISBN 0 521 42261 2 paperback

Transferred to digital printing 2004

We dedicate this book to the memory of our excellent colleague and very dear friend Jan Reiterman

Contents

Preface

The basic theme of our monograph is the syntax and semantics of a categorical theory of mathematical structures (as used in algebra, model theory, computer science, etc.). The semantics part is the study of properties of the categories of structures. We concentrate on two kinds of categories: locally presentable categories, which are rather close to quasivarieties of algebras, and the broader (and less "pleasant") accessible categories, which are close to classes of structures axiomatizable in first-order logic. The syntax part describes categories of structures by means of sketches: a sketch is a small category with specified limits and colimits, and a model of the sketch is a set-valued functor preserving the specified limits and colimits. Locally presentable categories are precisely the categories of models of limit-sketches (i.e., no colimit is specified), and accessible categories are precisely the categories of models of sketches. A different approach to syntax is by means of first-order logic: we characterize theories needed to axiomatize locally presentable and accessible categories.

The fundamentals of the theory of locally presentable categories are exhibited in Chapter 1, and those of accessible categories in Chapter 2. The rest of our monograph is devoted to some related topics: algebraic categories, injectivity, categories of models, and Vopěnka's large-cardinal principle. The book is completely self-contained: we expect the reader to be familiar with the basic category-theoretical concepts (such as adjoint, limit, etc.) which we mention briefly in the Preliminaries, but all the more advanced concepts are carefully explained in the text. Facts about large cardinal numbers, used in the last chapter, are presented in the Appendix.

Organization

Every chapter is concluded by a set of easy exercises which illustrate some of the features of the text. References appear in the historical remarks. Open problems are listed at the end of the book.

Acknowledgements

We are grateful to Hans-Eberhard Porst, Walter Tholen and Jiří Velebil for their suggestions on improvements of the text. Miroslav Dont undertook the extremely difficult task of turning our almost illegible manuscript into a perfect AMS-LaTeX file, which we greatly appreciate. For the diagrams he used the macro package LAMS-TeX of M. D. Spivak.

April 1993 J. A. & J. R.

Introduction

Locally presentable categories

The concept of a locally presentable category is one of the most fruitful concepts of category theory. The definition, generalizing the concept of an algebraic lattice, is natural and simple. The scope is very broad: varieties and quasivarieties of (many-sorted) algebras, Horn classes of relational structures, and functor-categories are all locally presentable. Furthermore, locally presentable categories enjoy a number of important properties: they are complete and cocomplete, wellpowered and co-wellpowered, and they have a strong generator.

The definition of a locally presentable category is due to P. Gabriel and F. Ulmer. Their lecture notes [Gabriel, Ulmer 1971], by now classical, are a profound, but by no means easily readable, treatise on the topic. One of the aims of our monograph is to make the fundamentals of the theory of locally presentable categories more accessible to readers who work in category theory, computer science, and related areas. We have collected these fundamentals in Chapter 1, where the basic properties of locally presentable categories are proved and several equivalent ways of introducing these categories are exhibited. For example, we show that locally presentable categories are precisely the categories sketchable by a limit sketch (i.e., the categories of all set-valued functors preserving specified limits).

Accessible categories

These generalize locally presentable categories by weakening cocompleteness to the existence of some directed colimits. The collection of all categories obtained by this generalization is much broader than that of all locally presentable categories, and it includes categories such as

fields and homomorphisms,

Hilbert spaces and linear contractions,

linearly ordered sets and order-preserving functions,

sets and one-to-one functions.

An important special case: for each sketch in the sense of C. Ehresmann the category of set-valued models of the sketch (i.e., set-valued functors preserving specified limits and specified colimits) is accessible. It turns out that this is actually no special case: each accessible category is sketchable, i.e., equivalent to the category of all models of some sketch. This fundamental relationship between accessible and sketchable categories was discovered by [Lair 1981], who called accessible categories "catégories modelables". The name "accessible" is due to [Makkai, Paré 1989], whose book is a comprehensive treatise devoted to accessible categories. Our prime aim in presenting (in Chapter 2) the fundamentals of the theory of accessible categories is to make this theory easy to grasp. Thus, for example, our proof of the equivalence of accessible and sketchable categories is conceptually simpler than any previously published proof, being based on the concept of a pure morphism (a concept "borrowed" from model theory). Unlike M. Makkai and R. Paré, we do not stress the 2-categorical aspects of the theory, although some of the basic results (e.g., on limits of accessible categories) are addressed.

Algebraic categories

Locally presentable categories are closely related to varieties and quasivarieties of many-sorted algebras. We devote Chapter 3 to this interrelationship. We present J. R. Isbell's characterization of quasivarieties (i.e., implicational classes) of algebras as precisely the locally presentable categories with a regularly projective regular generator. Varieties are precisely the quasivarieties with effective equivalence relations. We also introduce the Lawvere–Linton concept of algebraic theory, and prove that varieties are precisely the categories of models of algebraic theories (= product sketches). We finally study the concept of essentially algebraic theory due to P. Freyd, which is an equational theory of partial operations in which the domain of definition of each operation is determined by equations in the "preceding" operations. We prove that locally presentable categories are precisely the categories of models of essentially algebraic theories—this is a folklore result whose proof has not been published before.

Injectivity and generalizations of locally presentable categories

Some natural generalizations of locally presentable categories are studied in Chapter 4: weakly locally presentable categories, i.e., accessible categories with weak colimits (equivalently: with products) and locally multipresentable categories, i.e., accessible categories with multicolimits (equivalently: with connected limits). These concepts are closely related to those of orthogonality or injectivity w.r.t. a morphism, or w.r.t. a cone. Weakly locally presentable categories are precisely the full subcategories of locally presentable categories which are specified by injectivity w.r.t. a set of morphisms. And locally multipresentable categories are precisely the full subcategories of locally presentable categories specified by orthogonality w.r.t. a set of cones.

Categories of models

Chapter 5 deals with first-order logic. We call categories of models of first-order theories axiomatizable. In the finitary, many-sorted first-order logic L_ω we prove that categories axiomatizable by so-called limit theories are precisely the locally finitely presentable categories. This is a result in [Coste 1979]. More generally, in the λ-ary logic L_λ categories axiomatizable by limit theories are precisely the locally λ-presentable categories.

We also exhibit a full characterization of accessible categories by means of (more general) basic theories: a category is accessible iff it is axiomatizable by a basic theory in some of the logics L_λ. A crucial difference between the above two characterization results is that in the case of accessible categories we cannot restrict to a given λ: we show an example (1) of a finitely accessible category which cannot be axiomatized in L_ω and (2) of a basic theory in L_ω whose category is not finitely accessible.

Vopěnka's principle

Some results on locally presentable and accessible categories depend on the existence of certain large cardinal numbers. Notably, the large-cardinal Vopěnka's principle turns out to be equivalent to important properties of locally presentable categories. Vopěnka's principle implies e.g. that

(∗) a category is locally presentable iff it is cocomplete and has a colimit-dense set of objects.

This is quite surprising because it is thus possible to define "locally presentable" without a reference to the concept of presentable object. (The explanation is that, under Vopěnka's principle, in each category with a colimit-dense set all objects are presentable.) Conversely, the statement (∗) implies Vopěnka's principle. Thus, large cardinal numbers turn out to have a close link to locally presentable categories.

We devote Chapter 6 to the role of Vopěnka's principle in the theory of locally presentable and accessible categories. All concepts concerning large cardinal numbers which are needed in that chapter are explained in the Appendix.

0. Preliminaries

This monograph is devoted to a theory of locally presentable categories and accessible categories. We assume that the reader has basic knowledge of categories, functors, and adjoints, but we are careful to explain all the necessary concepts of model theory, logic, and set theory, as well as all the more advanced categorical notions in the text. We have concentrated all the required facts concerning cardinal numbers in the Appendix. We now recall some conventions and facts of category theory necessary for avoiding later misunderstandings. The proofs of the (standard) statements presented here can be found e.g. in [Adámek, Herrlich, Strecker 1990]*.

0.1 Set Theory. We distinguish, as in the Bernays–Gödel set theory, between *sets* and *classes*. Until Chapter 6 this is all that need be said—in other words, we just use naive set theory with a distinction between "small" and "large". But we use transfinite induction frequently; thus, the axiom of choice (for classes) is assumed without mention.

The first infinite cardinal is denoted by ω or $\aleph_0$, the next one by ω_1 or $\aleph_1$. Categories $\mathcal{K}$ are understood to be locally small, i.e., objects and morphisms form classes $\mathcal{K}^{\mathrm{obj}}$ and $\mathcal{K}^{\mathrm{mor}}$, respectively, whereas $\hom(A, B)$ is a set (for any pair A, B of objects). A class of objects of a category is called *essentially small* if it has a set of representatives w.r.t. isomorphism.

0.2 Composition is written from right to left, that is, if $f\colon A \to B$ and $g\colon B \to C$ are morphisms, then $g \cdot f$ [or gf] is their composite.

0.3 Comma-categories. For each object K of a category $\mathcal{K}$ we form the comma-category $K \downarrow \mathcal{K}$ of all arrows with the domain K, whose morphisms from $K \xrightarrow{a} A$ to $K \xrightarrow{b} B$ are the $\mathcal{K}$-morphisms $f\colon A \to B$ with $b = fa$. Dually, $\mathcal{K} \downarrow K$ denotes the comma-category of all arrows with the codomain K.

*References to the literature listed at the end of our book are denoted by square brackets.

0.4. By a **diagram** in a category $\mathcal{K}$ is meant a functor $D: \mathcal{D} \to \mathcal{K}$ from a small category $\mathcal{D}$ (called the scheme of the diagram D). The diagram D is said to be *finite* if $\mathcal{D}$ has finitely many morphisms. A category is called *(co)complete* provided that every diagram in it has a (co)limit.

Definition. Let $\mathcal{A}$ be a small, full subcategory of a category $\mathcal{K}$. For each object K in $\mathcal{K}$ the *canonical diagram* of K (w.r.t. $\mathcal{A}$) is the diagram of all arrows $A \to K$ where A lies in $\mathcal{A}$; more precisely, the canonical diagram is the natural forgetful functor $D: \mathcal{A} \downarrow K \to \mathcal{K}$. We say that K is a *canonical colimit of $\mathcal{A}$-objects* provided that the canonical diagram has a colimit with the colimit-object K and the colimit maps $D(A \xrightarrow{a} K) \xrightarrow{a} K$.

0.5 Hierarchy of Monomorphisms. A monomorphism $m: A \to B$ is called

(1) *regular* if m is an equalizer of some pair $f_1, f_2: B \to C$;

(2) *strong* if each commuting square

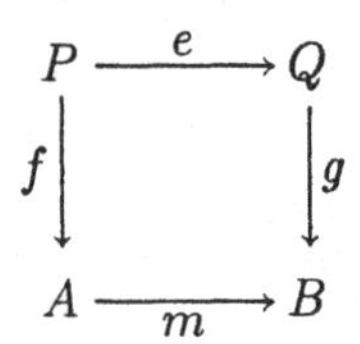

such that e is an epimorphism has a diagonal fill-in (i.e., a morphism $d: Q \to A$ with $f = d \cdot e$ and $g = m \cdot d$);

(3) *extremal* if every epimorphism $e: A \to A'$ through which m factorizes is an isomorphism.

Every regular monomorphism is strong, and the converse is true in each category with (epi, regular mono)-factorizations of morphisms. Every strong monomorphism is extremal, and the converse is true in each category with pushouts.

Every complete, wellpowered category has (epi, extremal mono)-factorizations as well as (extremal epi, mono)-factorizations of morphisms.

0.6 Generators. A set $\mathcal{G}$ of objects of a category is called a *generator* provided that for each pair $f_1, f_2: K \to K'$ of distinct morphisms there exists an object $G \in \mathcal{G}$ and a morphism $g: G \to K$ with $f_1 \cdot g \neq f_2 \cdot g$. The dual concept is cogenerator.

A generator $\mathcal{G}$ is called *strong* provided that for each object K and each proper subobject of K there exists a morphism $G \to K$ with $G \in \mathcal{G}$ which does not factorize through that subobject. A shorter definition is possible in a cocomplete category: $\mathcal{G}$ is a strong generator if every object is an extremal quotient of a coproduct of $\mathcal{G}$-objects. (It would be more reasonable, but unfortunately less standard, to call $\mathcal{G}$ an extremal generator.) Every category $\mathcal{K}$ with a strong generator $\mathcal{G}$ is *wellpowered*, i.e., each object has only a set of subobjects.

0.7 Adjoint Functors. A functor $F: \mathcal{K} \to \mathcal{L}$ is *right adjoint* to a functor $G: \mathcal{L} \to \mathcal{K}$ provided that there exists a natural isomorphism

$$\hom(G-, -) \cong \hom(-, F-).$$

Notation: $G \dashv F$. We often use *Freyd's adjoint functor theorem*: if $\mathcal{K}$ is a complete category, then a functor $F: \mathcal{K} \to \mathcal{L}$ is a right adjoint iff F preserves limits and satisfies the *solution-set condition* (which says that for each object L in $\mathcal{L}$ there exists a set of arrows $L \xrightarrow{f_i} FK_i$ in $\mathcal{L}$ such that every arrow $L \xrightarrow{f} FK$ factorizes as $f = Fk \cdot f_i$ for some i). We also have *Freyd's special adjoint functor theorem*: if $\mathcal{K}$ is a complete, wellpowered category with a cogenerator, then a functor $F: \mathcal{K} \to \mathcal{L}$ is a right adjoint iff F preserves limits.

0.8 Reflective Subcategories. A subcategory $\mathcal{A}$ of a category $\mathcal{K}$ is called

(1) *isomorphism-closed* provided that for each isomorphism $i: A \to A'$ in $\mathcal{K}$ with A in $\mathcal{A}$ we have A' and i in $\mathcal{A}$ too,

(2) *closed under limits* if every limit cone in $\mathcal{K}$ of a diagram in $\mathcal{A}$ lies in $\mathcal{A}$,

(3) *reflective* provided that the inclusion functor $\mathcal{A} \hookrightarrow \mathcal{K}$ is right adjoint.

The latter means that each object K of $\mathcal{K}$ has a *reflection map* $r_K: K \to A$, $A \in \mathcal{A}$, with the universal property that each morphism from K into an $\mathcal{A}$-object uniquely factorizes through r_K by an $\mathcal{A}$-morphism. If each r_K is an epimorphism, $\mathcal{A}$ is said to be *epireflective* in $\mathcal{K}$.

For complete, wellpowered, and co-wellpowered categories $\mathcal{K}$ the following holds: a full, isomorphism-closed subcategory of $\mathcal{K}$ is epireflective iff it is closed in $\mathcal{K}$ under products and extremal subobjects.

0.9 The Yoneda Lemma. For each small category $\mathcal{K}$ the *Yoneda embedding* is the functor $Y: \mathcal{K} \to \mathbf{Set}^{\mathcal{K}^{\mathrm{op}}}$ assigning to each object K of $\mathcal{K}$ the contravariant hom-functor $\hom(-, K): \mathcal{K}^{\mathrm{op}} \to \mathbf{Set}$, and to each morphism $f: K \to K'$ of $\mathcal{K}$ the natural transformation $\hom(-, f): \hom(-, K) \to$

$\hom(-,K')$ defined via composites with f. The fact that Y is a full embedding follows from the *Yoneda lemma*: for each functor F in $\mathbf{Set}^{\mathcal{K}^{op}}$ and each natural transformation $\tau\colon \hom(-,K) \to F$ there exists a unique element $x \in FK$ with $\tau_A(h) = Fh(x)$ for all $h \in \hom(A,K)$.

0.10 Cones. A set of morphisms with a common domain A is called a *cone* with domain A. Special cases: every morphism $f\colon A \to B$ is considered to be a cone with domain A, and every object A is considered to be the empty cone with domain A.

0.11 Cofinal Subdiagrams. A functor $H\colon \mathcal{D}_0 \to \mathcal{D}$ is said to be *cofinal* provided that for each object d in $\mathcal{D}$

(a) there exists a morphism $f\colon d \to Hd_0$ for some object d_0 in $\mathcal{D}_0$,

and

(b) given two such morphisms $f\colon d \to Hd_0$ and $f'\colon d \to Hd_0'$, there exist morphisms $g'\colon d_0 \to \bar{d}_0$ and $g\colon d_0' \to \bar{d}_0$ in $\mathcal{D}_0$ such that the square

$$\begin{array}{ccc} d & \xrightarrow{f} & Hd_0 \\ {\scriptstyle f'}\downarrow & & \downarrow{\scriptstyle Hg'} \\ Hd_0' & \xrightarrow[Hg]{} & H\bar{d}_0 \end{array}$$

commutes.

Observation. For each cofinal functor $H\colon \mathcal{D}_0 \to \mathcal{D}$ the categories $\mathcal{D}_0$ and $\mathcal{D}$ are "equivalent as diagram schemes" w.r.t. colimits: (1) a category has colimits over $\mathcal{D}$ iff it has colimits over $\mathcal{D}_0$, and (2) a functor preserves colimits over $\mathcal{D}$ iff it preserves colimits over $\mathcal{D}_0$.

In more detail: let $D\colon \mathcal{D} \to \mathcal{K}$ be a diagram. There is a bijective correspondence between compatible cocones of D and those of $D\cdot H$ (hence, a bijective correspondence between colimits of D and $D \cdot H$). In fact:

(1) Given a compatible cocone $(Dd \xrightarrow{c_d} C)_{d\in\mathcal{D}^{\mathrm{obj}}}$ for D, then the cocone

$$\left(DHd_0 \xrightarrow{c_{Hd_0}} C\right)_{d_0\in\mathcal{D}_0^{\mathrm{obj}}}$$

is compatible for $D \cdot H$.

(2) Given a compatible cocone $(DHd_0 \xrightarrow{c_{d_0}} C)_{d_0\in\mathcal{D}_0^{\mathrm{obj}}}$ for $D \cdot H$, then choose a morphism $f\colon d \to Hd_0$ for each d in $\mathcal{D}$. It is clear that $\hat{c}_d = c_{d_0} \cdot Df\colon Dd \to C$ is independent of the choice of f and d_0, and that the cocone $(Dd \xrightarrow{\hat{c}_d} C)_{d\in\mathcal{D}^{\mathrm{obj}}}$ is compatible for D.

Remark. In particular, a *cofinal subdiagram* of a diagram $D\colon \mathcal{D} \to \mathcal{K}$ is a subdiagram $D_0\colon \mathcal{D}_0 \to \mathcal{K}$ (i.e., $\mathcal{D}_0$ is a subcategory of $\mathcal{D}$ and $D_0 = D/_{\mathcal{D}_0}$) such that the inclusion functor $\mathcal{D}_0 \hookrightarrow \mathcal{D}$ is cofinal.

0.12 Equivalence of Categories is a full and faithful functor $E\colon \mathcal{K} \to \mathcal{L}$ which is isomorphism-dense, i.e., each object of $\mathcal{L}$ is isomorphic to an object of $E(\mathcal{K})$. We call $\mathcal{K}$ and $\mathcal{L}$ equivalent; the notation for this is $\mathcal{K} \approx \mathcal{L}$. The notation for isomorphic categories is $\mathcal{K} \cong \mathcal{L}$.

0.13 The Quasicategory of all Categories. We denote by **Cat** the category of all small categories and all functors. On two occasions we also refer to the quasicategory **CAT** of all categories and all functors. A quasicategory is defined as a category except that it lives beyond the universe of our set theory—thus, all objects may form a collection which fails to be a class, and the same is true about any $\hom(A, B)$. (We are vague here because the use is so infrequent and unimportant that an effort for axiomatization would be wasted. The reader may consult the monograph [Adámek, Herrlich, Strecker 1990].)

Chapter 1

Locally Presentable Categories

The first chapter is devoted to an important class of categories, the locally presentable categories, which is broad enough to encompass a great deal of mathematical life: varieties of algebras, implicational classes of relational structures, interesting cases of posets (domains, lattices), etc., and yet restricted enough to guarantee a number of completeness and smallness properties. Besides, locally presentable categories are closed under a number of categorical constructions (limits, comma-categories), see also Chapter 2. The basic concept, a finitely presentable object, can be regarded as a generalization of the concept of a finite (or compact) element in a Scott domain, i.e., an element a such that for each directed set $\{d_i \mid i \in I\}$ with $a \leq \bigvee_{i \in I} d_i$ it follows that $a \leq d_i$ for some $i \in I$. Now, an object A is finitely presentable if for each directed diagram $\{D_i \mid i \in I\}$ every morphism $A \to \operatorname{colim}_{i \in I} D_i$ factorizes (essentially uniquely) through D_i for some $i \in I$.

More generally, an object A is λ-presentable (for a cardinal λ) if every morphism from A to a λ-directed colimit $\operatorname{colim}_{i \in I} D_i$ factorizes (essentially uniquely) through some D_i. A category is locally λ-presentable iff it has colimits and is generated (in some strong sense) by a set of λ-presentable objects. We will see that there are many equivalent ways in which locally λ-presentable categories can be introduced: they are precisely

(1) the cocomplete categories in which every object is a λ-directed colimit of λ-presentable objects of a certain set (Definition 1.17);

(2) the cocomplete categories with a strongly generating set of λ-presentable objects (Theorem 1.20);

(3) the full, reflective subcategories of categories of relational structures closed under λ-directed colimits (Corollary 1.47);

(4) the categories of λ-continuous set-valued functors (Theorem 1.46);

(5) the λ-free cocompletions of small categories (Theorem 1.46);

(6) the λ-ary essentially algebraic categories (Theorem 3.36);

(7) the categories of models of λ-small limit-sketches (Corollary 1.52);

(8) the categories of models of λ-ary limit-theories (Theorem 5.30).

The role that the cardinal λ plays here is analogous to, say, an upper bound on the arity of operations in universal algebra. We begin with the important case of finitely presentable categories, i.e., with $\lambda = \aleph_0$.

The concept of a locally presentable category is closely related to that of a small-orthogonality class, i.e., the class $\mathcal{M}^\perp$ of all objects orthogonal to morphisms of a given set $\mathcal{M}$. We prove that (a) each locally presentable category is equivalent to a small-orthogonality class of some functor-category $\mathbf{Set}^{\mathcal{A}}$, and (b) every small-orthogonality class in a locally presentable category is locally presentable [Theorems 1.46 and 1.39].

A concept often easier to work with than λ-directed colimits is that of λ-directed unions. We show that locally presentable categories can equivalently be introduced by means of λ-directed unions—this is the local generation theorem 1.70 (which is, somewhat surprisingly, technically rather difficult).

1.A Locally Finitely Presentable Categories

The concept of a locally finitely presentable category can be viewed as a direct generalization of the concept of an algebraic lattice. Recall that a non-empty partially ordered set is called *directed* provided that each pair of elements has an upper bound. An element a of a partially ordered set $(K, \leq)$ is called *finite* (or compact) provided that for each directed set $D \subseteq K$ with $a \leq \bigvee D$ there exists $d \in D$ such that $a \leq d$. Now an *algebraic lattice* is a partially ordered set $(K, \leq)$ which is

(1) cocomplete, i.e., has all joins (and thus all meets)

and

(2) algebraic, i.e., every element is a directed join of finite elements.

When working with a category $\mathcal{K}$ rather than just a poset, directed joins have to be generalized to *directed colimits*, i.e., colimits of diagrams $D\colon (I, \leq) \to \mathcal{K}$, where $(I, \leq)$ is a directed poset (considered as a category).

Finitely Presentable Objects

1.1 Definition. An object K of a category $\mathcal{K}$ is called *finitely presentable* provided that its hom-functor

$$\hom(K, -): \mathcal{K} \to \mathbf{Set}$$

preserves directed colimits.

Explicitly: K is finitely presentable iff for each directed diagram

$$D: (I, \leq) \to \mathcal{K},$$

each colimit cocone $D_i \xrightarrow{c_i} C$ ($= \operatorname{colim} D$), $i \in I$, and each morphism $f: K \to C$ there exists i such that

(1) f factorizes through c_i, i.e., $f = c_i \cdot g \quad (i \in I)$ for some $g: K \to D_i$,

and

(2) the factorization is essentially unique in the sense that if $f = c_i \cdot g = c_i \cdot g'$, then $D(i \to j) \cdot g = D(i \to j) \cdot g'$ for some $j \geq i$.

1.2 Examples

(1) A set K is finitely presentable in the category **Set** of sets and functions iff K is finite. In fact, every set K is a colimit of the directed diagram of all finite subsets of K; if K is finitely presentable, then id_K must factorize through the inclusion of one of the finite subsets—thus, K is finite. Conversely, let K be a finite set, and let $\left(D_i \xrightarrow{c_i} C\right)_{i \in I}$ be a directed colimit in **Set**. For each function $f: K \to C$ and each element $x \in K$ there exists $i_x \in I$ such that $f(x)$ lies in the image of c_{i_x}. Since K is finite, and I is directed, there exists an upper bound $i \in I$ of all i_x ($x \in K$); thus, $f(K) \subseteq c_i(D_i)$. This implies that f factorizes through c_i. To show that the factorization is essentially unique, use the following property of directed colimits in **Set** (see Exercise 1.a): whenever elements $y, y' \in D_i$ fulfil $c_i(y) = c_i(y')$, then there exists $j \in I$ with $i \leq j$ such that $D(i \to j)(y) = D(i \to j)(y')$.

(2) For each set S (of *sorts*) let $\mathbf{Set}^S$ denote the category of *S-sorted sets*, i.e., collections $X = (X_s)_{s \in S}$ of sets X_s indexed by S, and *S-sorted functions* $f: X \to Y$, i.e., collections $f = (f_s)_{s \in S}$ of functions $f_s: X_s \to Y_s$ indexed by S. For each S-sorted set X we call the cardinal

$$\#X = \sum_{s \in S} \operatorname{card} X_s$$

the *power* of X. An S-sorted set is finitely presentable in $\mathbf{Set}^S$ iff it has finite power. The proof is analogous to (1) since (directed) colimits in $\mathbf{Set}^S$ are computed coordinate-wise.

(3) In the category **Pos** of posets (partially ordered sets) and order-preserving functions the finitely presentable objects are precisely the finite ones. The proof is analogous to (1) since directed colimits are computed on the level of **Set**.

Analogously in the category **Gra** of graphs (i.e., sets endowed with a binary relation) and homomorphisms (i.e., functions preserving the binary relation), the finitely presentable objects are precisely the finite graphs.

(4) Let S be a set of sorts, and let Σ be a (finitary, S-sorted, relational) *signature.* That is, with each symbol $\sigma \in \Sigma$ we are given an arity $\operatorname{ar}\sigma = (s_1, \ldots, s_n) \in S^n$. (If $n = 1$, then $\operatorname{ar}\sigma = s$ means that σ is a unary symbol of sort s, if $n = 2$, then σ is a binary symbol of sorts s_1, s_2; etc. The case $n = 0$ is denoted by $\operatorname{ar}\sigma = \emptyset$; this is a nullary symbol σ.)

A *relational structure* A of type Σ consists of an (*underlying*) S-sorted set $|A| = (A_s)_{s\in S}$ and, for each $\sigma \in \Sigma$, of a relation $\sigma_A \subseteq A_{s_1} \times A_{s_2} \times \cdots \times A_{s_n}$, where $\operatorname{ar}\sigma = (s_1, \ldots, s_n)$. (If $n = 0$, then $\sigma_A \subseteq A^{\emptyset}$, where $A^{\emptyset}$ is a terminal object. Thus, we just distinguish between two cases: $\sigma_A = \emptyset$ or $\sigma_A \neq \emptyset$.) Let $\mathbf{Rel}\,\Sigma$ denote the category of relational structures of type Σ, where morphisms $f\colon A \to B$ are the *homomorphisms,* i.e., S-sorted functions $f\colon |A| \to |B|$ such that for each $\sigma \in \Sigma$ of arity $(s_1, \ldots, s_n)$ with $n > 0$ we have that

$$(x_1, \ldots, x_n) \in \sigma_A \quad \text{implies} \quad (f_{s_1}(x_1), \ldots, f_{s_n}(x_n)) \in \sigma_B$$

and, for each σ of arity $\emptyset$, if $\sigma_A \neq \emptyset$ then $\sigma_B \neq \emptyset$. A relational structure A is finitely presentable in $\mathbf{Rel}\,\Sigma$ iff it has finitely many vertices (i.e., $|A|$ has finite power) and finitely many edges (i.e., $\sum_{\sigma\in\Sigma} \operatorname{card}\sigma_A$ is finite). The proof is analogous to (1) above.

(5) A group A is finitely presentable in **Grp**, the category of groups and homomorphisms, iff it can be presented by finitely many generators and finitely many equations in the usual algebraic sense. (That is, iff A is isomorphic to the quotient group of the free group $F\{x_i\}_{i=1}^n$ generated by $\{x_1, \ldots, x_n\}$ modulo a congruence generated by finitely many equations on $F\{x_i\}_{i=1}^n$.) For example, $(\mathbb{Z}, +)$ is finitely presentable and $(\mathbb{R}, +)$ is not.

In general, in each variety of finitary algebras, an algebra is finitely presentable iff it can be presented by finitely many generators and finitely many equations in the usual algebraic sense. A full proof will be presented later (see Theorem 3.12).

(6) Let **Aut** be the category of (deterministic, sequential) automata: objects are sixtuples $A = (Q, I, O, q_0, \delta, \beta)$ where

Q is a set of states,

I is a set of input symbols,

O is a set of output symbols,

$q_0 \in Q$ is the initial state,

$\delta\colon I \times Q \to Q$ is the next-state map,

and

$\beta\colon Q \to O$ is the output map.

Morphisms from A to $A' = (Q', I', O', q_0', \delta', \beta')$ are triples (f, i, o) of functions $f\colon Q \to Q'$, $i\colon I \to I'$, and $o\colon O \to O'$ satisfying

(i) $f(q_0) = q_0'$,

(ii) $f(\delta(q, x)) = \delta'(f(q), i(x))$,

(iii) $\beta'(f(q)) = \beta(o(q))$,

for all states $q \in Q$ and all inputs $x \in I$. Composition is defined coordinate-wise, and the identity morphisms are (id_Q, id_I, id_O).

An automaton is finitely presentable iff each of the sets Q, I, and O is finite. The proof is analogous to that in (1) since directed colimits in **Aut** are computed coordinate-wise.

(7) Let $\mathcal{A}$ be a small category. By the Yoneda lemma (see 0.9), every hom-functor is a finitely presentable object of $\mathbf{Set}^{\mathcal{A}}$.

(8) Let **CPO** denote the category of CPO's, i.e., *complete posets* (posets in which every directed set has a join) and *continuous functions* (i.e., functions preserving all directed joins). No non-empty object is finitely presentable in **CPO**. In fact, consider the following directed diagram D of inclusions of linearly ordered CPO's: $\{0\} \subseteq \{0, 1\} \subseteq \{0, 1, 2\} \subseteq \ldots$. A colimit in **CPO** can be described by the inclusions of those CPO's into $\omega^T = \{0, 1, \ldots, n, \ldots\} \cup \{T\}$. Now let K be a non-empty CPO and let $f\colon K \to C$ be the constant map of value T. This is a continuous function which does not factorize through any of the colimit maps of D.

(9) In the category **CSLat** of complete semilattices (= complete lattices) and join-preserving homomorphisms no object of more than one element is finitely presentable—the proof is analogous to that in (8) above.

(10) A topological space is finitely presentable in **Top**, the category of topological spaces and continuous functions, iff it is finite and discrete. In fact, any topological space A with a non-open subset $M \subseteq A$ fails to be finitely presentable: consider the sequence D_n of topological spaces for $n \in \mathbb{N} = \{0, 1, 2, \dots\}$, where D_n is the space on the disjoint union $A + \mathbb{N}$ of A and $\mathbb{N}$ such that a subset of D_n is open iff its intersection with A is open in A and its intersection with $\mathbb{N}$ is disjoint with $\{0, 1, \dots, n-1\}$. The colimit of $D = (D_n)_{n \in \mathbb{N}}$ is the indiscrete space on the set $A + \mathbb{N}$. The canonical injection of A into $\operatorname{colim} D$ does not factorize through any of the colimit morphisms $D_n \hookrightarrow \operatorname{colim} D$.

1.3 Proposition. *A finite colimit of finitely presentable objects is finitely presentable.*

PROOF. Let $D \colon \mathcal{D} \to \mathcal{K}$ be a finite diagram with colimit $(Dd \xrightarrow{k_d} K)_{d \in \mathcal{D}^{\mathrm{obj}}}$. Then we will prove that K is finitely presentable provided that each Dd is finitely presentable.

Suppose $D^* \colon (I, \leq) \to \mathcal{K}$ is a directed diagram with a colimit

$$(D_i^* \xrightarrow{c_i} C)_{i \in I}.$$

For each morphism $f \colon K \to C$ and every object d of $\mathcal{D}$ the morphism $f \cdot k_d$ factorizes through $c_{i(d)}$ for some $i(d) \in I$ (because Dd is finitely presentable). Since $\mathcal{D}$ has finitely many objects, there exists an upper bound $i_0 \in I$ of all $i(d)$, $d \in \mathcal{D}^{\mathrm{obj}}$. Then for each d there exists a factorization $f \cdot k_d = c_{i_0} \cdot g_d$ for some $g_d \colon Dd \to D_{i_0}^*$. Next, for each morphism $\delta \colon d \to d'$ in $\mathcal{D}$ we have two factorizations of $f \cdot k_d$: since $k_d = k_{d'} \cdot D\delta$ we get

$$f \cdot k_d = c_{i_0} \cdot g_d = c_{i_0} \cdot g_{d'} \cdot D\delta.$$

Thus, there exists $j(\delta) \geq i_0$ such that

$$D^*(i_0 \to j(\delta)) \cdot g_d = D^*(i_0 \to j(\delta)) \cdot g_{d'} \cdot D\delta.$$

Since $\mathcal{D}$ has finitely many morphisms, there exists an upper bound $i_1 \in I$ of all $j(\delta)$, $\delta \in \mathcal{D}^{\mathrm{mor}}$. The cocone

$$D(i_0 \to i_1) \cdot g_d \colon Dd \to D_{i_1}^* \qquad (d \in \mathcal{D}^{\mathrm{obj}})$$

is compatible for D. Thus, there exists $g\colon K \to D^*_{i_1}$ with $D(i_0 \to i_1) \cdot g_d = g \cdot k_d$ $(d \in \mathcal{D}^{\text{obj}})$. Consequently,

$$f \cdot k_d = c_{i_0} \cdot g_d = c_{i_1} \cdot D(i_0 \to i_1) \cdot g_d = c_{i_1} \cdot g \cdot k_d$$

for all $d \in \mathcal{D}^{\text{obj}}$, which implies $f = c_{i_1} \cdot g$.

To prove that f factorizes through c_{i_1} essentially uniquely, consider $g'\colon K \to D^*_{i_1}$ with $c_{i_1} \cdot g = c_{i_1} \cdot g'$. For each $d \in \mathcal{D}^{\text{obj}}$ we have two factorizations of $f \cdot k_d$:

$$f \cdot k_d = c_{i_1} \cdot g \cdot k_d = c_{i_1} \cdot g' \cdot k_d$$

which (since Dd is finitely presentable) implies that there exists $j(d)$ with $D(i_1 \to j(d)) \cdot g \cdot k_d = D(i_1 \to j(d)) \cdot g' \cdot k_d$. Finally, let j be an upper bound of all $j(d)$, $d \in \mathcal{D}^{\text{obj}}$, then $D(i_1 \to j) \cdot g \cdot k_d = D(i_1 \to j) \cdot g' \cdot k_d$ (for all d). This implies $D(i_1 \to j) \cdot g = D(i_1 \to j) \cdot g'$. □

Remark. Consequently, a split subobject (or a split quotient) of a finitely presentable object A is finitely presentable: we can express it by a coequalizer of two endomorphisms of A.

In contrast, a quotient (or, dually, a subobject) of a finitely presentable object need not be finitely presentable: consider an algebraic lattice as a category, then every object is a quotient of the initial object. Even a regular subobject can fail to be finitely presentable. For example, in the category of lattices the free lattice on three generators contains sublattices which are not finitely presentable, see [Whitman 1941].

Directed and Filtered Colimits

A number of authors prefer working with filtered rather than directed, colimits. (The obvious reason is that canonical diagrams are often filtered, but not directed.) In this subsection we will show that those two concepts are equivalent.

1.4 Definition. A category $\mathcal{D}$ is called *filtered* provided that every finite subcategory of $\mathcal{D}$ has a compatible cocone in $\mathcal{D}$. In other words,

(1) $\mathcal{D}$ is non-empty,

(2) for each pair D_1, D_2 of objects there exists an object D and morphisms $f_1\colon D_1 \to D$ and $f_2\colon D_2 \to D$ in $\mathcal{D}$,

(3) for each pair $g, g'\colon D_1 \to D_2$ of morphisms in $\mathcal{D}$ there exists a morphism $f\colon D_2 \to D$ in $\mathcal{D}$ with $f \cdot g = f \cdot g'$.

Observe that (2) and (3) imply that every pair of morphisms with a common domain can be completed to a commuting square.

Every directed poset, considered as a category, is filtered. The category with one object and two morphisms *id*, f satisfying $f \cdot f = f$ is filtered.

A *filtered diagram* in a category $\mathcal{K}$ is a diagram $D\colon \mathcal{D} \to \mathcal{K}$ whose scheme $\mathcal{D}$ is filtered. Colimits of such diagrams are called *filtered colimits*. We will show how to "reduce" every filtered diagram to a directed diagram. To make the concept of reduction precise, we use cofinality (see 0.11):

1.5 Theorem. *For every (small) filtered category $\mathcal{D}$ there exists a (small) directed poset $\mathcal{D}_0$ and a cofinal functor $H\colon \mathcal{D}_0 \to \mathcal{D}$.*

Remark. The following proof is somewhat more technical than the reader might expect at first sight. To realize the difficulty, consider the simple case where $\mathcal{D}$ has just one object d and two morphisms *id* and $f = f \cdot f$. Here $\mathcal{D}_0$ has to be an infinite directed category such as $d \xrightarrow{f} d \xrightarrow{f} d \xrightarrow{f} d \ldots$.

PROOF. I. Let us first suppose that $\mathcal{D}$ has the following property: every finite subcategory of $\mathcal{D}$ can be extended into a finite subcategory with a unique terminal object. Then the set I of all subcategories of $\mathcal{D}$ with a unique terminal object, ordered by inclusion, is obviously directed: given two such subcategories $\mathcal{A}_1, \mathcal{A}_2$, we extend $\mathcal{A}_1 \cup \mathcal{A}_2$ to a subcategory $\mathcal{A}$ with a unique terminal object, and we have an upper bound of $\mathcal{A}_1$, $\mathcal{A}_2$ in I. The functor $H\colon (I, \subseteq) \to \mathcal{D}$ defined by

$$H(\mathcal{A}) = \text{the terminal object of } \mathcal{A}$$
$$H(\mathcal{A} \to \mathcal{A}') = \text{the unique } \mathcal{A}'\text{-morphism from } H(\mathcal{A}) \text{ to } H(\mathcal{A}')$$

is cofinal. In fact, for each object d we have $id_d\colon d \to H\{d\}$. Given $f_i\colon d \to HA_i$ $(i = 1, 2)$, let $\mathcal{A} \in I$ contain $\mathcal{A}_1 \cup \mathcal{A}_2 \cup \{f_1, f_2\}$, then $H(\mathcal{A}_1 \to \mathcal{A}) \cdot f_1 = H(\mathcal{A}_2 \to \mathcal{A}) \cdot f_2$.

II. Let $\mathcal{D}$ be an arbitrary filtered category. Then the category $\mathcal{D}\times\omega$ (where ω is the linearly ordered category of natural numbers) is also filtered, and it has the property required in I above. In fact, if a subcategory $\mathcal{A}$ of $\mathcal{D}\times\omega$ has a compatible cocone with codomain (A, n), then the object $(A, n+1)$ is the unique terminal object of the following extension of $\mathcal{A}$: add to $\mathcal{A}$ the object $(A, n+1)$, its identity morphism and all composites of the given cocone and the canonical morphism $(A, n) \to (A, n+1)$. The projection functor $\mathcal{D} \times \omega \to \mathcal{D}$ is, obviously, cofinal. By I, we have a cofinal functor from a directed poset into $\mathcal{D} \times \omega$. It is clear that the composite of two cofinal functors is cofinal. □

Corollary. *A category has filtered colimits iff it has directed colimits. For such categories $\mathcal{K}$, a functor $F: \mathcal{K} \to \mathcal{L}$ preserves filtered colimits iff it preserves directed colimits.*

1.6. We have reduced filtered colimits to directed colimits. We will make a further step, and reduce directed colimits to colimits of chains (= well-ordered diagrams, or, diagrams whose schemes are ordinals).

Lemma. *Every infinite directed poset $(I, \leq)$ can be expressed as a union of a chain of directed subposets each of which has a smaller cardinality than* card I. *In more detail, if* card $I = \lambda$, *then there exist directed subposets $I_k \subseteq I$ $(k < \lambda)$ such that*

(i) $I = \bigcup_{k<\lambda} I_k$,

(ii) $I_k \subseteq I_{k'}$ *for* $k \leq k'$,

(iii) card $I_k < \lambda$ *for each* k,

and

(iv) $I_k = \bigcup_{k'<k} I_{k'}$ *for every limit ordinal* $k < \lambda$.

PROOF. Express I as
$$I = \{\, i_k \mid k < \lambda \,\}.$$
For each finite set $J \subseteq I$ choose an upper bound $j \in I$ and put $J^* = J \cup \{j\}$; for each infinite subset $J \subseteq I$ there exists a directed set $J^* \subseteq I$ of the same cardinality containing J. In fact, put $J^* = \bigcup_{n<\omega} J_n$, where $J_0 = J$ and J_{n+1} is obtained from J_n by adding, for each pair of elements in J_n, an upper bound. The following subposets I_k $(k < \lambda)$ of I have the required properties:

$$I_0 = \emptyset;$$
$$I_{k+1} = \left(I_k \cup \{\, i_k \,\}\right)^*;$$
$$I_k = \bigcup_{k'<k} I_{k'} \qquad \text{for limit ordinals } k < \lambda. \qquad \square$$

1.7 Corollary. *A category has directed colimits iff it has colimits of chains. For such categories $\mathcal{K}$, a functor $F: \mathcal{K} \to \mathcal{L}$ preserves directed colimits iff it preserves colimits of chains.*

PROOF. Let $\mathcal{K}$ be a category with colimits of chains. We will prove that $\mathcal{K}$ has colimits of diagrams D indexed by an arbitrary directed poset $(I, \leq)$. We proceed by transfinite induction on the cardinality $\lambda = \operatorname{card} I$.

First step: If λ is a finite cardinal, then I has a largest element, so there is nothing to prove.

Induction step: If the statement holds for all directed posets of cardinality less than λ, then we use Lemma 1.6: since $I = \bigcup_{k<\lambda} I_k$, a colimit of D can be simply constructed as a colimit of the λ-chain of $\operatorname{colim} D_k$, $k < \lambda$, where D_k is the diagram D restricted to I_k (and $\operatorname{colim} D_k$ exists by induction hypothesis).

Analogously for the statement on functors. □

Remark. Let us call a chain $D\colon \lambda \to \mathcal{K}$ (λ an infinite ordinal) *smooth* provided that for each limit ordinal $i < \lambda$ we have $D_i = \operatorname{colim}_{j<i} D_j$ with the colimit cocone formed by the D-arrows $D_j \to D_i$ ($j < i$). The above corollary can be sharpened by substituting colimits of chains by colimits of smooth chains. This follows from (iv) in Lemma 1.6.

1.8 Example. Unlike the situation for filtered and directed colimits, it is *not* true that every directed diagram is cofinal with a chain! For example, the directed poset of all finite sets of real numbers (ordered by inclusion) does not have a cofinal chain.

Locally Finitely Presentable Categories

1.9 Definition. A category $\mathcal{K}$ is called *locally finitely presentable* provided that it is cocomplete and has a set $\mathcal{A}$ of finitely presentable objects such that every object is a directed colimit of objects from $\mathcal{A}$.

Remark. The condition concerning $\mathcal{A}$ says that for every object K there exists a directed diagram in $\mathcal{A}$ (considered as a full subcategory of $\mathcal{K}$) such that K is a colimit object of that diagram.

The condition on the existence of $\mathcal{A}$ can be reformulated by the following two conditions (which are often easier to verify):

(1) every object is a directed colimit of finitely presentable objects,

and

(2) there exists, up-to isomorphism, only a set of finitely presentable objects.

In fact, assuming (1) and (2), any set $\mathcal{A}$ of representatives of finitely presentable objects has the property in the definition above. Conversely, given $\mathcal{A}$ as above, it is clear that (1) holds. To verify (2), observe first that $\mathcal{A}$ is a strong generator of $\mathcal{K}$; thus, $\mathcal{K}$ is a wellpowered category (see 0.6). It is sufficient to verify that every finitely presentable object K is a subobject of an object in $\mathcal{A}$. In fact, let D be a directed diagram in $\mathcal{A}$ with a colimit $(D_i \xrightarrow{k_i} K)_{i \in I}$. Since K is finitely presentable, id_A factorizes through some k_i, i.e., there exists $m: K \to D_i$ ($i \in I$) with $k_i \cdot m = id_K$. Thus, K is a split subobject of $D_i \in \mathcal{A}$.

1.10 Examples

(1) **Set** is locally finitely presentable. In fact (i) every set is a directed colimit of the diagram of all of its finite subsets (ordered by inclusion), and (ii) there exists, up to isomorphism, only a (countable) set of finite sets.

Analogously, **Pos**, **Rel** Σ, **Grp**, and **Aut** are locally finitely presentable categories.

(2) Every variety of finitary (many-sorted) algebras is locally finitely presentable, as will be proved in Chapter 3.

(3) **CPO** and **Top** are not locally finitely presentable.

(4) The category of finite sets is not locally finitely presentable since it is not cocomplete.

(5) A poset, considered as a category, is locally finitely presentable iff it is a complete lattice which is algebraic (i.e., each element is a directed join of finite elements).

A Criterion for Local Finite Presentability

In the definition of locally finitely presentable category $\mathcal{K}$ an important weakening is possible: instead of a set $\mathcal{A}$ of finitely presentable objects which "generates" all of $\mathcal{K}$ via directed colimits, it is sufficient to require that $\mathcal{A}$ be a strong generator (see 0.6):

1.11 Theorem. *A category is locally finitely presentable iff it is cocomplete, and has a strong generator formed by finitely presentable objects.*

PROOF. The necessity is clear. To prove the sufficiency, let $\mathcal{K}$ be a cocomplete category with a strong generator $\mathcal{A}$ formed by finitely presentable objects. Let $\overline{\mathcal{A}}$ be a closure of $\mathcal{A}$ under finite colimits (i.e., the smallest subcategory of $\mathcal{K}$ closed under finite colimits and containing $\mathcal{A}$). It is clear that $\overline{\mathcal{A}}$ is essentially small and, by Proposition 1.3, objects of $\overline{\mathcal{A}}$ are finitely presentable. (In fact, the collection of all finitely presentable objects is closed under finite colimits and, since it contains $\mathcal{A}$, it must contain $\overline{\mathcal{A}}$.) It is sufficient to prove that every object of $\mathcal{K}$ is a filtered colimit of objects of $\overline{\mathcal{A}}$.

For each object K we can form the canonical diagram D w.r.t. $\overline{\mathcal{A}}$ (see Definition 0.4). Since $\overline{\mathcal{A}}$ is closed under finite colimits, D is clearly filtered. Put $K^* = \operatorname{colim} D$ and for each morphism $f\colon A \to K$ with A in $\overline{\mathcal{A}}$ denote the corresponding colimit morphism by $f^*\colon A \to K^*$. Let $m\colon K^* \to K$ be the unique morphism with $f = m \cdot f^*$ for each f. We are going to show that m is an isomorphism. It is sufficient to verify that m is a monomorphism: since $\mathcal{A}$ is a strong generator and each morphism from an $\mathcal{A}$-object into K factorizes through m, it then follows that m is an isomorphism.

Given $p, q\colon B \to K^*$ with $m \cdot p = m \cdot q$, we will prove that $p = q$; it is sufficient to prove this in the case $B \in \mathcal{A}$, since the general case follows from the fact that $\mathcal{A}$ is a (strong) generator. The diagram D is filtered, and B is finitely presentable. Thus, there exists $f\colon A \to K$, for A in $\overline{\mathcal{A}}$, such that both p and q factorize through f^*. That is, we have $p', q'\colon B \to A$ with $p = f^* \cdot p'$ and $q = f^* \cdot q'$.

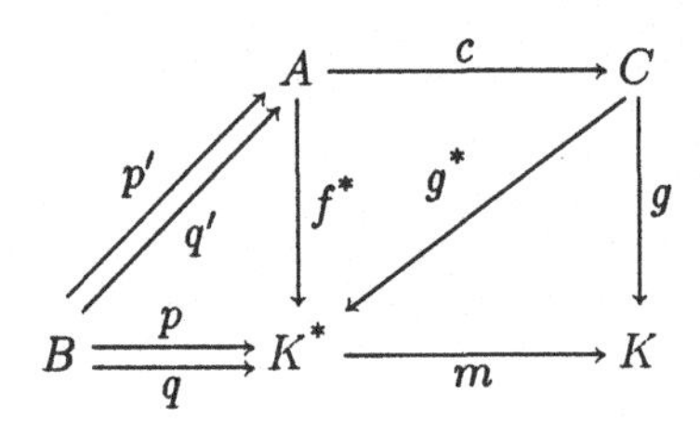

Let $c\colon A \to C$ be a coequalizer of p' and q'. Since A and B lie in $\overline{\mathcal{A}}$, it follows that C also lies in $\overline{\mathcal{A}}$, thus, the unique morphism $g\colon C \to K$ with $f = g \cdot c$ belongs to the diagram D. Since $f^* = (g \cdot c)^* = g^* \cdot c$, we have $p = g^* \cdot c \cdot p' = g^* \cdot c \cdot q' = q$. □

1.12 Example. For each small category $\mathcal{A}$ the category $\mathbf{Set}^{\mathcal{A}}$ of all functors from $\mathcal{A}$ to $\mathbf{Set}$ is locally finitely presentable: it is cocomplete, and the set of all hom-functors (which are finitely presentable objects) is a strong generator.

1.B Locally Presentable Categories

Analogously to the transition from finitary to infinitary algebras, we now generalize the concept of a locally finitely presentable category. We use a *regular cardinal* λ, i.e., an infinite cardinal which is not a sum of a smaller number of smaller cardinals. More precisely, λ is a regular cardinal iff it is infinite and cannot be expressed as $\lambda = \sum_{i<\alpha} \lambda_i$ where $\lambda_i < \lambda$ and $\alpha < \lambda$. For example, $\aleph_0$, $\aleph_1$ and every successor cardinal $\aleph_{i+1}$ are regular cardinals, whereas $\aleph_\omega$ $(= \sum_{i<\omega} \aleph_i)$ is not regular.

1.13 Definition. Let λ be a regular cardinal.

(1) A poset is called λ-*directed* provided that every subset of cardinality smaller than λ has an upper bound. A diagram whose scheme is a λ-directed poset is called a λ-directed diagram, and its colimit is called a λ-*directed colimit.*

(2) An object K of a category is called λ-*presentable* provided that its hom-functor $\hom(K, -)$ preserves λ-directed colimits. An object is called *presentable* if it is λ-presentable for some λ.

The case $\lambda = \aleph_0$ is the above concept of a directed colimit and a finitely presentable object. Observe that every λ-presentable object is λ'-presentable for each regular cardinal $\lambda' > \lambda$.

1.14 Examples

(1) A set is λ-presentable in **Set** iff it has cardinality smaller than λ.

An S-sorted set is λ-presentable in $\mathbf{Set}^S$ iff it has power smaller than λ.

(2) Analogously to finitary relational structures (see Example 1.2 (4)) we can introduce λ-ary relational structures: Let S be a set of sorts and let Σ be a λ-*ary* S-*sorted relational signature.* That is, with each symbol $\sigma \in \Sigma$ we are given an arity which is a collection $(s_i)_{i\in I}$ of sorts $s_i \in S$ with $\operatorname{card} I < \lambda$. A *relational structure* A of type Σ consists of an underlying S-sorted set $|A| = (A_s)_{s\in S}$ and of relations $\sigma_A \subseteq \prod_{i\in I} A_{s_i}$ (for each $\sigma \in \Sigma$ of arity $(s_i)_{i\in I}$). The category $\mathbf{Rel}\,\Sigma$ of relational structures of type Σ has as morphisms all *homomorphisms*, i.e., S-sorted maps preserving the corresponding relations.

For each λ-ary signature Σ, an object A of $\mathbf{Rel}\,\Sigma$ is λ-presentable iff it has less than λ vertices, i.e., $\#|A| < \lambda$, and less than λ edges, i.e., $\sum_{\sigma\in\Sigma} \operatorname{card} \sigma_A < \lambda$.

(3) By a *convergence space* on a set X is meant a relation $\to$ ("converges") between ω-sequences in X and elements of X such that

(a) for each $x \in X$ we have $\dot{x} \to x$ for the constant sequence $\dot{x} = (x, x, x, \dots)$;

(b) if $(x_n)_{n<\omega} \to x$ then $(x_{n_k})_{k<\omega} \to x$ for each subsequence (x_{n_k}) of (x_n).

The category **Con** of convergence spaces is a full subcategory of $\mathbf{Rel}\,\Sigma$ where Σ consists of one one-sorted ω-ary symbol σ (provided that $(x_0, x_1, x_2, \dots) \in \sigma$ is rewritten as $(x_1, x_2, x_3, \dots) \to x_0$). That is, morphisms of **Con** are the continuous functions, i.e., functions preserving convergence.

A convergence space is λ-presentable, where $\lambda \geq \omega_1$, iff its cardinality is smaller than λ.

(4) No non-empty *CPO* is presentable (in the category **CPO**); the proof is analogous to Example 1.2 (8). Analogously, no complete semilattice of more than one element is presentable in **CSLat** (cf. Example 1.2(9)).

Let $\boldsymbol{\omega}$ **CPO** denote the category of ωCPO's (i.e., posets in which every increasing ω-chain has a join) and ω-continuous maps (i.e., maps preserving joins of ω-chains). Then each finite ωCPO is ω_1-presentable, and an infinite ωCPO is λ-presentable iff it has cardinality smaller than λ.

(5) For each variety $\mathcal{V}$ of finitary algebras and each uncountable regular cardinal λ larger than the number of basic operations, an algebra is λ-presentable in $\mathcal{V}$ iff it has power less than λ. (See Corollary 3.13.)

In contrast, let $\mathcal{V}$ be the variety of algebras of one ω-ary operation with no equations. There exist arbitrarily large regular cardinals λ such that a λ-presentable algebra in $\mathcal{V}$ can have λ or more elements. (In fact, let α be a cardinal cofinal with ω, and let λ be a successor of α. Then the $\mathcal{V}$-free algebra A on α generators is λ-presentable, but since $\alpha^\omega > \alpha$, the cardinality of A is larger than α and, thus, larger or equal to λ.)

(6) No non-discrete space is presentable in **Top**; the proof is analogous to that in Example 1.2(10).

1.15 Definition. A diagram $D\colon \mathcal{D} \to \mathcal{K}$ is called λ-*small* provided that its scheme $\mathcal{D}$ has less than λ morphisms.

1.16 Proposition. *A colimit of a λ-small diagram of λ-presentable objects is λ-presentable.*

The proof is analogous to that of Proposition 1.3. □

Remark. Consequently, a split quotient of a λ-presentable object is λ-presentable. In contrast, neither a subobject nor a quotient of a λ-presentable object is λ-presentable in general, see Remark 1.3.

1.17 Definition. A category is called *locally λ-presentable* (λ a regular cardinal) provided that it is cocomplete, and has a set $\mathcal{A}$ of λ-presentable objects such that every object is a λ-directed colimit of objects from $\mathcal{A}$.

A category is *locally presentable* if it is locally λ-presentable for some regular cardinal λ.

1.18 Examples

(1) Locally ω-presentable categories are precisely the locally finitely presentable ones.

(2) $\boldsymbol{\omega}$ **CPO** is locally ω_1-presentable but not locally finitely presentable (see Example 1.14(3)).

(3) A poset, considered as a category, is locally presentable iff it is a complete lattice. The chain L_λ (λ regular) of all ordinals smaller or equal to λ is locally λ-presentable, but not locally λ'-presentable for any $\lambda' < \lambda$.

(4) **Rel** Σ is locally λ-presentable iff Σ is a λ-ary signature. **Con** is locally ω_1-presentable.

(5) The categories **CPO**, **CSLat**, and **Top** are not locally presentable.

1.19 Remark. Analogously to the case $\lambda = \aleph_0$ of locally finitely presentable categories (see Remark 1.9), the condition on $\mathcal{A}$ can be reformulated to the following conditions:

(1) every object is a λ-directed colimit of λ-presentable objects,

and

(2) there exists, up to isomorphism, only a set of λ-presentable objects.

Notation. In a locally λ-presentable category $\mathcal{K}$ all λ-presentable objects have a set of representatives (w.r.t. isomorphism). Any such set is denoted by

$$\mathbf{Pres}_\lambda\, \mathcal{K}$$

and is considered as a (small) full subcategory of $\mathcal{K}$.

1.20 Theorem. *A category is locally λ-presentable iff it is cocomplete, and has a strong generator formed by λ-presentable objects.*

The proof is analogous to that of Theorem 1.11. □

Remark. Observe that a locally λ-presentable category $\mathcal{K}$ is locally λ'-presentable for all regular cardinals $\lambda' \geq \lambda$, and that every object of $\mathcal{K}$ is presentable. (See Proposition 1.16.)

1.21 Remark. Analogously to Theorem 1.5, λ-directed diagrams can be substituted by *λ-filtered diagrams* in the definition of λ-presentability. A category $\mathcal{D}$ is called λ-filtered provided that each subcategory of less than λ morphisms has a compatible cocone in $\mathcal{D}$. This means that

(1) $\mathcal{D}$ is non-empty,

(2) for each collection D_i, $i \in I$, of less than λ objects of $\mathcal{D}$ there exists an object D and morphisms $f_i \colon D_i \to D$, $i \in I$, in $\mathcal{D}$,

(3) for each collection $g_i \colon D_1 \to D_2$, $i \in I$, of less than λ morphisms in $\mathcal{D}$ there exists a morphism $f \colon D_2 \to D$ in $\mathcal{D}$ with $f \cdot g_i$ independent of i.

For every (small) λ-filtered category $\mathcal{D}$ there exists a (small) λ-directed poset $\mathcal{D}_0$ and a cofinal functor $H : \mathcal{D}_0 \to \mathcal{D}$. Thus λ-filtered and λ-directed colimits are "equivalent", see 0.11. In particular, an object is λ-presentable iff its hom-functor preserves λ-filtered colimits. All this is quite analogous to Corollary 1.5. In contrast, the step from directed colimits to colimits of chains (see Corollary 1.7) does not generalize to λ: a category with ω_1-directed colimits of chains need not have ω_1-directed colimits in general, see Exercise 1.c.

The reason why λ-filtered diagrams are preferred by some authors to λ-directed ones is obvious: whereas in the definition of locally λ-presentable category each object is presented by some (unspecified) λ-directed colimit of λ-presentable objects, one can be more specific and use the canonical colimit (see 0.4), as we shall see now. However, the canonical colimit is λ-filtered, not λ-directed.

1.22 Proposition. *For each object K of a locally λ-presentable category the canonical diagram w.r.t. $\mathbf{Pres}_\lambda \mathcal{K}$ is λ-filtered, and K is its canonical colimit.*

PROOF. The first statement is an immediate corollary of Proposition 1.16: for $\mathcal{A} = \mathbf{Pres}_\lambda \mathcal{K}$, any subcategory $\mathcal{D}$ of $\mathcal{A} \downarrow K$ of less than λ morphisms has a compatible cocone, e.g., a colimit of the forgetful functor $\mathcal{D} \to \mathcal{K}$.

Next, for each object K there exists a λ-directed diagram D of λ-presentable objects with a colimit $(D_i \xrightarrow{k_i} K)_{i \in I}$. To prove that the canonical diagram w.r.t. $\mathcal{A}$ has the canonical colimit K, it is sufficient to observe that the objects $D_i \xrightarrow{k_i} K$ of $\mathcal{A} \downarrow K$ form a cofinal subdiagram (see 0.11) of the canonical diagram: each arrow $A \xrightarrow{a} K$ in $\mathcal{A} \downarrow K$ factorizes essentially uniquely through some k_i (since $\hom(A, -)$ preserves the λ-directed colimit of D). Now use Exercise 1.o(3). □

1.23 Definition. A small, full subcategory $\mathcal{A}$ of a category $\mathcal{K}$ is called *dense* provided that every object of $\mathcal{K}$ is a canonical colimit of $\mathcal{A}$-objects (see 0.4).

Remark. Every dense subcategory is a strong generator. The converse does not hold, e.g., the vector space $\mathbb{R}$ forms a (singleton) strong generator in **Vec** which is not dense (see the Example 1.24(4)).

1.24 Examples

(1) In a locally λ-presentable category we have seen that $\mathbf{Pres}_\lambda\, \mathcal{K}$ is a dense subcategory.

(2) In the category **Pos** the single-object subcategory consisting of the two-element chain is dense.

(3) No single-object subcategory of **Gra** is dense: given a graph $A = (|A|, \sigma)$, then either A is discrete (i.e., $\sigma = \emptyset$), hence no non-discrete object is a colimit of copies of A, or A is non-discrete, and then no non-empty discrete object is a colimit of copies of A.

The full subcategory consisting of

$$A = (\{0\}, \emptyset) \qquad \text{and} \qquad B = (\{0,1\}, \{(0,1)\})$$

is dense in **Gra**.

(4) The difference between canonical and arbitrary colimits is well illustrated by the vector space $\mathbb{R}$ in the category **Vec** of real vector spaces: although every object is a coproduct of copies of $\mathbb{R}$, it is not true that $\mathbb{R}$ is dense. To verify this, consider a map $f\colon A \to B$ between vector spaces which is homogenous (i.e., $f(rx) = rf(x)$ for all $r \in \mathbb{R}$ and $x \in A$) but not additive. The passage

$$\mathbb{R} \xrightarrow{a} A \quad \longmapsto \quad \mathbb{R} \xrightarrow{fa} B$$

is a natural transformation of the canonical diagrams. (In fact, if a is linear, then each $f \cdot a$ is homogenous and, thus, linear.) However, since f is not linear, the natural transformation above is not induced by any morphism from A to B, which demonstrates that A is not a colimit of its canonical diagram w.r.t. $\mathbb{R}$.

On the other hand, the space $\mathbb{R} \times \mathbb{R}$ is dense in **Vec**.

(5) In a variety of (one-sorted) algebras where all arities are smaller or equal to n the free algebra on n generators forms a singleton dense subcategory.

(6) In the category of complete lattices and complete homomorphisms no small subcategory is dense. (This follows easily from the fact that for each cardinal λ there exists a λ-complete lattice which is not complete.)

(7) The category **Top** of topological spaces and continuous maps also has no dense subcategory. This follows easily from the fact that for each cardinal λ there exists a topological space which is not discrete (i.e., not every subset is open) but whose subspaces of cardinality smaller than λ are all discrete.

1.25 Notation. Given a small, full subcategory $\mathcal{A}$ of a category $\mathcal{K}$ we define the *canonical functor*

$$E: \mathcal{K} \to \mathbf{Set}^{\mathcal{A}^{\mathrm{op}}}$$

analogously to the Yoneda embedding (0.9): E assigns to each object K the restriction

$$EK = \hom(-, K)/_{\mathcal{A}^{\mathrm{op}}}$$

of its hom-functor to $\mathcal{A}^{\mathrm{op}}$, and to each morphism $f: K \to K'$ the domain-codomain restriction of the Yoneda transformation $\hom(-, f)$.

1.26 Proposition. *Let $\mathcal{A}$ be a small, full subcategory of a category $\mathcal{K}$. The canonical functor $E: \mathcal{K} \to \mathbf{Set}^{\mathcal{A}^{\mathrm{op}}}$*

(i) *is full and faithful iff $\mathcal{A}$ is dense,*

(ii) *preserves λ-directed colimits iff every object of $\mathcal{A}$ is λ-presentable in $\mathcal{K}$.*

PROOF. (i) is trivial since morphisms from EK to EK' in $\mathbf{Set}^{\mathcal{A}^{\mathrm{op}}}$ are precisely the compatible cocones of the canonical diagram of K with the codomain K'.

(ii) Let E preserve λ-directed colimits. Given a morphism $f\colon A \to K$ with A in $\mathcal{A}$ and given a λ-directed colimit $(K_i \xrightarrow{k_i} K)_{i\in I}$ in $\mathcal{K}$, then since $f \in (EK)_A$ and EK is a λ-directed colimit of EK_i, there exists $i \in I$ with $f \in (Ek_i)_A$; i.e., f factorizes through k_i. The essential uniqueness of such a factorization follows from the description of directed colimits in **Set** (see Exercise 1.a).

Conversely, let each object of $\mathcal{A}$ be λ-presentable. Given a λ-directed colimit $(K_i \xrightarrow{k_i} K)_{i\in I}$ in $\mathcal{K}$ and given an element $f \in (EK)A = \hom(A, K)$, there exists an (essentially unique) $i \in I$ such that f factorizes through k_i, i.e., $f \in (Ek_i)_A[(EK_i)A]$. This proves that the cocone $\big((EK_i)A \xrightarrow{(Ek_i)_A} (EK)A\big)_{i\in I}$ is a (λ-directed) colimit in **Set**, see Exercise 1.a. Since colimits in $\mathbf{Set}^{\mathcal{A}}$ are computed object-wise, we conclude that $(EK_i \xrightarrow{Ek_i} EK)_{i\in I}$ is a colimit in $\mathbf{Set}^{\mathcal{A}}$. □

1.27 Proposition. *For each small, full subcategory $\mathcal{A}$ of a cocomplete category $\mathcal{K}$ the canonical functor $E\colon \mathcal{K} \to \mathbf{Set}^{\mathcal{A}^{\mathrm{op}}}$ is a right adjoint.*

Proof. Given an object $F\colon \mathcal{A}^{\mathrm{op}} \to \mathbf{Set}$ of $\mathbf{Set}^{\mathcal{A}^{\mathrm{op}}}$, let $\mathcal{D}$ be the category of pairs (A, a) where A is an object of $\mathcal{A}$ and $a \in FA$, with morphism $f\colon (A, a) \to (A', a')$ those $\mathcal{K}$-morphisms $f\colon A \to A'$ which fulfil $a = Ff(a')$. The diagram given by the forgetful functor $D\colon \mathcal{D} \to \mathcal{K}$ has a colimit $(A \xrightarrow{k_{A,a}} K)_{A\in\mathcal{A}, a\in FA}$. This defines a natural transformation $r\colon F \to E(K)$ by

$$r_A(a) = k_{A,a} \in \hom(A, K) \qquad \text{for} \quad A \in \mathcal{A},\ a \in FA.$$

r is universal since, given a natural transformation $r'\colon F \to E(K')$, then the cone $(A \xrightarrow{r'_A(a)} K')_{A\in\mathcal{A}, a\in FA}$ is compatible with D, and the unique morphism $k\colon K \to K'$ with $r'_A(a) = k \cdot r_A(a)$ $(A \in \mathcal{A},\ a \in FA)$ is the unique $\mathcal{K}$-morphism with $r' = E(k) \cdot r$. □

1.28 Corollary. *Every locally presentable category is complete.*

In fact, by Proposition 1.26 every locally presentable category is equivalent to a full subcategory of $\mathbf{Set}^{\mathcal{A}^{\mathrm{op}}}$, and by Proposition 1.27 this subcategory is reflective (thus, complete). □

1.29 Corollary. *For each category $\mathcal{K}$ the following are equivalent:*

(i) *$\mathcal{K}$ is equivalent to a full, reflective subcategory of a locally presentable category,*

(ii) *$\mathcal{K}$ is equivalent to a full, reflective subcategory of* $\mathbf{Set}^{\mathcal{A}}$ *for some small category* $\mathcal{A}$,

(iii) *$\mathcal{K}$ is cocomplete and has a dense subcategory.*

In fact, (iii) $\Rightarrow$ (ii) follows from Propositions 1.26 and 1.27, (ii) $\Rightarrow$ (i) is clear, and to prove (i) $\Rightarrow$ (iii), observe that if a category $\mathcal{L}$ has a dense subcategory $\mathcal{L}_0$, then for each full, reflective subcategory $\mathcal{K}$ of $\mathcal{L}$ the reflections of $\mathcal{L}_0$-objects in $\mathcal{K}$ form a dense subcategory of $\mathcal{K}$. □

1.30 Remark. Let $\mathcal{K}$ be a locally λ-presentable category. For each regular cardinal $\mu \geq \lambda$ we know that

(1) a μ-small colimit of λ-presentable objects is μ-presentable (see Proposition 1.16).

The converse is also true:

(2) every μ-presentable object is a μ-small colimit of λ-presentable objects.

The proof (which is rather technical) can be found in [Makkai, Paré 1989], pp. 35–37. However, the following weaker statement is trivial: each μ-presentable object K is a split quotient of a μ-small colimit of λ-presentable objects.

In fact, express K as a λ-directed colimit $(D_i \xrightarrow{k_i} K)_{i\in I}$ of λ-presentable objects D_i. Consider the poset I^* of all subsets $J \subseteq I$ with $\operatorname{card} J < \mu$, ordered by inclusion. For each J form a full subdiagram $(D_i)_{i\in J}$ of D, and let D^*_J be a colimit of this subdiagram. We obtain a natural μ-directed diagram D^* of all D^*_J's and all the canonical arrows $D^*_J \to D^*_{J'}$ for $J \subseteq J'$. It is clear that K is a colimit of this diagram, and since K is μ-presentable, id_K factorizes through some of the colimit maps $d_J\colon D^*_J \to K$. Thus, d_J is a split epimorphism. □

1.31 Example. For each small category $\mathcal{A}$ and each regular cardinal $\lambda > \operatorname{card} \mathcal{A}^{\mathrm{obj}}$, a functor F is λ-presentable in $\mathbf{Set}^{\mathcal{A}}$ iff $\sum_{X\in\mathcal{A}^{\mathrm{obj}}} \operatorname{card} FX < \lambda$. In fact, if F satisfies that inequality, then it is a colimit of less than λ hom-functors (by the Yoneda lemma 0.9) and since $\lambda > \operatorname{card} \mathcal{A}^{\mathrm{obj}}$, that diagram is λ-small—thus, F is λ-presentable by Proposition 1.16. Conversely, if F is λ-presentable, then it belongs to the (iterated) closure of hom-functors under λ-small colimits, see Remark 1.30. Each hom-functor satisfies the above inequality, and a λ-small colimit of functors satisfying that inequality satisfies it too.

1.C Representation Theorem

The aim of the present section is to prove that the following classes of categories coincide:

(1) locally λ-presentable categories;

(2) full, reflective subcategories of $\mathbf{Set}^{\mathcal{A}}$ (or of $\mathbf{Rel}\,\Sigma$) closed under λ-directed colimits;

(3) categories $\mathbf{Cont}_\lambda\,\mathcal{A}$ of λ-continuous set-valued functors defined on a small category $\mathcal{A}$.

(The list will be continued in Section 1.D and Chapter 3, where we will also show that locally presentable categories are precisely the sketchable categories and the essentially algebraic categories.)

We begin with the concept of a small-orthogonality class, since this is the basic technical tool used to prove the representation theorem below.

Orthogonality Classes

1.32 Definition

(1) An object K is said to be *orthogonal* to a morphism $m\colon A \to A'$ provided that for each morphism $f\colon A \to K$ there exists a unique morphism $f'\colon A' \to K$ such that the triangle

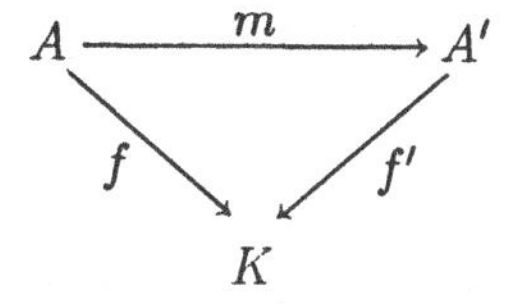

commutes.

(2) For each class $\mathcal{M}$ of morphisms in a category $\mathcal{K}$ we denote by $\mathcal{M}^\perp$ the full subcategory of $\mathcal{K}$ of all objects orthogonal to each $A \xrightarrow{m} A'$ in $\mathcal{M}$.

Conversely, a full subcategory of $\mathcal{K}$ is called a *(small-)orthogonality class* provided that it has the form $\mathcal{M}^\perp$ for a (small) collection $\mathcal{M}$ of morphisms of $\mathcal{K}$.

1.33 Examples

(1) Every full, reflective, isomorphism-closed subcategory of a category $\mathcal{K}$ is an orthogonality class in $\mathcal{K}$. In fact, given such a subcategory $\mathcal{A}$, we

choose, for each object K in $\mathcal{K}$, a reflection map $m_K : K \to K'$ in $\mathcal{A}$. Then

$$\mathcal{A} = \{\, m_K \,\}^{\perp}_{K \in \mathcal{K}^{\text{obj}}} .$$

In fact, every object of $\mathcal{A}$ is, obviously, orthogonal to each m_K. Conversely, every object K orthogonal to its reflection map m_K lies in $\mathcal{A}$: since id_K factorizes through m_K we know that m_K is a split monomorphism, hence, an isomorphism (see Exercise 1.g), and $\mathcal{A}$ is closed under isomorphisms.

The fullness is substantial here, see Exercise 1.h.

(2) Let **Frm** denote the category of *frames* (i.e., complete lattices in which joins distribute over finite meets: $\left(\bigvee_{i\in I} a_i\right) \wedge b = \bigvee_{i\in I}(a_i \wedge b)$) and frame homomorphisms (i.e., maps preserving joins and finite meets). Complete Boolean algebras form a small-orthogonality class of **Frm**: they are precisely the frames orthogonal to the following inclusion:

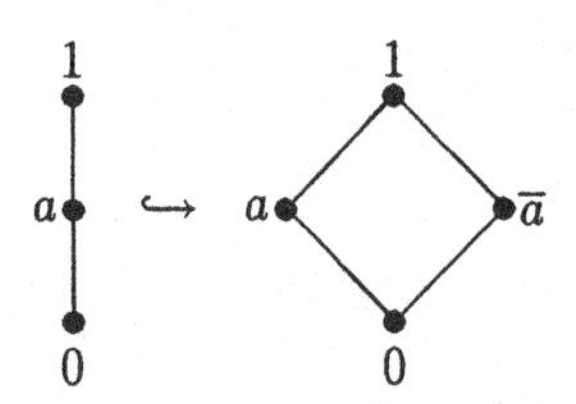

This is an example of a small-orthogonality class which is not reflective (in fact, **Frm** is a cocomplete category and the category of complete Boolean algebras is not, see Exercise 1.q).

(3) Let **Alg** 2 be the category of algebras of one binary operation, let $A = F\{x,y,z\}$ be the algebra of all terms in variables x, y, z, and let A' be the quotient algebra $F\{x,y,z\}/\!\sim$ under the smallest congruence $\sim$ with $(xy)z \sim x(yz)$. An algebra is orthogonal to the quotient map $m\colon A \to A'$ iff it is associative. Thus, $\{m\}^{\perp}$ is the class of all semigroups.

(4) **Pos** is a small-orthogonality class of **Gra**, viz, $\mathbf{Pos} = \{m_1, m_2, m_3\}^{\perp}$ for the following morphisms:

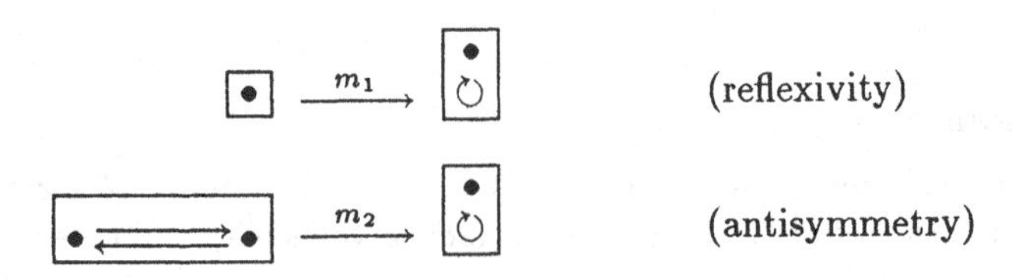

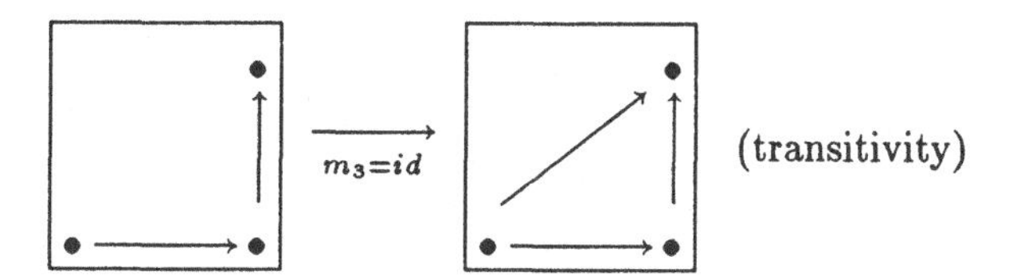

(5) An example of an orthogonality class which is not a small-orthogonality class: complete join-semilattices in the category **Pos*** of all posets and all functions preserving (all existing) joins. Recall that each poset P has a reflection in the subcategory **CSLat** of complete join-semilattices: the reflection is the poset $I(P)$ of all ideals (i.e., downwards closed sets $I \subseteq P$ closed under all existing joins) ordered by inclusion. The reflection map $P \to I(P)$ assigns to each element p the prime ideal $\downarrow p = \{x \in P \mid x \leq p\}$.

By (1), **CSLat** is an orthogonality class of **Pos***. However, **CSLat** $\neq \mathcal{M}^{\perp}$ for any small collection $\mathcal{M}$ of morphisms in **Pos***, see Exercise 1.h.

(6) In the category **Gra** the objects orthogonal to the inclusion

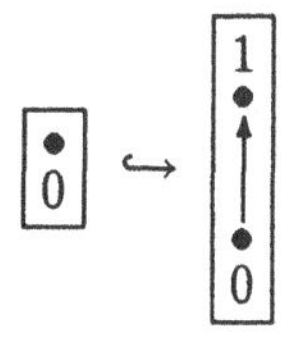

are precisely the unary algebras (on one operation).

(7) Orthogonality in $\mathbf{Set}^{\mathcal{A}}$ can serve to express the preservation of specified limits by functors from $\mathcal{A}$ to **Set**. For example, let $(A_1 \times A_2 \xrightarrow{\pi_i} A_i)$ be a product in $\mathcal{A}$. A functor $F \colon \mathcal{A} \to \mathbf{Set}$ preserves this product iff for each element of $F(A_1 \times A_2)$, i.e., each natural transformation $\varphi \colon \hom(A_1 \times A_2, -) \to F$ in $\mathbf{Set}^{\mathcal{A}}$, there exists a unique pair of elements in $FA_1 \times FA_2$, i.e., a unique natural transformation $\varphi' \colon \hom(A_1, -) \times \hom(A_2, -) \to F$, such that the following triangle

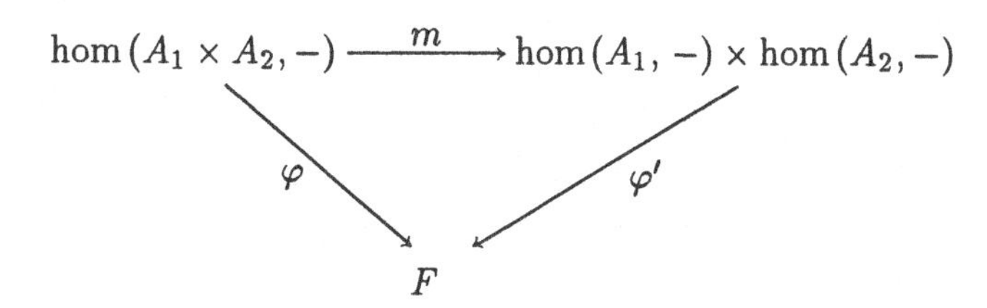

commutes. Here m is the canonical natural transformation whose components $\hom(A_1 \times A_2, -) \to \hom(A_i, -)$ correspond to the projections $\pi_i \colon A_1 \times A_2 \to A_i$. Thus

$$\{m\}^\perp = \text{the full subcategory of } \mathbf{Set}^{\mathcal{A}} \text{ formed by functors preserving the product } (A_1 \times A_2 \xrightarrow{\pi_i} A_i).$$

(8) More generally, given a collection of limits $\lim D_i$ $(i \in I)$ in a small category $\mathcal{A}$, the full subcategory of $\mathbf{Set}^{\mathcal{A}}$ formed by all functors preserving those limits is the orthogonality class $\mathcal{M}^\perp$, where $\mathcal{M} = \{ m_i \}_{i \in I}$ consists of the canonical maps

$$m_i \colon \hom(\lim_d D_i d, -) \to \lim_d \big[\hom(D_i d, -)\big].$$

A further generalization is straightforward: given a collection of diagrams D_i with a compatible cone $\sigma(D_i)$ for each i, the full subcategory of $\mathbf{Set}^{\mathcal{A}}$ formed by all functors F with $F\big(\sigma(D_i)\big) = \lim FD_i$ for each i is an orthogonality class in $\mathbf{Set}^{\mathcal{A}}$.

1.34 Observation. Each orthogonality class is closed under limits. In fact, if $\mathcal{M}$ is a class of morphisms in $\mathcal{K}$ and if a diagram $D \colon \mathcal{D} \to \mathcal{M}^\perp$ has a limit $(L \xrightarrow{\pi_d} Dd)_{d \in \mathcal{D}^{\text{obj}}}$ in $\mathcal{K}$, then $L \in \mathcal{M}^\perp$: for each morphism $f \colon A \to L$ and each $m \colon A \to A'$ in $\mathcal{M}$ we have the unique cone $f_d \colon A' \to Dd$ with

$$\pi_d \cdot f = f_d \cdot m \quad (d \in \mathcal{D}^{\text{obj}})$$

(since $Dd \in \mathcal{M}^\perp$). The unicity clearly guarantees that this cone is compatible. Thus, there is a unique $f' \colon A' \to L$ with

$$\pi_d \cdot f' = f_d \quad (d \in \mathcal{D}^{\text{obj}}).$$

The last condition is equivalent to $f' \cdot m = f$. □

The question of whether a given orthogonality class is a reflective subcategory or not is sometimes called the *orthogonal subcategory problem.* In a locally presentable category the general answer is dependent on set theory, as we prove in Chapter 6, see Corollary 6.24. However, small-orthogonality classes are always "well-behaved", as we prove now: they are reflective and, moreover, as categories they are locally presentable.

Orthogonal-reflection Construction

1.35 Definition. A *λ-orthogonality class* is a class of the form $\mathcal{M}^\perp$ such that every morphism in $\mathcal{M}$ has a λ-presentable domain and a λ-presentable codomain.

Remark. In a locally presentable category (a) each λ-orthogonality class is a small-orthogonality class and (b) for each small-orthogonality class there exists a regular cardinal λ such that this is a λ-orthogonality class. This follows from the fact that each object is presentable, and that for each regular cardinal λ there exists by Remark 1.19 essentially a set of λ-presentable objects.

Proposition. *Each λ-orthogonality class is closed under λ-directed colimits.*

PROOF. Let $\mathcal{M} = \{A_i \xrightarrow{m_i} A_i'\}_{i \in I}$ be a set of morphisms in a category $\mathcal{K}$ such that each A_i and A_i' $(i \in I)$ is λ-presentable. It follows that $\mathcal{M}^{\perp}$ is closed under λ-directed colimits. In fact, let $(K_t \xrightarrow{k_t} K)_{t \in T}$ be a colimit of a λ-directed diagram in $\mathcal{M}^{\perp}$. For each morphism $f\colon A_i \to K$ there exists $t \in T$ and a factorization $f = k_t \cdot f_1$ (since A_i is λ-presentable), and there exists $f_1'\colon A_i' \to K_t$ with $f = f_1' \cdot m_i$ (since $K_t \in \mathcal{M}^{\perp}$). Thus $f' = k_t \cdot f_1'\colon A_i' \to K$ fulfils $f = f' \cdot m_i$. To prove that f' is unique, let $f''\colon A_i' \to K$ also fulfil $f = f'' \cdot m_i$. Then there exists $t_0 \in T$ such that both f' and f'' factorize through k_{t_0}, say, $f' = k_{t_0} \cdot f_1$ and $f'' = k_{t_0} \cdot f_2$. Since A_i' is λ-presentable and since $f_1, f_2\colon A_i' \to K_{t_0}$ fulfil $k_{t_0} \cdot f_1 = k_{t_0} \cdot f_2$, there exists $t \geq t_0$ such that the connecting arrow $k_{t_0,t}\colon K_{t_0} \to K_t$ satisfies $k_{t_0,t} \cdot f_1 = k_{t_0,t} \cdot f_2$. Then $f' = k_t \cdot k_{t_0,t} \cdot f_1 = f''$. □

1.36 Orthogonal Reflection. One of the crucial results for the development of the theory of locally presentable categories is that every small-orthogonality class in a locally presentable category $\mathcal{K}$ is a reflective subcategory of $\mathcal{K}$. This can be proved abstractly, e.g., using the concept of pure subobject—we will exhibit such a proof in Chapter 2 (Theorem 2.48). In the present chapter we exhibit a constructive proof which, although less elegant than the abstract one, has the advantage of making the structure of the desired reflection quite lucid. If $\mathcal{M}$ consists of a single arrow $A \xrightarrow{m} A'$ then for each object X we could proceed as follows:

(1) given a morphism $f\colon A \to X$ which does not factorize through m, form a pushout

$$\begin{array}{ccc} A & \xrightarrow{m} & A' \\ {\scriptstyle f}\downarrow & & \downarrow{\scriptstyle f'} \\ X & \xrightarrow[r_0]{} & X' \end{array}$$

and consider r_0 as the initial step of the desired reflection (i.e., in further steps work with X' in place of X);

(2) given a morphism $f\colon A \to X$ which factorizes non-uniquely through m, say, $f = p \cdot m = q \cdot m$, form the coequalizer $r_0\colon X \to X'$ of p, q and consider it as the initial step of the desired reflection.

The idea of the construction below is an iterative application of such steps. We proceed by transfinite induction; the limit steps are performed by forming the appropriate colimit. If $\mathcal{K}$ is locally λ-presentable and if A and A' are λ-presentable objects, we obtain the desired reflection in λ steps.

1.37 Orthogonal-reflection Construction. Let $\mathcal{K}$ be a cocomplete category, and let $\mathcal{M}$ be a set of morphisms of $\mathcal{K}$. For each object K of $\mathcal{K}$ we form a chain $x_{i,j}\colon X_i \to X_j$ $(i, j$ ordinals, $i \leq j)$ by transfinite induction as follows:

I. First step: $X_0 = K$.

II. Isolated step: Given X_i, form a diagram

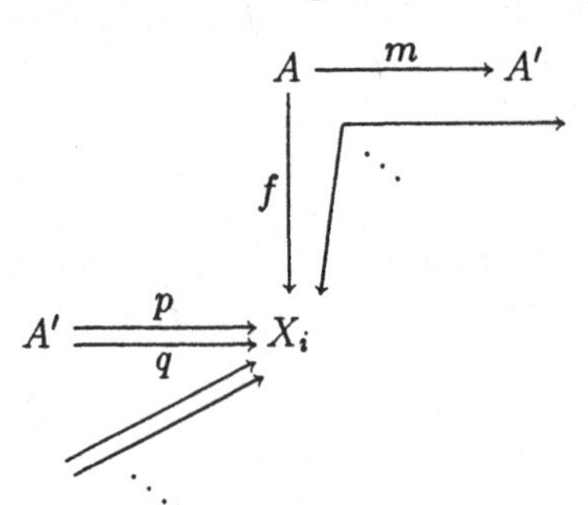

indexed by (i) all spans $X_i \xleftarrow{f} A \xrightarrow{m} A'$ with $m \in \mathcal{M}$ and (ii) all pairs $p, q\colon A' \to X_i$ of morphisms for which there exists $m\colon A \to A'$ in $\mathcal{M}$ with $p \cdot m = q \cdot m$. Denote by X_{i+1} the colimit object of a colimit of that diagram; the colimit maps are denoted as follows:

$$\begin{array}{ccc} A & \xrightarrow{\quad m \quad} & A' \\ {\scriptstyle f}\big\downarrow & & \big\downarrow{\scriptstyle f'} \\ X_i & \xrightarrow[x_{i,i+1}]{} & X_{i+1} \end{array} \qquad \text{for each } X_i \xleftarrow{f} A \xrightarrow{m} A',\ m \in \mathcal{M}.$$

Thus, $x_{i,i+1}$ is universal w.r.t. (i) commutative squares for all the spans above and (ii) the property that

$$p \cdot m = q \cdot m \quad \text{implies} \quad x_{i,i+1} \cdot p = x_{i,i+1} \cdot q$$

(for all $m \in \mathcal{M}$ and all parallel pairs p, q with the codomain X_i).

III. Limit step: Form a colimit X_i of the previously defined chain and call the colimit maps $x_{j,i}: X_j \to X_i \quad (j < i)$.

Proposition. *The above construction stops after i_0 steps (i.e., x_{i_0,i_0+1} is an isomorphism) iff the object X_{i_0} is orthogonal to $\mathcal{M}$. Then $x_{0,i_0}: K \to X_{i_0}$ is a reflection of K in $\mathcal{M}^\perp$.*

PROOF. (1) If x_{i_0,i_0+1} is an isomorphism, then $X_{i_0} \in \mathcal{M}^\perp$. In fact, for each $m: A \to A'$ in $\mathcal{M}$ and each $f: A \to X_{i_0}$ we have the commutative square above, which yields

$$f = x_{i_0,i_0+1}^{-1} \cdot f' \cdot m.$$

To show that this factorization is unique, consider $f = p \cdot m = q \cdot m$. Then one of the pairs coequalized by $x_{i,i+1}$ is p, q. Since $x_{i,i+1}$ is an isomorphism, we have $p = q$.

(2) If X_{i_0} is orthogonal to $\mathcal{M}$, then the diagram defining X_{i_0+1} has a compatible cocone formed by finding, for each span $X_{i_0} \xleftarrow{f} A \xrightarrow{m} A'$ with $m \in \mathcal{M}$ the unique $f^*: A' \to X_{i_0}$ with $f = f^* \cdot m$. Let $g: X_{i_0+1} \to X_{i_0}$ be the unique morphism with $g \cdot x_{i_0,i_0+1} = id$ and $g \cdot f' = f^*$ for each span above. Then $x_{i_0,i_0+1} \cdot g = id$ because for each span we have $(x_{i_0,i_0+1} \cdot g) \cdot f' = f'$; this follows from the orthogonality of X_{i_0}, since $\left[(x_{i_0,i_0+1} \cdot g) \cdot f'\right] \cdot m = x_{i_0,i_0+1} \cdot f^* \cdot m = x_{i_0,i_0+1} \cdot f = f' \cdot m$. Thus, $x_{i_0,i_0+1} = g^{-1}$.

(3) If the construction stops after i_0 steps, we will show that x_{0,i_0} is a reflection map. Let $h: K \to L$ be a morphism with $L \in \mathcal{M}^\perp$. Then we define a compatible cocone $h_i: X_i \to L$ by the following transfinite induction:

First step: $h_0 = h$.

Isolated step: we construct a compatible cocone of the diagram defining X_{i+1} as follows: for each span $X_i \xleftarrow{f} A \xrightarrow{m} A'$, $m \in \mathcal{M}$, there exists a unique $f^*: A' \to L$ with $h_i \cdot f = f^* \cdot m$, and given $p \cdot m = q \cdot m$, then $(h_i \cdot p) \cdot m = (h_i \cdot q) \cdot m$ implies $h_i \cdot p = h_i \cdot q$. Thus, there exists a unique $h_{i+1}: X_{i+1} \to L$ with $h_{i+1} \cdot x_{i,i+1} = h_i$ and $h_{i+1} \cdot f' = f^*$ for each span as above.

Limit step: obvious.

Thus, if the construction stops after i_0 steps, then each morphism $h: K \to L$ with $L \in \mathcal{M}^\perp$ factorizes as $h = h_{i_0} \cdot x_{0,i_0}$. The uniqueness of the factorization is easy to verify: if $h = h' \cdot x_{0,i_0}$, then $h_i = h' \cdot x_{i,i_0}$ for all $i \leq i_0$ (which follows by transfinite induction on i). □

1.38 Theorem. *For each λ-orthogonality class in a cocomplete category the reflection construction always stops in λ steps.*

PROOF. We will prove that X_λ lies in $\mathcal{M}^\perp$. In fact, for each morphism $m: A \to A'$ in $\mathcal{M}$, since A is λ-presentable, all morphisms $f: A \to X_\lambda$ factorize through m. This follows from the existence of a factorization $f = x_{i,\lambda} \cdot \overline{f}$ for some $i < \lambda$: we have $x_{i,i+1} \cdot \overline{f} = \overline{f}' \cdot m$, thus

$$f = x_{i+1,\lambda} \cdot x_{i,i+1} \cdot \overline{f} = (x_{i+1,\lambda} \cdot \overline{f}') \cdot m.$$

To verify that the factorization is unique, consider $p, q: A' \to X_\lambda$ with $f = p \cdot m = q \cdot m$. Since A' is λ-presentable, there exists $j < \lambda$ such that p, q both factorize through $x_{j,\lambda}$, say, $p = x_{j,\lambda} \cdot \overline{p}$ and $q = x_{j,\lambda} \cdot \overline{q}$. Then $f = x_{j,\lambda} \cdot \overline{p} \cdot m = x_{j,\lambda} \cdot \overline{q} \cdot m$ which implies (since A is λ-presentable) that there exists $j_1 > j$ with $x_{j,j_1} \cdot \overline{p} \cdot m = x_{j,j_1} \cdot \overline{q} \cdot m$. Since m equalizes $x_{j,j_1} \cdot \overline{p}$ and $x_{j,j_1} \cdot \overline{q}$, it follows that x_{j_1,j_1+1} coequalizes $x_{j,j_1} \cdot \overline{p}$ and $x_{j,j_1} \cdot \overline{q}$; hence $x_{j_1,\lambda}$ also coequalizes them. In other words, $p = q$. Consequently, $X_\lambda \in \mathcal{M}^\perp$. Now apply Proposition 1.37. □

1.39 Theorem. *Let $\mathcal{K}$ be a locally λ-presentable category. The following conditions on a full subcategory $\mathcal{A}$ of $\mathcal{K}$ are equivalent:*

(i) *$\mathcal{A}$ is a λ-orthogonality class in $\mathcal{K}$;*

(ii) *$\mathcal{A}$ is a reflective subcategory of $\mathcal{K}$ closed under λ-directed colimits.*

Furthermore, they imply that $\mathcal{A}$ is locally λ-presentable.

Remark. In the proof we will see that (ii) implies that every object λ-presentable in $\mathcal{K}$ has a reflection $r_K: K \to K^*$ such that K^* is λ-presentable in $\mathcal{A}$ and

$$\mathcal{A} = \{ r_K \mid K \in \mathbf{Pres}_\lambda \mathcal{K} \}^\perp.$$

PROOF. (i) ⇒ (ii) by Proposition 1.35 and Theorem 1.38.

(ii) ⇒ (i). For each λ-presentable object K of $\mathcal{K}$ let $r_K: K \to K^*$ be a reflection. Then K^* is λ-presentable in $\mathcal{A}$ because for each λ-directed colimit $(A_i \xrightarrow{a_i} A)$ in $\mathcal{A}$ and each morphism $f: K^* \to A$, since $\mathcal{A}$ is closed under λ-directed colimits in $\mathcal{K}$, the morphism $f \cdot r_K$ factorizes as $f \cdot r_K = a_i \cdot f'$ for an (essentially unique) i and $f': K \to A_i$ factorizes (uniquely) as $f' = f'' \cdot r_K$. The collection $\mathcal{M} = \{ r_K \mid K \in \mathbf{Pres}_\lambda \mathcal{K} \}$ is small, and $\mathcal{A} = \mathcal{M}^\perp$. In fact, $\mathcal{A} \subseteq \mathcal{M}^\perp$ is clear. Conversely, given an $\mathcal{M}$-orthogonal object K, express K as a λ-directed colimit of λ-presentable objects $(K_i \xrightarrow{k_i} K)_{i \in I}$. For each i we have a reflection $r_i: K_i \to K_i^*$ in $\mathcal{A}$, and since $K \in \mathcal{M}^\perp$ implies that $\mathcal{K}$ is orthogonal to r_i, there is a unique $k_i^*: K_i^* \to K$ with $k_i = k_i^* \cdot r_i$. It is easy to see that the diagram in $\mathcal{A}$ obtained by reflections from the above λ-directed diagram has a colimit

$(K_i^* \xrightarrow{k_i^*} K)_{i \in I}$. Since $\mathcal{A}$ is closed under λ-directed colimits, we conclude that K lies in $\mathcal{A}$.

Furthermore, (ii) implies that $\mathcal{A}$ is locally λ-presentable: the reflections of λ-presentable objects of $\mathcal{K}$ in $\mathcal{A}$ form an (essentially small) collection of λ-presentable objects in $\mathcal{A}$, and each object of $\mathcal{A}$ is a λ-directed colimit of these reflections. □

1.40 Corollary. *Every small-orthogonality class of a locally presentable category is locally presentable.*

1.41 Example. For every small category $\mathcal{A}$, there is a binary signature Σ such that the category $\mathbf{Set}^{\mathcal{A}}$ is equivalent to an ω-orthogonality class in $\mathbf{Rel}\,\Sigma$ (thus, to a full, reflective subcategory of $\mathbf{Rel}\,\Sigma$, closed under directed colimits).

In fact, let $\mathcal{A}$-objects be sorts and $\mathcal{A}$-morphisms be relation symbols, i.e.,

$$S = \mathcal{A}^{\mathrm{obj}} \quad \text{and} \quad \Sigma = \mathcal{A}^{\mathrm{mor}},$$

where the arity of each $f\colon a \to b$ in Σ is $a \times b$. We will find a set $\mathcal{M}$ of morphisms with finitely presentable domains and codomains in $\mathbf{Rel}\,\Sigma$ such that $\mathcal{M}^{\perp}$ is equivalent to $\mathbf{Set}^{\mathcal{A}}$. For each functor $F\colon \mathcal{A} \to \mathbf{Set}$ consider the Σ-structure whose a-sorted underlying set is Fa ($a \in \mathcal{A}^{\mathrm{obj}}$) and whose relation corresponding to $f\colon a \to b$ is the subset of $Fa \times Fb$ which is the graph of $Ff\colon Fa \to Fb$. Under this identification, $\mathbf{Set}^{\mathcal{A}}$ becomes a full subcategory of $\mathbf{Rel}\,\Sigma$. To define the appropriate set $\mathcal{M}$, we introduce the following simple Σ-structures:

P_a, where $a \in \mathcal{A}^{\mathrm{obj}}$, has all underlying sets empty except the a-sort set $\{0\}$, and all relations are empty.

Q_f, where $f\colon a \to b$ is a morphism, has all underlying sets empty except the a-sort set $\{0\}$ and the b-sort set $\{1\}$ (in the case $a \neq b$), or the a-sort set $\{0, 1\}$ (in the case $a = b$), and all relations are empty except f which consists of $(0, 1)$ alone.

Now $\mathbf{Set}^{\mathcal{A}}$ is presented in $\mathbf{Rel}\,\Sigma$ by orthogonality to the following morphisms (with finitely presentable domains and codomains):

(1) for each morphism $f\colon a \to b$ of $\mathcal{A}$ the inclusion

$$P_a \hookrightarrow Q_f$$

(which guarantees that the relations are graphs of functions);

(2) for each object a of $\mathcal{A}$ the quotient morphism

$$Q_{id_a} \to Q_{id_a}/\sim$$

where $0 \sim 1$ (which guarantees the preservation of identity maps);

(3) for each commutative triangle

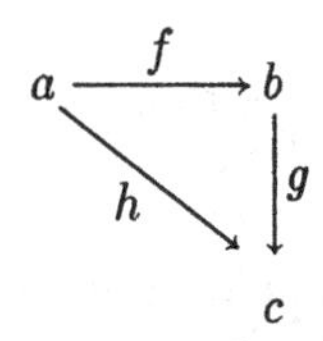

in $\mathcal{A}$ the inclusion

$$(Q_f + Q_g)/\sim \hookrightarrow (Q_f + Q_g + Q_h)/\approx$$

where $\sim$ is the smallest equivalence merging 0 in Q_g with 1 in Q_f, and $\approx$ is the smallest equivalence merging

0 in Q_f with 0 in Q_h,

0 in Q_g with 1 in Q_f, and

1 in Q_g with 1 in Q_h.

(This guarantees the preservation of composition.)

Free Cocompletions

Recall that a diagram whose scheme has less than λ morphisms is called λ-small.

1.42 Notation. For each small category $\mathcal{A}$ and each regular cardinal λ we denote by

$$\mathbf{Cont}_\lambda\, \mathcal{A}$$

the category of all set-valued functors on $\mathcal{A}$ preserving all λ-small limits existing in $\mathcal{A}$. (This is a full subcategory of $\mathbf{Set}^{\mathcal{A}}$.)

1.43 Remark. Since hom-functors preserve limits, we have a restriction of the Yoneda embedding (see 0.9)

$$Y : \mathcal{A} \to \mathbf{Cont}_\lambda\, \mathcal{A}^{\mathrm{op}}.$$

It is clear that Y is a full embedding. We are going to show that this extension of $\mathcal{A}$ is a free cocompletion preserving λ-small colimits.

A functor is called *cocontinuous* if it preserves (small) colimits.

1.44 Definition. Let $\mathcal{A}$ be a category.

(1) By a *free cocompletion* of $\mathcal{A}$ is meant a full embedding $E\colon \mathcal{A} \to \mathcal{A}^*$ such that

a. $\mathcal{A}^*$ is a cocomplete category,

b. every functor $F\colon \mathcal{A} \to \mathcal{B}$ with $\mathcal{B}$ cocomplete has an extension to a cocontinuous functor $F^*\colon \mathcal{A}^* \to \mathcal{B}$, unique up to natural isomorphism.

(2) Let λ be a regular cardinal. By a *λ-free cocompletion* of $\mathcal{A}$ is meant a full embedding $E\colon \mathcal{A} \to \mathcal{A}^*$ such that

a. $\mathcal{A}^*$ is a cocomplete category, and E preserves λ-small colimits (which exist) in $\mathcal{A}$,

b. every functor $F\colon \mathcal{A} \to \mathcal{B}$ with $\mathcal{B}$ cocomplete which preserves λ-small colimits has an extension to a cocontinuous functor

$$F^*: \mathcal{A}^* \to \mathcal{B},$$

unique up to natural isomorphism.

1.45 Proposition. *Let $\mathcal{A}$ be a small category.*

(i) *The Yoneda embedding $Y: \mathcal{A} \to \mathbf{Set}^{\mathcal{A}^{\mathrm{op}}}$ is a free cocompletion of $\mathcal{A}$.*

(ii) *For each regular cardinal λ, the Yoneda embedding $Y: \mathcal{A} \to \mathbf{Cont}_\lambda\, \mathcal{A}^{\mathrm{op}}$ is a λ-free cocompletion of $\mathcal{A}$.*

Remark. Explicitly, (i) states that for each functor $F\colon \mathcal{A} \to \mathcal{B}$, with $\mathcal{B}$ cocomplete, there exists a cocontinuous functor $F^*\colon \mathbf{Set}^{\mathcal{A}^{\mathrm{op}}} \to \mathcal{B}$ with $F = F^* \cdot Y$. We call F^* (determined up to a natural isomorphism) a *left Kan extension* of the functor F.

PROOF. (i). Let $F\colon \mathcal{A} \to \mathcal{B}$ be a functor with $\mathcal{B}$ cocomplete. We extend F to $F^*: \mathbf{Set}^{\mathcal{A}^{\mathrm{op}}} \to \mathcal{B}$ as follows. For each object H of $\mathbf{Set}^{\mathcal{A}^{\mathrm{op}}}$ we denote by $\mathcal{D}_H$ the category of all pairs (A, a) where A is an object of $\mathcal{A}$ and $a \in HA$. The morphisms $f\colon (A, a) \to (A', a')$ of $\mathcal{D}_H$ are those morphisms $f\colon A \to A'$ of $\mathcal{A}$ which fulfil $(Hf)(a') = a$. The forgetful functor $D_H : \mathcal{D}_H \to \mathcal{A}$ defines a diagram in $\mathcal{A}$, and we put

$$F^*H = \operatorname{colim} F \cdot D_H.$$

Every morphism $h\colon H \to H'$ in $\mathbf{Set}^{\mathcal{A}^{\mathrm{op}}}$ leads to a compatible cocone of the diagram $F \cdot D_H$ with codomain F^*H'; we define $F^*h\colon F^*H \to F^*H'$ as the unique factorization of that cocone.

(a) F^* is cocontinuous. In fact, we prove more: F^* is a left adjoint whose right adjoint is the functor $F_*: \mathcal{B} \to \mathbf{Set}^{\mathcal{A}^{\mathrm{op}}}$ defined by

$$\begin{aligned} F_*B &= \hom(-, B) \cdot F^{\mathrm{op}} : \mathcal{A}^{\mathrm{op}} \to \mathbf{Set} && \text{(for objects } B) \\ F_*b &= \hom(-, b) \cdot F^{\mathrm{op}} : F_*B \to F_*B' && \text{(for morphisms} \\ &&& b: B \to B'). \end{aligned}$$

It is easy to verify that for each object B of $\mathcal{B}$ and each functor

$$H: \mathcal{A}^{\mathrm{op}} \to \mathbf{Set}$$

we have a bijection

$$\frac{F^*H \to B}{H \to F_*B}$$

(natural in H and B) since each of these arrows expresses a collection of morphisms $d_{A,a}: FA \to B$ ($A \in \mathcal{A}^{\mathrm{obj}}$, $a \in HA$) such that for each $\mathcal{A}$-morphism $f: A' \to A$ we have $d_{A,a} = d_{A',Hf(a)} \cdot Ff$.

(b) F^* is unique up to natural isomorphism. To verify this, we use the fact that each functor $H : \mathcal{A}^{\mathrm{op}} \to \mathbf{Set}$ is a canonical colimit of hom-functors. Since the canonical diagram of H w.r.t. hom-functors is $Y \cdot D_H : \mathcal{D}_H \to \mathbf{Set}^{\mathcal{A}^{\mathrm{op}}}$, we have, for each cocontinuous functor F^+ extending F (i.e., with $F = F^+ \cdot Y$),

$$F^+H \cong F^+(\operatorname{colim} Y \cdot D_H) \cong \operatorname{colim} F^+ \cdot Y \cdot D_H = \operatorname{colim} F \cdot D_H = F^*H.$$

(ii). The proof is analogous to (i), we just have to verify that Y preserves λ-small colimits, and that if F preserves λ-small colimits, then F_* maps $\mathcal{B}$ into $\mathbf{Cont}_\lambda\, \mathcal{A}^{\mathrm{op}}$. The latter is obvious: $F^{\mathrm{op}}: \mathcal{A}^{\mathrm{op}} \to \mathcal{B}^{\mathrm{op}}$ preserves λ-small limits, thus, so does $F_*B = \hom(-, B) \cdot F^{\mathrm{op}}$. To prove the former, let $D: \mathcal{D} \to \mathcal{A}$ be a λ-small diagram with a colimit $(Dd \xrightarrow{c_d} C)_{d \in \mathcal{D}^{\mathrm{obj}}}$ in $\mathcal{A}$. Let $f_d: \hom(-, Dd) \to H$ ($d \in \mathcal{D}^{\mathrm{obj}}$) be a compatible cocone in $\mathbf{Cont}_\lambda\, \mathcal{A}^{\mathrm{op}}$. Since H preserves λ-small limits in $\mathcal{A}^{\mathrm{op}}$, we know that $(HC \xrightarrow{Hc_d} HDd)$ is a limit of $H \cdot D^{\mathrm{op}}$ in $\mathbf{Set}$. The elements $x_d = (f_d)_{Dd}(id)$ form a compatible collection of $H \cdot D^{\mathrm{op}}$, thus, there exists a unique $x \in HC$ with $x_d = Hc_d(x)$ for each d. In other words, there exists a unique $f: YC \to H$ with $f_d = f \cdot Yc_d$. Thus, $(YDd \xrightarrow{Yc_d} YC)$ is a colimit of $Y \cdot D$ in $\mathbf{Cont}_\lambda\, \mathcal{A}^{\mathrm{op}}$. □

1.46 Representation Theorem. *Let λ be a regular cardinal. For each category $\mathcal{K}$ the following statements are equivalent:*

(i) *$\mathcal{K}$ is locally λ-presentable;*

(ii) $\mathcal{K}$ *is equivalent to* $\mathbf{Cont}_\lambda\, \mathcal{A}$ *for some small category* $\mathcal{A}$;

(iii) $\mathcal{K}$ *is equivalent to a* λ*-orthogonality class in* $\mathbf{Set}^{\mathcal{A}}$ *for some small category* $\mathcal{A}$;

(iv) $\mathcal{K}$ *is equivalent to a full, reflective subcategory of* $\mathbf{Set}^{\mathcal{A}}$ *closed under* λ*-directed colimits for some small category* $\mathcal{A}$;

(v) $\mathcal{K}$ *is a* λ*-free cocompletion of a small category* $\mathcal{A}$.

Remark. The category $\mathcal{A}$ in conditions (ii)–(v) can be chosen as the dual of $\mathbf{Pres}_\lambda\, \mathcal{K}$. (Thus, (v) generalizes the well-known fact that every algebraic lattice is a free completion of its subposet of all finite elements.)

Consequently, the category

$$\mathbf{c}_\lambda\text{-}\mathbf{Cat}$$

of all small categories with λ-small limits and functors preserving λ-small limits is dually equivalent to the quasicategory (see 0.13)

$$\mathbf{lp}_\lambda\text{-}\mathbf{CAT}$$

of all locally λ-presentable categories and functors preserving limits and λ-directed colimits. In fact, the functor $R\colon (\mathbf{c}_\lambda\text{-}\mathbf{Cat})^{\mathrm{op}} \to \mathbf{lp}_\lambda\text{-}\mathbf{CAT}$ defined by

$$R\mathcal{A} = \mathbf{Cont}_\lambda\, \mathcal{A}^{\mathrm{op}} \quad \text{and} \quad RH = F \mapsto F \cdot H^{\mathrm{op}}$$

is an equivalence (see Exercise 1.s for details).

PROOF. (i) $\Rightarrow$ (ii). Let $\mathcal{K}$ be a locally λ-presentable category, and let $\mathcal{A} = \mathbf{Pres}_\lambda\, \mathcal{K}$. Then $\mathcal{A}$ is dense (see Proposition 1.22), and, by Proposition 1.26 the canonical functor $E\colon \mathcal{K} \to \mathbf{Set}^{\mathcal{A}^{\mathrm{op}}}$ preserves λ-directed colimits, and $\mathcal{K}$ is equivalent to $E(\mathcal{K})$. Let $\mathcal{K}'$ be the category of all functors in $\mathbf{Set}^{\mathcal{A}^{\mathrm{op}}}$ naturally isomorphic to some EK, $K \in \mathcal{K}^{\mathrm{obj}}$. We will prove that $\mathcal{K}'$, which is equivalent to $\mathcal{K}$, is precisely $\mathbf{Cont}_\lambda\, \mathcal{A}^{\mathrm{op}}$.

Every functor in $\mathcal{K}'$ preserves λ-small limits. In fact, for every object K the functor $EK = \hom(-, K)/_{\mathcal{A}^{\mathrm{op}}}$ preserves λ-small limits in $\mathcal{A}^{\mathrm{op}}$ because, by Proposition 1.16, $\mathcal{A}^{\mathrm{op}}$ is closed under λ-small limits in $\mathcal{K}^{\mathrm{op}}$ (and $\hom(-, K)$ preserves limits in $\mathcal{K}^{\mathrm{op}}$).

Conversely, we prove that if a functor $H\colon \mathcal{A}^{\mathrm{op}} \to \mathbf{Set}$ preserves λ-small limits, then it is a λ-filtered colimit of functors in $\mathcal{K}$; then H lies in $\mathcal{K}'$ because $\mathcal{K}'$ is closed under λ-filtered colimits (since E preserves λ-directed colimits, thus (by Remark 1.21), λ-filtered colimits). The Yoneda embedding $Y\colon \mathcal{A} \to \mathbf{Set}^{\mathcal{A}^{\mathrm{op}}}$ fulfils $Y(\mathcal{A}) \subseteq \mathcal{K}'$, and H is a canonical colimit of objects of $Y(\mathcal{A})$. It remains to prove that this colimit is λ-filtered. By the Yoneda lemma (see 0.9), each λ-small subcategory of $Y(\mathcal{A}) \downarrow H$ has the form $YD(\mathcal{D}) \downarrow H$ for some λ-small diagram $D\colon \mathcal{D} \to \mathcal{A}$. If $(Dd \xrightarrow{c_d} C)$ is a colimit of D in $\mathcal{A}$, then $(HC \xrightarrow{Hc_d} HDd)$ is a limit of $HD^{\mathrm{op}}\colon \mathcal{D}^{\mathrm{op}} \to \mathcal{A}^{\mathrm{op}}$ in **Set**. Thus, the elements $x_d \in HDd$ representing the given maps $\hom(-, Dd) \to H$ lead to a unique $x \in HC$ with $Hc_d(x) = x_d$. Then the subcategory $YD(\mathcal{D}) \downarrow H$ has a compatible cocone with the codomain $YC \to H$ (given by $x \in HC$) formed by c_d. Thus, $Y(\mathcal{A}) \downarrow H$ is λ-filtered.

(ii) $\Rightarrow$ (iii). In Example 1.33(8) we have seen how $\mathbf{Cont}_\lambda\, \mathcal{A}^{\mathrm{op}}$ can be presented as the orthogonality class $\mathcal{M}^\perp$ in $\mathbf{Set}^{\mathcal{A}^{\mathrm{op}}}$, where $\mathcal{M}$ consists of the canonical natural transformations

$$m\colon \hom(\lim_d Dd, -) \to \lim_d[\hom(Dd, -)]$$

for λ-small diagrams D in $\mathcal{A}^{\mathrm{op}}$. Since the domain of each m is finitely presentable (see Example 1.2(7)) and the codomain is λ-presentable (see Proposition 1.16), it follows that $\mathcal{M}^\perp$ is a λ-orthogonality class in $\mathbf{Set}^{\mathcal{A}^{\mathrm{op}}}$.

(iii) $\Rightarrow$ (iv) $\Rightarrow$ (i). See Theorem 1.39.

(v) and (ii) are equivalent by Proposition 1.45. □

1.47 Corollary. *A category is locally λ-presentable iff it is equivalent to a full, reflective subcategory of* $\mathbf{Rel}\,\Sigma$ *(for some finitary signature Σ) closed under λ-directed colimits.*

This follows from the representation theorem and Example 1.41. □

1.48 Example. The category **Ban** of (complex) Banach spaces and linear contractions is locally ω_1-presentable.

To verify this, we introduce a larger category **Tot** of *totally convex spaces*. A totally convex space is an ω-ary algebra A (i.e., an underlying set $|A|$ together with operations which assign results to sequences in $|A|$) whose operations are indexed by all sequences $(\alpha_n)_{n=0}^\infty$ of complex numbers satisfying $\sum_{n=0}^\infty |\alpha_n| \le 1$, and which are denoted by

$$(x_n)_{n=0}^\infty \mapsto \sum_{n=0}^\infty \alpha_n x_n \qquad \text{for sequences } (x_n)_{n=0}^\infty \text{ in } A,$$

subject to the following equations:

$$\text{(1)} \qquad \sum_{n=0}^\infty \delta_n^k x_n = x_k \quad \text{where } \delta_k^k = 1 \text{ and } \delta_n^k = 0 \text{ for all } n \ne k;$$

$$\text{(2)}\qquad \sum_{n=0}^{\infty}\alpha_n\Big(\sum_{k=0}^{\infty}\beta_{kn}x_k\Big)=\sum_{k=0}^{\infty}\Big(\sum_{n=0}^{\infty}\alpha_n\beta_{kn}\Big)x_k.$$

A homomorphism $f\colon X \to Y$ of totally convex spaces is a function satisfying $f(\sum \alpha_n x_n) = \sum \alpha_n f(x_n)$ for each (α_n) and each (x_n). The category **Tot** of totally convex spaces and homomorphisms is a variety of ω-ary operations; we will see in Chapter 3 that this implies that **Tot** is locally ω_1-presentable.

Ban is equivalent to a full subcategory of **Tot**: the unit ball $U(B)$ of each Banach space B is a totally convex space. That subcategory is reflective since, given a totally convex space A, we have the following equivalence on the set $\mathbb{C} \times A$, where $\mathbb{C}$ is the set of all complex numbers:

$$(\alpha, x) \sim (\alpha', x') \quad \text{iff there exists } \beta > \max(|\alpha|, |\alpha'|) \text{ with } \frac{\alpha}{\beta}x = \frac{\alpha'}{\beta}x'.$$

Then the quotient algebra $(\mathbb{C} \times A)/\!\sim$, together with the function $A \to (\mathbb{C} \times A)/\!\sim$ assigning to each a the equivalence class of $(1, a)$, is a reflection of A in the subcategory $U(\mathbf{Ban})$. The subcategory $U(\mathbf{Ban})$ is clearly closed under ω_1-directed colimits. Thus, **Ban** is a locally ω_1-presentable category by Theorem 1.39.

Limit Sketches

1.49. We have seen above that locally presentable categories can be described as categories of set-valued functors preserving λ-small limits. We will now show that, more generally, set-valued functors preserving specified limits (or, still more generally, turning specified cones to limits) always form a locally presentable category. Such categories are known as categories of models of limit sketches. Let us elaborate on this concept by specifying a collection of cones and working with set-valued functors turning the specified cones into limits.

Definition

(1) By a *limit sketch* is meant a triple $\mathscr{S} = (\mathcal{A}, \mathbf{L}, \sigma)$ consisting of a small category $\mathcal{A}$, a set $\mathbf{L}$ of diagrams in $\mathcal{A}$, and an assignment σ of a compatible cone $\sigma(D)$ in $\mathcal{A}$ to each diagram $D \in \mathbf{L}$.

(2) By a *model* of the limit sketch $\mathscr{S}$ is meant a functor $F\colon \mathcal{A} \to \mathbf{Set}$ such that for each diagram D in $\mathbf{L}$ the F-image of the cone $\sigma(D)$ is a limit of $F \cdot D$ in **Set**. The full subcategory of $\mathbf{Set}^{\mathcal{A}}$ consisting of models of $\mathscr{S}$ is denoted by

$$\mathbf{Mod}\,\mathscr{S}.$$

1.50 Examples

(1) $\mathbf{Cont}_\lambda \mathcal{A}$ is the special case of $\mathbf{Mod}\,\mathscr{S}$ for the limit sketch $\mathscr{S}$ where $\mathbf{L}$ is a set of diagrams representing all λ-small diagrams in $\mathcal{A}$, and σ assigns to each diagram a limit cone.

(2) The category $\mathbf{Alg}\,2$ of algebras on one binary operation $(o\colon X \times X \to X)$ can be expressed as the category of models of the following sketch $\mathscr{S}$:

$\mathcal{A}$ is the category of two objects a and a^2 and, besides the identity morphisms, precisely three morphisms $o, p_1, p_2\colon a^2 \to a$;

$\mathbf{L}$ is the singleton set containing the discrete diagram which consists of two copies of a;

σ is the following cone:

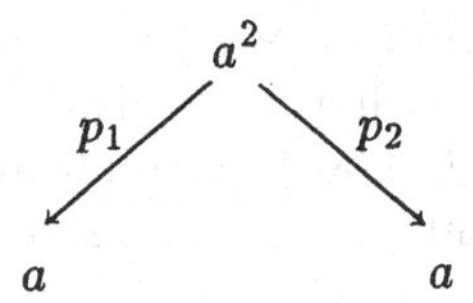

We have $\mathbf{Alg}\,2 \approx \mathbf{Mod}\,\mathscr{S}$. In fact, a model of the sketch $\mathscr{S}$ is a functor $F\colon \mathcal{A} \to \mathbf{Set}$ such that $Fa^2 = Fa \times Fa$ with projections Fp_1, Fp_2. Thus, a model is given by a set $X = Fa$ and a binary operation $Fo\colon X \times X \to X$. Morphisms in $\mathbf{Mod}\,\mathscr{S}$ are precisely the usual algebraic homomorphisms.

(3) To express the category of *commutative* binary algebras (i.e., the subcategory of $\mathbf{Alg}\,2$ given by the equation $o(x,y) = o(y,x)$), we extend the above category $\mathcal{A}$ by a new morphism $s\colon a^2 \to a^2$ with the following composition

$$s \cdot s = id$$
$$p_1 \cdot s = p_2$$
$$p_2 \cdot s = p_1$$
$$o \cdot s = o.$$

(4) Associativity can also be expressed by a finite-product sketch, see Exercise 1.r.

(5) The category $\mathbf{Gra}$ of graphs is the model category of the following sketch $\mathscr{S}$:

$\mathcal{A}$ has three objects a, a^2, r and, besides the identity morphisms, three morphisms: $p_1, p_2\colon a^2 \to a$ (projections) and $i\colon r \to a^2$ (expressing the inclusion of the binary relations).

L contains two diagrams: the discrete diagram consisting of two copies of a, and the span $r \xrightarrow{i} a^2 \xleftarrow{i} r$.

σ assigns the following cones:

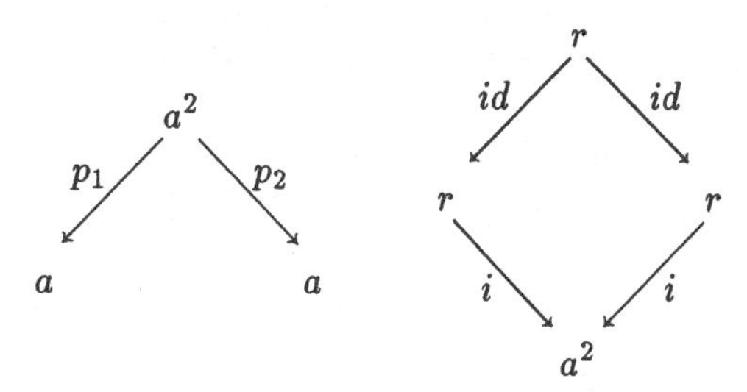

A model of this sketch $\mathscr{S}$ is a functor $F\colon \mathcal{A} \to \mathbf{Set}$ such that $Fa^2 = Fa \times Fa$ (with projections Fp_i) and Fi is a subobject of $Fa \times Fa$, since the square

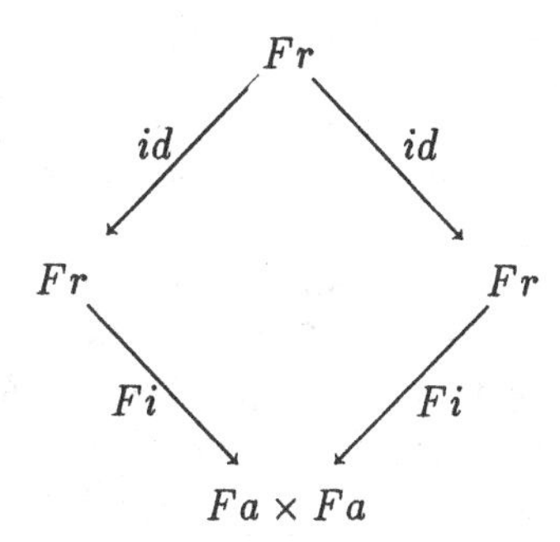

is a pullback. Thus, a model is given by a set $X = Fa$ and a subset $\sigma \subseteq X \times X$ representing the subobject $Fi\colon Fr \to X \times X$. Morphisms in $\mathbf{Mod}\,\mathscr{S}$ are precisely the graph homomorphisms.

(6) The full subcategory of **Gra** consisting of reflexive graphs (i.e., such that the relation contains the diagonal) can be obtained from the previous sketch by adding a new morphism $j\colon a \to r$ to $\mathcal{A}$ satisfying $p_1 \cdot i \cdot j = p_2 \cdot i \cdot j = id_a$.

Analogously, antisymmetry and transitivity can be sketched in order to express **Pos** as a category of models.

1.51 Proposition. *For each limit-sketch $\mathscr{S}$ the category* $\mathbf{Mod}\,\mathscr{S}$ *is locally presentable.*

PROOF. This follows from the fact, established in Example 1.33(8), that $\mathbf{Mod}\,\mathscr{S}$ is an orthogonality class in the (locally finitely presentable) category $\mathbf{Set}^{\mathcal{A}}$. Thus we can apply Theorem 1.39. □

1.52 Corollary. *A category is locally presentable iff it is equivalent to the category* $\mathbf{Mod}\,\mathscr{S}$ *of models of a limit sketch* $\mathscr{S}$.

Remark. In more detail, a category is locally λ-presentable iff it is equivalent to $\mathbf{Mod}\,\mathscr{S}$ for a limit sketch $\mathscr{S} = (\mathcal{A}, \mathbf{L}, \sigma)$ such that all diagrams in $\mathbf{L}$ are λ-small. In fact, if diagrams in $\mathbf{L}$ are λ-small, then $\mathbf{Mod}\,\mathscr{S}$ is, as shown in Example 1.33(8), a λ-orthogonality class of $\mathbf{Set}^{\mathcal{A}}$.

1.53. For each sketch $\mathscr{S} = (\mathcal{A}, \mathbf{L}, \sigma)$ and each category $\mathcal{K}$ we can introduce *models of $\mathscr{S}$ in the category $\mathcal{K}$*: these are functors $F: \mathcal{A} \to \mathcal{K}$ such that for each diagram D in $\mathbf{L}$ the F-image of the cone $\sigma(D)$ is a limit of $F \cdot D$ in $\mathcal{K}$. The full subcategory of $\mathcal{K}^{\mathcal{A}}$ consisting of all models of $\mathscr{S}$ is denoted by

$$\mathbf{Mod}(\mathscr{S}, \mathcal{K}).$$

Proposition. *If $\mathcal{K}$ is a locally λ-presentable category, then* $\mathbf{Mod}(\mathscr{S}, \mathcal{K})$ *is also locally λ-presentable for each limit sketch $\mathscr{S}$ with λ-small diagrams.*

PROOF. By representation theorem 1.46 we can assume that $\mathcal{K} = \mathbf{Cont}_\lambda\, \mathcal{B}$ for a suitable small category $\mathcal{B}$. To prove that $\mathbf{Mod}(\mathscr{S}, \mathcal{K})$ is locally λ-presentable, we observe that it is equivalent to $\mathbf{Mod}\,\mathscr{S}^*$ where $\mathscr{S}^* = (\mathcal{A} \times \mathcal{B}, \mathbf{L}^*, \sigma^*)$ is the following sketch: $\mathbf{L}^*$ consists of all diagrams $D \times D'$ where D ranges through $\mathbf{L}$, D' ranges through a set of representatives of λ-small diagrams in $\mathcal{B}$, and $\sigma^*(D \times D') = \sigma(D^*) \times \lim D'$. □

1.54 Corollary. *For each locally λ-presentable category $\mathcal{K}$ all functor-categories $\mathcal{K}^{\mathcal{A}}$ ($\mathcal{A}$ small) are locally λ-presentable.*

1.55 Examples

(1) Let $\mathbf{2}$ be the two-chain $0 \leq 1$, considered as a category. Then $\mathcal{K}^{\mathbf{2}}$ is the category of $\mathcal{K}$-morphisms. Thus, if $\mathcal{K}$ is locally λ-presentable, then $\mathcal{K}^{\mathbf{2}}$ is locally λ-presentable (and λ-presentable objects of $\mathcal{K}^{\mathbf{2}}$ are, obviously, precisely the $\mathcal{K}$-morphisms with a λ-presentable domain and co-domain).

(2) Consider the limit-sketch $\mathscr{S}$ consisting of the single pullback

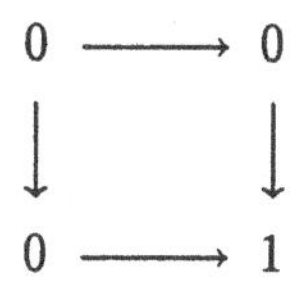

in **2**. Then $\mathbf{Mod}(\mathscr{S}, \mathcal{K})$ is the category of all $\mathcal{K}$-monomorphisms.

1.D Properties of Locally Presentable Categories

1.56 Remark. We know that every locally presentable category is

(1) complete and wellpowered (since it is equivalent to a full, reflective subcategory of the complete and wellpowered category $\mathbf{Set}^{\mathcal{A}}$, see Theorem 1.46),

(2) cocomplete (by definition),

and we are now to going to prove the (non-trivial) fact that it is

(3) co-wellpowered.

The last result can be derived from properties of accessible categories, as will be seen in Chapter 2 (Theorem 2.49), but we now present a direct proof. We first prove the local presentability of all comma-categories $K \downarrow \mathcal{K}$ and $\mathcal{K} \downarrow K$ (see 0.3):

1.57 Proposition. *If $\mathcal{K}$ is a locally λ-presentable category, then for each object K the comma-categories $K \downarrow \mathcal{K}$ and $\mathcal{K} \downarrow K$ are locally λ-presentable.*

PROOF. (1) The comma-category $\mathcal{K} \downarrow K$ is locally λ-presentable by Theorem 1.20: $\mathcal{K} \downarrow K$ is, obviously, cocomplete, and its (directed) colimits are inherited from $\mathcal{K}$. If $\mathcal{A}$ is a strong generator formed by λ-presentable objects of $\mathcal{K}$, then $\bigcup_{A \in \mathcal{A}} \hom(A, K)$ is a strong generator formed by λ-presentable objects of $\mathcal{K} \downarrow K$.

(2) To prove that $K \downarrow \mathcal{K}$ is locally λ-presentable, let us consider $\mathcal{K}$ as a full, reflective subcategory of $\mathbf{Rel}\,\Sigma$ closed under λ-directed colimits (see Corollary 1.47). Then $K \downarrow \mathcal{K}$ is, obviously, a full, reflective subcategory of $K \downarrow \mathbf{Rel}\,\Sigma$ closed under λ-directed colimits; thus, it is sufficient to prove that $K \downarrow \mathbf{Rel}\,\Sigma$ is locally finitely presentable (see Theorem 1.39).

If $|K| = (K_s)_{s \in S}$ are the underlying sets of K, define a signature $\Sigma_1 = \Sigma + \coprod_{s \in S} K_s$ where each $a \in K_s$ is a unary symbol of sort s. Consider the full subcategory $\mathcal{A}$ of $\mathbf{Rel}\,\Sigma_1$ consisting of precisely those Σ_1-structures A such that the relations corresponding to the elements of $|K|$ all have cardinality 1, and the corresponding S-sorted function is a Σ-homomorphism from K to A. Then $\mathcal{A}$ is clearly isomorphic to $K \downarrow \mathbf{Rel}\,\Sigma$, and since $\mathcal{A}$ is closed in $\mathbf{Rel}\,\Sigma_1$ under

(i) products,

(ii) extremal subobjects (thus, $\mathcal{A}$ is epireflective),

(iii) directed colimits,

it follows that $\mathcal{A}$ is locally finitely presentable—see Theorem 1.39. □

1.58 Theorem. *Every locally presentable category is co-wellpowered.*

PROOF. For each object A of a locally presentable category $\mathcal{K}$ we first choose an uncountable regular cardinal λ such that $\mathcal{K}$ is locally λ-presentable and A is a λ-presentable object. We are going to find a set of representatives of all epimorphisms with domain A.

For each morphism $f: A \to B$ in $\mathcal{K}$ we denote by

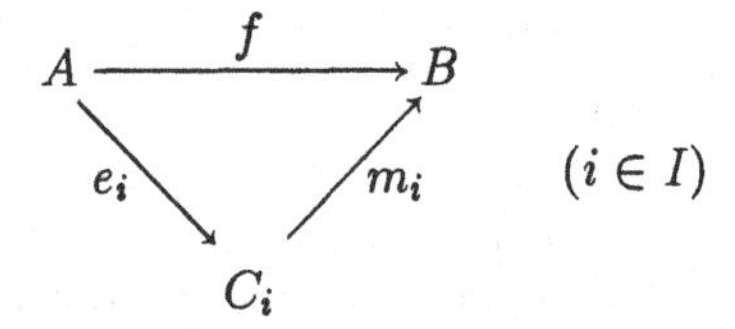

the set of all factorizations of f through an object C_i of $\mathbf{Pres}_\lambda\,\mathcal{K}$. Let $\mathcal{D}(f)$ be the category whose objects are C_i $(i \in I)$ and whose morphisms are those morphisms $h: C_i \to C_j$ for which $e_j = he_i$ and $m_i = m_j h$. Since $\mathbf{Pres}_\lambda\,\mathcal{K}$ is closed under λ-small colimits (see Proposition 1.16), $\mathcal{D}(f)$ is λ-filtered. From the fact that $\mathcal{K}$ is locally λ-presentable it follows that $B = \operatorname{colim}_{\mathcal{D}(f)} C_i$ in $\mathcal{K}$, and since A is λ-presentable, we also have

(1) $$f = \operatorname*{colim}_{\mathcal{D}(f)} e_i \quad \text{in} \quad A \downarrow \mathcal{K}.$$

It is sufficient to prove that for every epimorphism f the full subcategory $\overline{\mathcal{D}(f)}$ of $D(f)$ formed by all factorizations in which e_i is an epimorphism is cofinal in $\mathcal{D}(f)$ (see 0.11). In fact, we then know that

(2) $$f = \operatorname*{colim}_{\overline{\mathcal{D}(f)}} e_i,$$

and it is clear that the collection of all possible categories $\overline{\mathcal{D}(f)}$ is a set: if e_i is an epimorphism, then it determines the factorization $f = m_i \cdot e_i$

uniquely, thus, $\overline{\mathcal{D}(f)}$ can be considered as a subcategory of the (small) comma-category $A \downarrow \mathbf{Pres}_\lambda \mathcal{K}$.

Thus, for each $i \in I$ we want to find $j \in I$ such that e_j is an epimorphism and a morphism $C_i \to C_j$ exists in $\mathcal{D}(f)$ (see Exercise 1.o(3)). First, let P_i denote a pushout of e_i, e_i, and P a pushout of f, f:

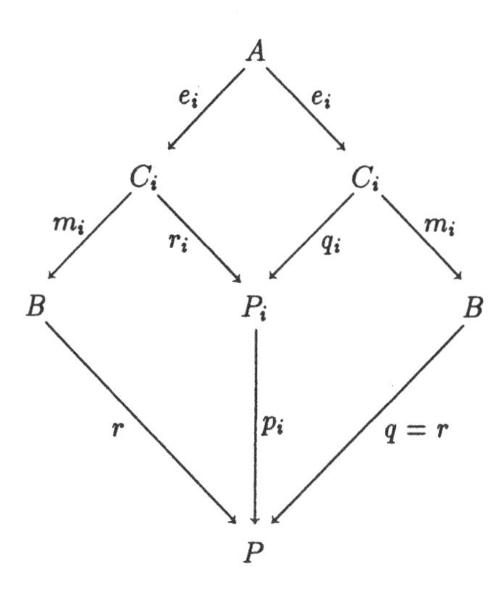

Since f is an epimorphism, we have $r = q$ (and, given $i \in I$, e_i is an epimorphism iff $r_i = q_i$). For each morphism $h\colon C_i \to C_j$ of $\mathcal{D}(f)$ there exists a unique morphism

$$(3) \qquad h^*\colon P_i \to P_j \quad \text{with} \quad h^* \cdot r_i = r_j \cdot h \quad \text{and} \quad h^* \cdot q_i = q_j \cdot h.$$

These morphisms form a λ-filtered diagram whose colimit is $(P_i \xrightarrow{p_i} P)_{i \in I}$. Since for each $i \in I$ the object C_i is λ-presentable, and since $p_i \cdot r_i = p_i \cdot q_i$ ($= r \cdot m_i$), we conclude that there exists $i^* \in I$ and a morphism $c_i\colon C_i \to C_{i^*}$ of $\mathcal{D}(f)$ with

$$(4) \qquad c_i^* \cdot r_i = c_i^* \cdot q_i.$$

We are ready to prove that for each $i_0 \in I$ there is a morphism from C_{i_0} to C_j such that e_j is an epimorphism. Consider the following ω-chain in $\mathcal{D}(f)$,

$$C_{i_0} \xrightarrow{c_{i_0}} C_{i_0^*} = C_{i_1} \xrightarrow{c_{i_1}} C_{i_1^*} = C_{i_2} \xrightarrow{c_{i_2}} C_{i_2^*} \cdots,$$

and let $(C_{i_n} \xrightarrow{k_n} \overline{C})_{n<\omega}$ be a colimit of this chain in $\mathcal{K}$. Since λ is uncountable, we know that $\overline{C}$ is λ-presentable (see Proposition 1.16), thus, we can assume that $\overline{C} \in \mathbf{Pres}_\lambda \mathcal{K}$. The unique morphism $\overline{m}\colon \overline{C} \to B$ with

$\overline{m} \cdot k_n = m_{i_n}$ $(n < \omega)$ yields a factorization $f = \overline{m} \cdot k_0 \cdot e_{i_0}$ which is an object of $\mathcal{D}(f)$, say, $(k_0 \cdot e_{i_0}, \overline{m}) = (k_j, m_j)$ for $j \in I$. To prove that e_j is an epimorphism, i.e., $r_j = q_j$, we use the fact that $k_n = k_{n+1} \cdot c_{i_n}$ (thus, $k_n^* = k_{n+1}^* \cdot c_{i_n}^*$); hence, from (3) and (4) we get

$$r_j \cdot k_n = k_{n+1}^* \cdot c_{i_n}^* \cdot r_{i_n} = k_{n+1}^* \cdot c_{i_n}^* \cdot q_{i_n} = q_j \cdot k_n$$

for each $n < \omega$.

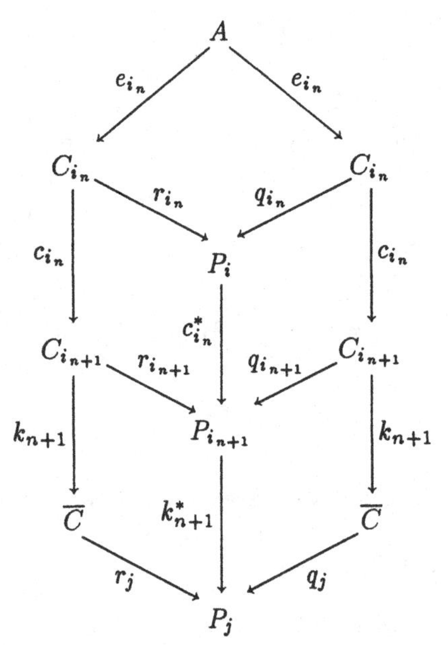

Thus, $r_j = q_j$, which implies that $k_0 \colon C_{i_0} \to C_j$ is the required morphism. □

1.59 Proposition. *In each locally λ-presentable category $\mathcal{K}$ all λ-directed colimits commute with λ-small limits. That is, given a diagram*

$$D \colon I \times J \to \mathcal{K} \quad \textit{with } (I, \leq) \ \lambda\textit{-directed, } J \ \lambda\textit{-small,}$$

then the canonical morphism

$$c \colon \operatorname*{colim}_I \lim_J D_{i,j} \to \lim_J \operatorname*{colim}_I D_{i,j}$$

is an isomorphism.

PROOF. It is sufficient to prove that in **Set** λ-directed colimits commute with λ-small limits for each λ. It then follows immediately that each category $\mathbf{Set}^{\mathcal{B}}$ has this property too. Consequently, each full subcategory of $\mathbf{Set}^{\mathcal{B}}$ closed under limits and λ-directed colimits also has that property. It follows from the representation theorem, 1.46, that every locally λ-presentable category is equivalent to such a subcategory of $\mathbf{Set}^{\mathcal{B}}$.

Let $D\colon I \times J \to \mathbf{Set}$ be a diagram, where $(I, \leq)$ is λ-directed and J is λ-small. Let us describe $\operatorname{colim}_I \lim_J D_{i,j}$ first. For each $i \in I$, $\lim_J D_{i,j}$ can be described as the subset of $\prod_{j \in J} D_{i,j}$ formed by those $\langle x_{i,j} \rangle_{j \in J}$ which for each J-morphism $p\colon j \to j'$ fulfil

$$D(id, p)(x_{i,j}) = x_{i,j'}.$$

(The limit maps are the usual projections.) Given $i \leq i'$ in I, denote by $f_{i,i'}\colon \lim_J D_{i,j} \to \lim_J D_{i',j}$ the function defined by

$$f_{i,i'}(\langle x_{i,j} \rangle) = \langle D(i \to i', id)(x_{i,j}) \rangle .$$

Then $\operatorname{colim}_I \lim_J D_{i,j}$ can be described as the quotient set of $\coprod_{i \in I} \lim_J D_{i,j}$ under the following equivalence $\sim$:

$$\langle x_{i,j} \rangle \sim \langle y_{i',j} \rangle \quad \text{iff} \quad f_{i,i_0}(\langle x_{i,j} \rangle) = f_{i',i_0}(\langle y_{i',j} \rangle)$$

for some $i_0 \in I$ with $i \leq i_0$, $i' \leq i_0$. The colimit cocone takes $\langle x_{i,j} \rangle$ to the equivalence class $[\langle x_{ij} \rangle]$ containing it.

Next, we describe $\lim_J \operatorname{colim}_I D_{i,j}$. For each $j \in J$ we have $\operatorname{colim}_I D_{i,j} = \coprod_{i \in I} D_{i,j} / \approx_j$ where, given $x \in D_{i,j}$ and $y \in D_{i',j}$, then $x \approx_j y$ iff

$$D(i \to i_0, id)(x) = D(i' \to i_0, id)(y) \text{ for some } i_0 \in I \text{ with } i \leq i_0, i' \leq i_0.$$

The colimit cocone assigns to each $x \in D_{i,j}$ the class $[x]_j$ containing x. It is obvious that for each morphism $p\colon j \to j'$ in J we have: $x \approx_j y$ implies $D(id, p)(x) \approx_{j'} D(id, p)(y)$. It follows that we can define a function $p^*\colon \operatorname{colim}_I D_{i,j} \to \operatorname{colim}_I D_{i,j'}$ by $p^*([x_j]) = [D(id, p)(x)]_{j'}$. Then $\lim_J \operatorname{colim}_I D_{ij}$ can be described as the subset of $\prod_{j \in J} \operatorname{colim}_I D_{i,j}$ formed by those $\langle [x_{i_j,j}]_j \rangle$ which for each J-morphism $p\colon j \to j'$ fulfil $p^*([x_{i_j,j}]_j) = [x_{i_{j'},j'}]_{j'}$. (The limit maps are the usual projections, again.) The above canonical morphism is given by $c\colon [\langle x_{i,j} \rangle] \mapsto \langle [x_{i,j}]_j \rangle$. This map is:

(a) One-to-one, because whenever $x_{i,j} \approx_j y_{i,j}$ for each $j \in J$, then $\langle x_{i,j} \rangle \sim \langle y_{i,j} \rangle$. This follows from the fact that, since I is λ-directed and J is λ-small, we can choose $i_0 \in I$ with

$$D(i \to i_0, id)(x_{i,j}) = D(i \to i_0, id)(y_{i,j})$$

independently of j.

(b) Onto, because given $\langle [x_{i_j,j}]_j \rangle$ in $\lim_J \operatorname{colim}_I D_{ij}$, there exists $i \in I$ with $i_j \leq i$ for each $j \in J$ (again since I is λ-directed), and then c maps the element $[\langle f_{i_j,i}(x_{i_j}) \rangle]$ of $\operatorname{colim}_I \lim_J D_{i,j}$ onto $\langle [x_{i_j,j}]_j \rangle$. □

1.60 Corollary. *In a locally λ-presentable category $\mathcal{K}$ both the class of all monomorphisms and the class of all regular monomorphisms are stable under λ-directed colimits. That is, given λ-directed diagrams $D, D' \colon \mathcal{D} \to \mathcal{K}$ and a natural (regular-)monotransformation $\delta \colon D \to D'$, then*

$$\operatorname{colim} \delta \colon \operatorname{colim} D \to \operatorname{colim} D'$$

is a (regular) monomorphism.

1.61 Proposition. *Every locally presentable category has both (strong epi, mono)- and (epi, strong mono)-factorizations of morphisms.*

PROOF. In a locally presentable category $\mathcal{K}$ the existence of pushouts guarantees that strong and extremal monomorphisms coincide, thus, since $\mathcal{K}$ is complete and wellpowered, it has (epi, strong mono)-factorizations, see 0.5. Dually, since $\mathcal{K}$ is cocomplete, co-wellpowered, and has pullbacks, it has (strong epi, mono)-factorizations. □

1.62 Proposition. *In a locally λ-presentable category each λ-directed colimit of (regular) monomorphisms has the property that* (i) *every colimit cocone consists of (regular) monomorphisms, and* (ii) *for every compatible cocone of (regular) monomorphisms the factorizing morphism is a (regular) monomorphism too.*

PROOF. It is easy to verify that each of the categories $\mathbf{Set}^{\mathcal{A}}$ has the mentioned property. The proposition then follows from Theorem 1.46 (since the inclusion of a reflective subcategory preserves (regular) monomorphisms). □

1.63. Recall that an object A is a *union* of subobjects $m_i \colon A_i \to A$ $(i \in I)$ provided that A does not have a proper subobject containing all m_i $(i \in I)$.

Corollary. *In a locally λ-presentable category λ-directed unions are λ-directed colimits. More precisely, if A is a λ-directed union of subobjects $m_i = A_i \to A$, $i \in I$, then the diagram of all A_i's and the factorizations of m_i through m_j for $i \leq j$ has a colimit $(A_i \xrightarrow{m_i} A)_{i \in I}$.*

PROOF. The factorizing morphism $\operatorname{colim} A_i \to A$ is a monomorphism by Proposition 1.62, and it is an extremal epimorphism by the definition of union. □

1.64 Theorem. *If $\mathcal{K}$ and $\mathcal{K}^{op}$ are both locally presentable categories, then $\mathcal{K}$ is equivalent to a complete lattice.*

PROOF. We are to verify that $\hom(K, L)$ has at most one element for each pair of objects K, L of $\mathcal{K}$. It is sufficient to show that $\hom(K, K) = \{ id_K \}$: if we apply this to $K' = K + K$, we conclude that the two injections of K to K' are equal, hence, there exist no two distinct morphisms from K to L.

Let λ be a regular cardinal such that K is λ-presentable, and both $\mathcal{K}$ and $\mathcal{K}^{op}$ are locally λ-presentable. For each morphism $f: K \to K$ we will prove that $f = id_K$. Let I be a set of cardinality λ, choose $i_0 \in I$, and denote by I^* the λ-directed set of all subsets of I of cardinality smaller than λ and containing i_0 (ordered by inclusion). Consider the λ-directed diagram D whose objects are all powers K^J, where $J \in I^*$, and whose morphism from K^J to $K^{J'}$ $(J \subseteq J')$ is the unique morphism $K^J \to K^{J'}$ that composes with the jth projection as π_j, if $j \in J$, or as $f \cdot \pi_{i_0}$, if $j \in J' - J$ (here π_j denotes the jth projection of K^J). We define a compatible cocone $m_J : K^J \to K^I$ of the diagram D as follows: the composite of m_J with the jth projection is π_j $(j \in J)$ or $f \cdot \pi_{i_0}$ $(j \in I - J)$. Each m_J is a split monomorphism, since the projection $p_J: K^I \to K^J$ fulfils $p_J \cdot m_J = id$. By Corollary 1.60, $m = \operatorname{colim} m_J : \operatorname{colim} D \to K^I$ is a regular monomorphism. Furthermore, since $p_J \cdot m_J = id$, the morphism $p_J \cdot m$ is a (split) epimorphism. The canonical λ-directed limit $(p_J : K^I \to K^J)_{J \in I^*}$ fulfils $m = \lim_J (p_J \cdot m)$. By applying Corollary 1.60 to $\mathcal{K}^{op}$, we conclude that m is an epimorphism, thus, an isomorphism. It follows that $(K^J \xrightarrow{m_J} K^I)_{J \in I^*}$ is a colimit of D.

Since K is λ-presentable, the diagonal $\Delta_I : K \to K^I$ factorizes through some m_J. Hence $\Delta_I = m_J \cdot \Delta_J$, and by composing this with the jth projection for some $j \in I - J$, we get $id_K = f$. □

1.65 Examples

(1) Compact Hausdorff 0-dimensional spaces do not form a locally presentable category because the dual category is **Bool** (Stone duality) which is locally presentable.

(2) Unlike Banach spaces (see Example 1.48), Hilbert spaces do not form a locally presentable category. This follows from the observation that the category **Hil** of Hilbert spaces and linear contractions is self-dual.

In fact, the functor E: **Hil** $\to$ **Hil**^{op} which leaves objects unchanged and assigns to each linear contraction $f: A \to B$ the unique adjoint contraction $f^* : B \to A$ defined by

$$(fx, y) = (x, f^* y) \qquad \text{for all } x \in A \text{ and } y \in B$$

is an equivalence of categories.

1.66 Adjoint Functor Theorem. *A functor between locally presentable categories is a right adjoint iff it preserves limits and λ-directed colimits for some regular cardinal λ.*

PROOF. Let $F\colon \mathcal{K} \to \mathcal{L}$ be a functor, and let λ_0 be a regular cardinal such that $\mathcal{K}$ and $\mathcal{L}$ are locally λ_0-presentable.

(1) If F is continuous and preserves λ_1-directed colimits, we will show that the solution-set condition of Freyd's adjoint functor theorem (0.7) is satisfied. Given an object L in $\mathcal{L}$, let λ be a regular cardinal larger or equal to both λ_1 and λ_0, such that L is λ-presentable. It is sufficient to show that each arrow $f\colon L \to FK$ factorizes through an arrow $f'\colon L \to FK'$ such that K' is λ-presentable in $\mathcal{K}$ (i.e., $f = Fh \cdot f'$ for some $h\colon K' \to K$). In fact, since $\mathcal{K}$ is locally λ-presentable, there is a λ-directed colimit $(K_i \xrightarrow{h_i} K)$ such that each K_i is λ-presentable in $\mathcal{K}$. Since $\lambda \geq \lambda_1$, we have a λ-directed colimit $(FK_i \xrightarrow{Fh_i} FK)$ in $\mathcal{L}$. Consequently, each arrow $f\colon L \to FK$ factorizes through some Fh_i.

(2) Let F be a right adjoint with a left adjoint G. Let λ be a regular cardinal such that for each λ_0-presentable object L in $\mathcal{L}$ the object GL is λ-presentable in $\mathcal{K}$. Then F preserves λ-directed colimits. In fact, let $(D_i \xrightarrow{k_i} K)_{i\in I}$ be a colimit of a λ-directed diagram D in $\mathcal{K}$. Since $\mathcal{L}$ is locally λ_0-presentable, in order to prove that FD has a colimit $(FD_i \xrightarrow{Fk_i} FK)_{i\in I}$, it is sufficient, by Exercise 1.o(1), to verify that each λ_0-presentable object L of $\mathcal{L}$ has the following properties:

(i) each morphism $L \to FK$ factorizes through some Fk_i;

(ii) if a pair of morphisms $h, h'\colon L \to FK_i$ satisfies $Fk_i \cdot h = Fk_i \cdot h'$, then there exists $j \in I$ with $i \leq j$ such that $FD(i \to j)\cdot h = FD(i \to j)\cdot h'$.

The property (i) follows from the adjunction

$$\frac{L \to FK}{GL \to K}$$

since GL is λ-presentable in $\mathcal{K}$ and K is a λ-directed colimit of D. Analogously the property (ii) follows from the adjunction

$$\frac{L \to FD_i}{GL \to D_i}$$

again via the λ-presentability of GL. □

1.E Locally Generated Categories

An important variant of the definition of locally presentable category works with λ-directed unions (or, equivalently, λ-directed colimits of monomorphisms, see Corollary 1.63) instead of general λ-directed colimits.

1.67 Definition. Let λ be a regular cardinal. An object A of a category is called *λ-generated* provided that $\hom(A, -)$ preserves λ-directed colimits of monomorphisms. An object is called *generated* if it is λ-generated for some regular cardinal λ.

1.68 Examples

(1) In the categories **Set**, $\mathbf{Set}^S$, **Pos**, and $\mathbf{Rel}\,\Sigma$ the concepts of λ-presentable and λ-generated object coincide.

(2) An algebra in a finitary variety is λ-generated iff it has a set of less than λ-generators in the usual algebraic sense, as we will prove in Chapter 3. Thus, for example, in the category of algebras given by infinitely many constant symbols no finite algebra of more than one element is finitely presentable, but it can be finitely generated.

(3) No topological space which is not discrete is generated in **Top** (the proof is as in Example 1.2).

1.69 Proposition. *In a locally λ-presentable category*

(i) *each strong quotient of a λ-generated object is λ-generated,*

(ii) *each λ-generated object is a strong quotient of a λ-presentable object.*

PROOF. (i). This follows from the fact that the colimit cocone of a λ-directed diagram of monomorphisms is formed by monomorphisms (see Proposition 1.62). Thus, if $e\colon A \to B$ is a strong epimorphism and A is λ-generated, then for each $f\colon B \to \operatorname{colim} D$, where D is a λ-directed diagram of monomorphisms, we factorize $f \cdot e$ through some of the colimit maps of D, and use the diagonalization property of strong epimorphisms.

(ii). Let A be λ-generated, and let D be a λ-directed diagram of λ-presentable objects with a colimit $(D_i \xrightarrow{a_i} A)_{i\in I}$. Factorize each a_i as a strong epimorphism $e_i\colon D_i \to D_i'$ followed by a monomorphism $m_i\colon D_i' \to A$. For $i \le j$ in I we see that m_i factorizes through m_j, thus, the objects D_i' form a λ-directed diagram D' of monomorphisms with a colimit $(D_i' \xrightarrow{m_i} A)_{i\in I}$. Since A is λ-generated, id_A factorizes through some m_i, therefore, m_i is an isomorphism. Then A is a strong quotient of D_i. ☐

Corollary. *In a locally presentable category there exists, up to isomorphism, only a set of λ-generated objects for each regular cardinal λ.*

This follows from Theorem 1.58 and Proposition 1.69. □

1.70 Local Generation Theorem. *A category is locally presentable iff it is cocomplete and has, for some regular cardinal λ, a set $\mathcal{A}$ of λ-generated objects such that every object is a λ-directed colimit of its subobjects from $\mathcal{A}$.**

Remark. We do not claim that the category is locally λ-presentable for the given cardinal λ; in fact, a counterexample with $\lambda = \omega$ is presented below.

PROOF. I. Necessity. Let $\mathcal{K}$ be a locally λ-presentable category, and let $\mathcal{A}$ be a set of representatives for all λ-generated objects. Each object K in $\mathcal{K}$ is a colimit of a λ-directed diagram D of λ-presentable objects. Factorize each of the colimit maps $d_i : D_i \to K$ as a strong epimorphism $e_i : D_i \to D'_i$ with $D'_i \in \mathcal{A}$ followed by a monomorphism $m_i : D'_i \to K$ (see Propositions 1.61 and 1.69). Then K is a colimit of the obvious λ-directed diagram of monomorphisms between the D'_i.

II. Sufficiency. Let $\mathcal{K}$ be a cocomplete category, and let $\mathcal{A}$ be a set as in the theorem.

(i) We prove that the set $\mathcal{A}$ is dense in $\mathcal{K}$, and that each λ-generated object is isomorphic to an object in $\mathcal{A}$. The latter is clear from the given property of $\mathcal{A}$. Let K be an arbitrary object of $\mathcal{K}$, and let D be a λ-directed diagram of subobjects $m_i : D_i \to K$ $(i \in I)$ with $D_i \in \mathcal{A}$ and with a colimit $(D_i \xrightarrow{m_i} K)_{i \in I}$. Then K is a colimit of the canonical diagram w.r.t. $\mathcal{A}$ because D is cofinal in it (see 0.11): each morphism $f : A \to K$, $A \in \mathcal{A}$, factorizes through some m_i (since A is λ-generated).

(ii) $\mathcal{K}$ is equivalent to a full, reflective subcategory of $\mathbf{Set}^{\mathcal{A}^{\mathrm{op}}}$ closed under λ-directed colimits of monomorphisms. In fact, the canonical functor $E : \mathcal{K} \to \mathbf{Set}^{\mathcal{A}^{\mathrm{op}}}$ is full and faithful (see Proposition 1.26) and it is a right adjoint (by Proposition 1.27). It also preserves λ-directed colimits of monomorphisms—this follows from the fact that for each $A \in \mathcal{A}$, $\hom(A, -)$ preserves these colimits (the proof is analogous to that of Proposition 1.26). $\mathcal{K}$ is equivalent to $E(\mathcal{K})$, and $E(\mathcal{K})$ has all the required properties.

(iii) By (i) and (ii) it is sufficient to prove that for each locally finitely presentable category $\mathcal{L}$ (here $\mathcal{L} = \mathbf{Set}^{\mathcal{A}^{\mathrm{op}}}$), if $\mathcal{K}$ is a full, reflective subcategory of $\mathcal{L}$ closed under λ-directed colimits of monomorphisms in $\mathcal{L}$, then

*More precisely: of subobjects representable by monomorphisms with domains in $\mathcal{A}$.

$\mathcal{K}$ is locally presentable. Observe that since $\mathcal{A}$ is a set of representatives of all λ-generated objects of $\mathcal{K}$, every object of $\mathcal{A}$ has only a set of strong quotients L in $\mathcal{K}$: by Proposition 1.69 each such quotient is represented by an arrow with a codomain in $\mathcal{A}$.

Let $\mathcal{M}$ denote the collection of all reflection arrows of objects X of $\mathcal{L}$ in $\mathcal{K}$ where X is

(a) either a colimit of a finite diagram of $\mathcal{A}$-objects in $\mathcal{L}$,

or

(b) the codomain of a multiple pushout (co-intersection) in $\mathcal{L}$ of a cone of strong epimorphisms in $\mathcal{K}$ with a domain in $\mathcal{A}$.

Since $\mathcal{L}$ is co-wellpowered (see Theorem 1.58), $\mathcal{M}$ is essentially small. By Theorem 1.39, $\mathcal{M}^{\perp}$ is a locally presentable category, closed in $\mathcal{L}$ under λ_0-directed colimits for some regular cardinal λ_0. $\mathcal{K}$ is, obviously, a full, reflective subcategory of $\mathcal{M}^{\perp}$ closed under colimits of $\overline{\lambda}$-directed diagrams of monomorphisms for some regular cardinal $\overline{\lambda} \geq \lambda$. Besides, the choice of $\mathcal{M}$ guarantees that $\mathcal{K}$ is closed in $\mathcal{M}^{\perp}$ under finite colimits of $\mathcal{A}$-objects and under co-intersections of strong quotients of $\mathcal{A}$-objects.

(iv) We will prove that every morphism $f: A \to X$ in the category $\mathcal{M}^{\perp}$ with $A \in \mathcal{A}$ factorizes as a strong epimorphism $e_f: A \to \overline{A}$ in $\mathcal{K}$ followed by a monomorphism $m_f: \overline{A} \to X$ in $\mathcal{M}^{\perp}$. In fact, let $e_f: A \to \overline{A}$ be the co-intersection of all strong epimorphisms in $\mathcal{K}$ through which f factorizes. Then $f = m_f \cdot e_f$ for some $m_f: \overline{A} \to X$. Since $\mathcal{M}^{\perp}$ has (strong epi, mono)-factorizations (see Proposition 1.61), a co-intersection of strong epimorphisms is a strong epimorphism. To show that m_f is a monomorphism, consider $p_1, p_2: P \to \overline{A}$ in $\mathcal{M}^{\perp}$ with $m_f \cdot p_1 = m_f \cdot p_2$. Since $\mathcal{K}$ is reflective in $\mathcal{M}^{\perp}$, we can restrict ourselves to P in $\mathcal{K}$. The coequalizer c of p_1 and p_2 in $\mathcal{M}^{\perp}$ (or in $\mathcal{K}$) yields a strong epimorphism $c \cdot e_f$ in $\mathcal{K}$ through which f factorizes—thus, $c \cdot e_f$ factorizes through e_f, which implies that c is an isomorphism.

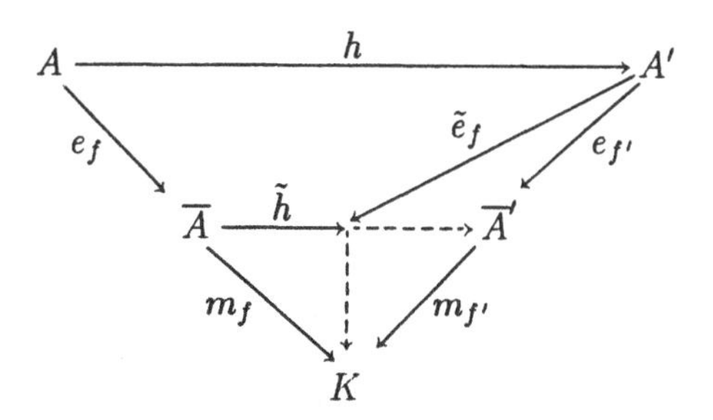

The above factorization has the following "coherence" property: given $f: A \to X$ and $f': A' \to X$ in the category $\mathcal{M}^\perp$ with $A, A' \in \mathcal{A}$, if f factorizes through f', then m_f factorizes through $m_{f'}$. In fact, given $f = f' \cdot h$, let us form the pushout of e_f and h in $\mathcal{M}^\perp$ (or in $\mathcal{K}$), $\tilde{h} \cdot e_f = \tilde{e}_f \cdot h$. Since extremal and strong epimorphisms coincide in $\mathcal{K}$ (see 0.5), it is easy to see that $\tilde{e}_f$ is a strong epimorphism in $\mathcal{K}$. Now, f' factorizes through $\tilde{e}_f$; hence, $e_{f'}$ factorizes through $\tilde{e}_f$, and it follows easily that m_f factorizes through $m_{f'}$.

(v) We will prove that $\mathcal{K}$ is a small-orthogonality class in $\mathcal{M}^\perp$, thereby verifying that $\mathcal{K}$ is locally presentable (Theorem 1.39). Let $\mathcal{N}$ be the (essentially small) collection of all reflections of $\overline{\lambda}$-generated objects of $\mathcal{M}^\perp$ in $\mathcal{K}$. We will prove that

$$\mathcal{K} = \mathcal{N}^\perp.$$

$\mathcal{K} \subseteq \mathcal{N}^\perp$ is obvious, thus, we just prove that each $X \in \mathcal{N}^\perp$ lies in $\mathcal{K}$. Since $\mathcal{M}^\perp$ is locally presentable, we know (from the necessity part of the present proof) that there exists a regular cardinal $\lambda_1 \geq \overline{\lambda}$ such that each object of $\mathcal{M}^\perp$ is a λ_1-directed colimit of its λ_1-generated subobjects. Let D be such a diagram for the given object X, with a colimit cocone $(D_i \xrightarrow{d_i} X)_{i \in I}$. Let $r_i: D_i \to D_i'$ be a reflection of D_i in $\mathcal{K}$, then $X \in \mathcal{N}^\perp$ implies that d_i factorizes as $d_i = d_i' \cdot r_i$ for a unique $d_i': D_i' \to X$. Since D_i is λ_1-generated in $\mathcal{M}^\perp$, and since both $\mathcal{K}$ and $\mathcal{M}^\perp$ are closed under λ_1-directed colimits of monomorphisms in $\mathcal{L}$, it follows that D_i' is λ_1-generated in $\mathcal{K}$. By (ii) above we can assume $D_i' \in \mathcal{A}$. Then we have a factorization $d_i' = m_i \cdot e_i$, for $e_i: D_i' \to D_i''$ as in (iv), and due to the above coherence property, the objects D_i'' form a λ_1-directed diagram of monomorphisms. This diagram obviously has a colimit $(D_i'' \xrightarrow{m_i} X)_{i \in I}$. Since $\mathcal{K}$ is closed under λ-directed colimits of monomorphisms in $\mathcal{L}$, this proves that X lies in $\mathcal{K}$. □

1.71 Example of a category which satisfies the condition of the local generation theorem with $\lambda = \omega$, yet, it is not locally finitely presentable.

Let Σ be a one-sorted signature of countably many unary relation symbols $\sigma_0, \sigma_1, \sigma_2, \ldots$. Let $\mathcal{K}$ be the full subcategory of $\mathbf{Rel}\,\Sigma$ of all structures satisfying the implication

$$(\sigma_1 = \sigma_n \quad \text{for all } n \geq 1) \implies (\exists! x)\sigma_0(x).$$

Then

(1) $\mathcal{K}$ is a reflective subcategory of $\mathbf{Rel}\,\Sigma$: a reflection of a Σ-structure A outside of $\mathcal{K}$ is obtained by (a) adding a new element to $(\sigma_0)_A$ in the case $(\sigma_0)_A = \emptyset$, or (b) merging all elements of $(\sigma_0)_A$. Furthermore,

$\mathcal{K}$ is closed under ω_1-directed colimits. Thus, $\mathcal{K}$ is locally ω_1-presentable by Theorem 1.39.

(2) $\mathcal{K}$ is also closed under directed unions in $\mathbf{Rel}\,\Sigma$, thus, every finite structure in $\mathcal{K}$ is finitely generated in $\mathcal{K}$. It follows easily that $\mathcal{K}$ satisfies the condition of Theorem 1.70 with $\lambda = \omega$.

(3) $\mathcal{K}$ is not locally finitely presentable because the following Σ-structure A contained in $\mathcal{K}$,

$$|A| = \{0, 1\}, \quad \sigma_0 = \{1\}, \quad \sigma_n = \{0\} \text{ for } n > 0,$$

is not finitely presentable in $\mathcal{K}$. In fact, consider the following directed colimit of structures K_n in $\mathcal{K}$: K_0 is the structure with

$$|K_0| = \omega, \quad \sigma_0 = \emptyset \quad \text{and} \quad \sigma_n = \{n\} \text{ for } n > 0,$$

and K_n is the quotient of K_0 under the smallest equivalence merging $0, 1, \ldots, n$. The ω-chain of quotient maps $K_0 \to K_1 \to K_2 \ldots$ in $\mathcal{K}$ has a colimit $(K_n \xrightarrow{k_n} A)_{n\in\omega}$ where $k_n(i) = 0$ for all $i < n$. It is obvious that A is not a directed colimit of finitely presentable objects of $\mathcal{K}$.

1.72 Remarks

(1) P. Gabriel and F. Ulmer introduce in [Gabriel, Ulmer 1971] the following concept of a *locally generated category*: a cocomplete category with a strong generator formed by λ-generated objects and such that every λ-generated object has only a set of strong quotients. They prove that a category is locally presentable iff it is locally generated. It is an open problem whether the last condition (equivalent to co-wellpoweredness w.r.t. strong epimorphisms) can be deleted from their definition of locally generated categories.

(2) The difference between our local generation theorem and the result of Gabriel and Ulmer is that we pass from λ-directed colimits to λ-directed unions in *both* the places where they appear in the definition of locally λ-presentable category, whereas Gabriel and Ulmer do so only once (and they "pay" by the co-wellpoweredness condition). More precisely, a cocomplete category $\mathcal{K}$ is locally presentable iff there exists a regular cardinal λ such that

(a) $\mathcal{K}$ has a set of λ-presentable objects whose closure under λ-directed colimits is all of $\mathcal{K}$ (by definition), or

(b) $\mathcal{K}$ has a set of λ-generated objects whose closure under λ-directed colimits of monomorphisms is all of $\mathcal{K}$ (by our local generation theorem),

or

(c) $\mathcal{K}$ has a set of λ-generated objects whose closure under λ-directed colimits is all of $\mathcal{K}$, provided that $\mathcal{K}$ is co-wellpowered w.r.t. strong epimorphisms (by Gabriel and Ulmer).

Exercises

1.a Directed Colimits

(1) Prove that a colimit of a directed diagram $D\colon (I, \leq) \to \mathbf{Set}$ can be described as follows. Let C be the quotient set of $\coprod_{i\in I} D_i = \bigcup_{i\in I} D_i \times \{ i \}$ under the following equivalence (where $d_{i,j}$ are the connecting maps):

$$(x, i) \sim (x', i') \quad \text{iff there exists } j \in I \text{ with } d_{i,j}(x) = d_{i',j}(x')$$

and let $c_i : D_i \to C$ assign to each $x \in D_i$ the equivalence class of (x, i). Then $(D_i \xrightarrow{c_i} C)_{i\in I}$ is a colimit of D.

(2) Let D be a directed diagram in the category **Set**. Prove that a compatible cocone $\left(D_i \xrightarrow{c_i} C\right)_{i\in I}$ of D is a colimit iff

(i) it is a collective epimorphism, i.e., $C = \bigcup_{i\in I} c_i(D_i)$,

(ii) whenever $c_i(y) = c_{i'}(y')$ for $y \in D_i$, $y' \in D_{i'}$, then there exists an upper bound j of i and i' with $d_{i,j}(y) = d_{i',j}(y')$.

(3) Prove that directed colimits in **Pos** are formed on the level of sets (with the ordering given by the union of the images of the orderings of D_i by the colimit maps).

(4) Prove that (directed) colimits on $\mathbf{Set}^{\mathcal{A}}$ are computed object-wise.

(5) Describe directed colimits in $\mathbf{Rel}\,\Sigma$.

1.b Non-regular Cardinals

(1) Prove that for each infinite cardinal λ there exists a smallest cardinal λ_0 such that λ is a sum of λ_0 cardinals smaller than λ. Prove that if λ is considered as the linearly ordered set of all smaller ordinals, then λ_0 is a cofinal subset of λ, and λ_0 is a regular cardinal—it is called the *cofinality* of λ and denoted by $\operatorname{cf}\lambda$.

(2) Prove that if the concept of a λ-directed colimit is extended from regular cardinals to all infinite cardinals λ, then the following holds: if λ_0 is the cofinality of λ, then λ-directed colimits can always be reduced to λ_0-directed ones. In contrast, if $\lambda_0 < \lambda$ are two *regular* cardinals, verify that λ-directed colimits cannot be reduced to λ_0-directed ones: find a category with λ-directed colimits which fails to have λ_0-directed colimits.

(3) Suppose we delete the requirement that λ be regular from the definition of a locally λ-presentable category. Prove that then a category is locally λ-presentable iff it is locally λ_0-presentable, where λ_0 is the cofinality of λ.

1.c Directed Joins and Joins of Chains

(1) Verify "Iwamura's lemma": a poset with joins of chains is a CPO. (Hint: see Corollary 1.7.)

(2) In contrast, verify that the following poset P has joins of chains of cofinality$\geq \omega_1$ without having all ω_1-directed joins. Let

$$P_n = \{ i \mid i \text{ is an ordinal}, i \leq \aleph_n \} \qquad \text{for } n = 1, 2, 3, \ldots$$

be the linearly ordered set of all ordinals smaller than or equal to the nth infinite cardinal $\aleph_n$, and let P be the subposet of $\prod_{1 \leq n < \omega} P_n$ formed by those (x_n) for which there exist finitely many n's only with $x_n = \aleph_n$.

(a) Prove that $\prod_{1 \leq n < \omega} P_n$ is a complete lattice, and that P is closed in it under joins of chains of cofinality larger than ω.

(b) Verify that the ω_1-directed set $\prod_{1 \leq n < \omega}(P_n - \{\aleph_n\})$ has no upper bound in P.

(3) Generalize (2) from ω_1 to any uncountable regular cardinal.

(4) Find a category with ω_1-directed colimits in which an object need not be ω_1-presentable although its hom-functor preserves colimits of chains of cofinality $\geq \omega_1$. (Hint: add a greatest element to the poset P in (2).)

1.d Sets and Classes of Presentable Objects

(1) Find a category $\mathcal{K}$ which is cocomplete and in which every object is a directed colimit of finitely presentable objects, although $\mathcal{K}$ is not locally presentable.

(2) Prove directly from the definition of finitely presentable category $\mathcal{K}$ that

(a) $\mathcal{K}$ is wellpowered. (Hint: the functor $U\colon \mathcal{K} \to \mathbf{Set}^{\mathcal{A}}$ with $UK = \hom(-, K)/_{\mathcal{A}}$ preserves monomorphisms.)

(b) $\mathcal{K}$ has, up-to isomorphism, only a set of finitely presentable objects. (Hint: every finitely presentable object K is a colimit of a directed diagram D in $\mathcal{A}$, then id_K factorizes through some of the colimit maps $D_i \to K$, thus, K is a subobject of D_i.)

1.e Strong and Extremal Epimorphisms

(1) Prove that every strong epimorphism is extremal.

(2) Prove that in a category with pullbacks every extremal epimorphism is strong.

(3) Prove that in a category with (strong epi, mono)-factorizations of morphisms, extremal and strong epimorphisms coincide.

(4) Find a (finite) category with an extremal epimorphism which is not strong.

1.f Generators

(1) Prove that any category with a strong generator is wellpowered. Is it necessarily co-wellpowered?

(2) Find a one-object generator in **Gra**. Does there exist a one-object strong generator in **Gra**?

(3) Prove that **Top** and **CSLat** do not have a strong generator.

(4) Prove that the category of compact Hausdorff spaces has "many" strong generators (any set containing at least one nonempty space is a strong generator), yet, it has no dense small subcategory.

1.g Reflection Maps. Let $\mathcal{A}$ be a full, reflective subcategory of $\mathcal{K}$.

(1) If all reflection maps are monomorphisms, prove that they all are epimorphisms. Does it follow that they are isomorphisms?

(2) Prove that every reflection map which is a split monomorphism is an isomorphism.

1.h Complete Semilattices as an Orthogonality Class

(1) Prove that **CSLat** is a reflective subcategory of **Pos**. Put $P = \omega \cup \{T\}$ with $0 < 1 < \cdots < T$, and describe a reflection in **CSLat** for P as an object of **Pos** or of **Pos***, see Example 1.33(5). (Hint: for each poset P consider the extension of P to the poset $J(P)$ of all downwards closed subsets of P where we identify each $p \in P$ with $\downarrow p = \{\, x \in P \mid x \leq p \,\}$.)

(2) Verify that **CSLat** is not an orthogonality class in **Pos**. (Hint: see Observation 1.34.)

(3) Verify that the class of complete semilattices is not closed in the category **Pos*** under λ-directed joins for any cardinal λ. Conclude that **CSLat** is not a small-orthogonality class in **Pos***.

1.i Categories of Continuous Functors

Describe $\mathbf{Cont}_{\omega}\,\mathcal{A}$ and $\mathbf{Cont}_{\omega_1}\,\mathcal{A}$ for the following categories:

(1) $\mathcal{A}$ is a small discrete category.

(2) $\mathcal{A} = (\omega^T)^{\mathrm{op}}$ where ω^T is the poset of natural numbers $0 \leq 1 \leq 2 \leq \cdots$ with a top element T ($\geq n$ for all n).

(3) $\mathcal{A}$ is a commutative square.

1.j Orthogonality Classes

(1) Prove that an intersection of orthogonality classes is an orthogonality class.

(2) Prove that each orthogonality class is an intersection of small-orthogonality classes.

(3) Verify that the class of all graphs (X, σ) with a single loop (i.e., $x \in X$ with $(x, x) \in \sigma$) is a small-orthogonality class in **Gra**.
Is the class of all graphs with at least one loop an orthogonality class in **Gra**?

(4) Verify that the class of all graphs such that any vertex is a source of exactly one arrow is a small-orthogonality class in **Gra**.

(5) Show that every small-orthogonality class in the category **Ab** of abelian groups is a "single-orthogonality class", i.e., has the form $\{m\}^{\perp}$ for a single morphism. (Hint: use coproducts.)

1.k Self-dual Categories

A category equivalent to its dual is called *self-dual.*

(1) Verify in detail that the category **Hil** of Hilbert spaces is self-dual. Why is it clear that the category **Ban** of Banach spaces is not self-dual?

(2) Verify that the category whose objects are sets and morphisms are partial injective maps is self-dual.

(3) Verify that the category **CSLat** is self-dual.

1.l Limit Sketches

(1) Find a limit sketch whose category of models is equivalent to **Pos**.

(2) Given limit sketches $\mathscr{S}_1$ and $\mathscr{S}_2$, their *tensor product* $\mathscr{S}_1 \otimes \mathscr{S}_2$ is the limit sketch whose underlying category is the product of the underlying categories of $\mathscr{S}_1$ and $\mathscr{S}_2$, whose diagrams are $D_1 \times D_2$ for D_i ranging through the diagrams of $\mathscr{S}_i$ for $i = 1, 2$, and whose cones are the obvious pairs formed from the cones of $\sigma_1(D_1)$ and $\sigma_2(D_2)$. Prove that for $\mathcal{K} = \mathbf{Mod}\,\mathscr{S}_2$, the models of $\mathscr{S}_1$ in $\mathcal{K}$ are, up to equivalence of categories, precisely the models of $\mathscr{S}_1 \otimes \mathscr{S}_2$ in **Set**, i.e.,

$$\mathbf{Mod}(\mathscr{S}_1, \mathbf{Mod}\,\mathscr{S}_2) \cong \mathbf{Mod}\,\mathscr{S}_1 \otimes \mathscr{S}_2.$$

1.m Adjoints and Directed Colimits

(1) Find an example of a right adjoint between cocomplete categories which does not preserve λ-directed colimits for any λ.

(2) Find an example of a functor preserving limits and directed colimits which is not a right adjoint.

1.n Orthogonal-reflection Construction

(1) Verify that for every ordinal i there is an instance of the orthogonal-reflection construction 1.37 which stops first after i steps.

(2) Let $\mathcal{K}$ be a cocomplete, co-wellpowered category. Show that for each class $\mathcal{M}$ of epimorphisms, $\mathcal{M}^{\perp}$ is an epireflective subcategory. (Hint: the orthogonal-reflection construction 1.37 can be extended to this case due to the co-wellpoweredness.)

1.o λ-Directed Colimits

(1) Let $\mathcal{K}$ be a locally λ-presentable category, and let $D\colon (I, \leq) \to \mathcal{K}$ be a λ-directed diagram. Prove that a compatible cocone $(D_i \xrightarrow{c_i} C)_{i \in I}$ is a colimit of D iff the following holds for each λ-presentable object K of $\mathcal{K}$:

(i) every morphism $K \to C$ factorizes through some c_i;

(ii) if a pair $h, h'\colon K \to C_i$ satisfies $c_i \cdot h = c_i \cdot h'$ then there exists $j \geq i$ with $D(i \to j) \cdot h = D(i \to j) \cdot h'$.

(2) Verify that in (1) we did not use the fact that $\mathcal{K}$ is cocomplete; it is sufficient to assume that $\mathcal{K}$ has λ-directed colimits.

(3) Prove that a cofinal full subdiagram (see 0.11) of a λ-filtered diagram is λ-filtered and that condition (b) of 0.11 is superfluous (i.e., a full subcategory $\mathcal{D}_0$ of a λ-filtered category $\mathcal{D}$ is cofinal iff for each object d of $\mathcal{D}$ there exists a morphism $d \to d_0$ with d_0 in $\mathcal{D}_0$).

More generally, (b) can be omitted in 0.11 whenever $\mathcal{D}_0$ is directed and the functor H has the following property: given $Hh = Hh' \cdot f$, then $f = Hh''$ for some h''.

1.p Kan Extension. Let $\mathcal{A}$ be a small category, and $\mathcal{B}$ a cocomplete category.

(1) For each natural transformation $f\colon F \to G$ in $\mathcal{B}^{\mathcal{A}}$ prove that there exists a unique extension to a natural transformation $f^*: F^* \to G^*$.

(2) Conclude that Kan extension defines a functor K from the category $\mathcal{B}^{\mathcal{A}}$ to the quasicategory $\mathcal{B}^{\mathbf{Set}^{\mathcal{A}^{\mathrm{op}}}}$.

(3) Prove that Kan extension preserves colimits, i.e., given a diagram $D\colon \mathcal{D} \to \mathcal{B}^{\mathcal{A}}$ with a colimit $(Dd \xrightarrow{c_d} F)$ in $\mathcal{B}^{\mathcal{A}}$, then F^* is a colimit of the diagram $KD\colon \mathcal{D} \to \mathcal{B}^{\mathbf{Set}^{\mathcal{A}^{\mathrm{op}}}}$.

1.q Complete Boolean Algebras and complete homomorphisms form a category which does not have countable coproducts. Prove this. (Hint: a free object on countably many generators does not exist, see [Hales 1964].)

1.r Prove that the category of semigroups (i.e., associative binary algebras) is equivalent to **Mod** $\mathscr{S}$ for the following sketch $\mathscr{S} = (\mathcal{A}, \mathbf{L}, \sigma)$. Let $\mathcal{A}$ be the category of three objects a, a^2, a^3, whose morphisms are generated by the following morphisms:

$p_1, p_2 \colon a^2 \to a$	(projections of a^2 as a product $a \times a$);
$q_1, q_2, q_3 \colon a^3 \to a$	(projections of a^3 as a product $a \times a \times a$);
$q_{1,2}, q_{2,3} \colon a^3 \to a^2$	(projections);
$o \colon a^2 \to a$	(operation);
$o \times id, id \times o \colon a^3 \to a^2$	(here $o \times id$ and $id \times o$ are just notations for two morphisms).

The composition is defined freely subject to equations expressed by the commutativity of the following diagrams:

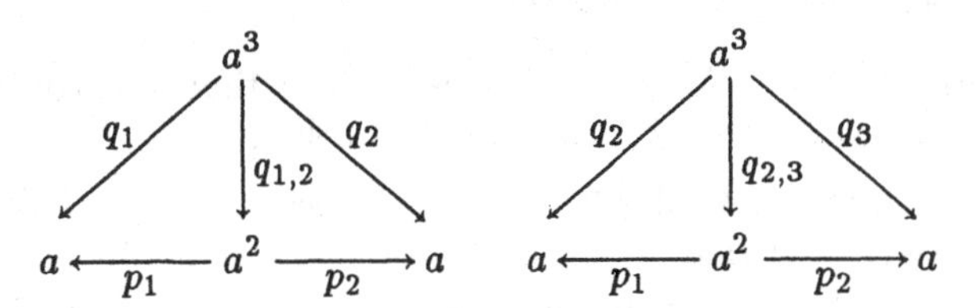

(expressing the role of the two projections $q_{1,2}$ and $q_{2,3}$),

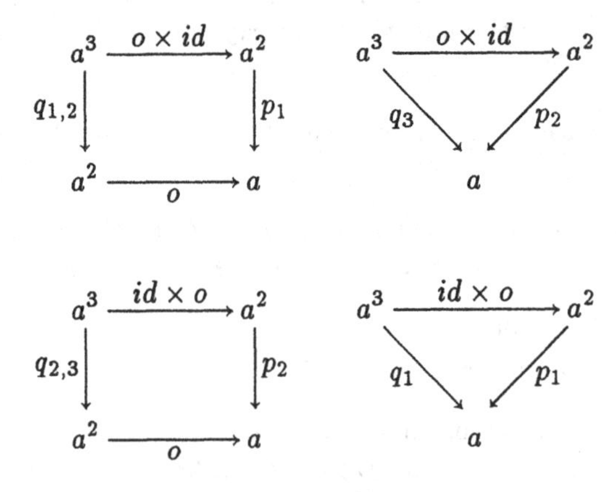

(expressing the role of $o \times id$ and $id \times o$), and

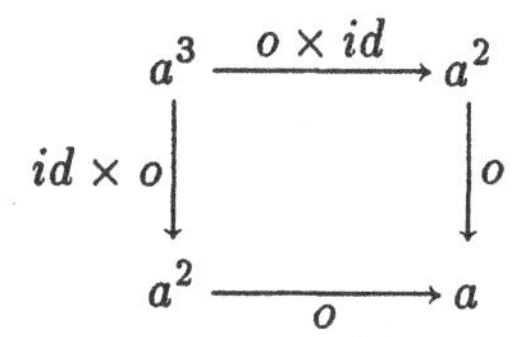

(expressing the associativity).

Further, let $\mathbf{L}$ contain two discrete diagrams: one of two copies of a and one of three copies of a. Finally, σ assigns the following cones to them:

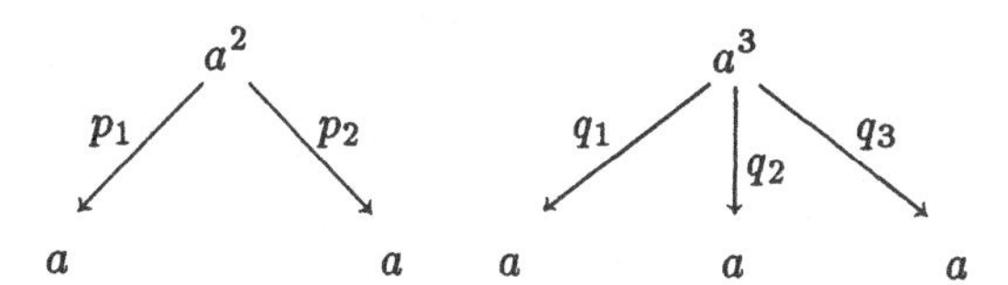

1.s Adjoint Functors

(1) Prove that for each functor $F: \mathcal{K} \to \mathcal{L}$ preserving limits and λ-directed colimits, with $\mathcal{K}$ and $\mathcal{L}$ locally λ-presentable, there exists a left adjoint $G: \mathcal{L} \to \mathcal{K}$ with $G(\mathbf{Pres}_\lambda\, \mathcal{L}) \subseteq \mathbf{Pres}_\lambda\, \mathcal{K}$. (Hint: F is a right adjoint by Theorem 1.66, and any left adjoint of F preserves λ-presentability of objects.)

(2) Define a functor $Q: \mathbf{lp}_\lambda\text{-}\mathbf{CAT} \to (\mathbf{c}_\lambda\text{-}\mathbf{Cat})^{\mathrm{op}}$ (see Remark 1.46) on objects by $Q\mathcal{K} = (\mathbf{Pres}_\lambda\, \mathcal{K})^{\mathrm{op}}$ and on morphisms $F: \mathcal{K} \to \mathcal{L}$ by choosing G as in (1) and denoting by QF the domain-codomain restriction of G^{op}. Verify that F is well-defined. (Hint: $\mathbf{Pres}_\lambda\, \mathcal{K}$ is skeletal.)

(3) If $\mathcal{A}$ is a small category with split idempotents, prove that λ-presentable objects of $\mathbf{Cont}_\lambda\, \mathcal{A}^{\mathrm{op}}$ are precisely the representable functors. Conclude that $QR \cong Id$. (Hint: express a λ-presentable object K as a λ-directed colimit of hom-functors. Then K is a split subobject of one of them, and since $\mathcal{A}$ has split idempotents, K is then representable.)

(4) Prove that $RQ \cong Id$. (Hint: by Theorem and Remark 1.46, every λ-presentable category $\mathcal{K}$ is a λ-free cocompletion of $(\mathbf{Pres}_\lambda\, \mathcal{K})^{\mathrm{op}}$, i.e., $\mathcal{K}$ is canonically isomorphic to $RQ\mathcal{K}$.)

Historical Remarks

The material of Chapter 1 is essentially contained in the lecture notes [Gabriel, Ulmer 1971]. We have found this text a continuous source of inspiration, in spite of the difficulties we had in discovering what it contains. The definition of locally λ-presentable categories we have chosen as basic seems to us the most fundamental both per se and also in connection with the λ-accessible categories in Chapter 2, whereas Gabriel and Ulmer chose the condition of our Theorem 1.20 for the definition.

The concepts of a presentable and a generated object have been introduced independently by [Artin, Grothendieck, Verdier 1972] and [Gabriel, Ulmer 1971].

Lemma 1.6 is, essentially, the proof in [Skornyakov 1964] of the famous Iwamura' lemma, 1c(1). The original source is [Iwamura 1944].

The idea of an iterative construction of orthogonal reflection is also due to Gabriel and Ulmer, and it was later developed by [Kelly 1980] from which our concrete form of the construction essentially stems. The reflectivity of orthogonal subcategories is discussed in [Freyd, Kelly 1972]. Limit sketches appeared first in the thesis [Lawvere 1963] where varieties of finitary algebras were shown to be sketchable by finite-product sketches; this was generalized to infinitary varieties and product sketches in [Linton 1966]. A full account of this is presented in Chapter 3 below. Limit sketches in general were introduced by [Ehresmann 1966] and studied by his school and, independently, by [Kennison 1968] and [Gabriel, Ulmer 1971].

The equivalence of directed and filtered colimits is well-known folklore; our treatment essentially follows [Artin, Grothendieck, Verdier 1972].

The example of Banach spaces as special totally convex spaces is due to [Pumplün, Röhrl 1984].

The local generation theorem is new. It has been published in [Adámek, Rosický 1991], but the proof was much inspired by Gabriel and Ulmer.

Chapter 2

Accessible Categories

We saw in Chapter 1 that locally presentable categories encompass a considerable collection of "everyday" categories. There are, however, other categories which are well-related to λ-directed colimits, but are not (co)complete, e.g., the categories of fields, Hilbert spaces, linearly ordered sets, etc. These are the accessible categories we investigate in the present chapter.

Accessible categories can be introduced in several equivalent ways: they are precisely

(1) the categories with λ-directed colimits in which every object is a λ-directed colimit of λ-presentable objects of a certain set (Definition 2.1);

(2) the full subcategories of functor categories $\mathbf{Set}^{\mathcal{A}}$ closed under λ-directed colimits and λ-pure subobjects (Corollary 2.36);

(3) the free cocompletions of small categories w.r.t. λ-directed colimits (Theorem 2.26);

(4) the categories sketchable by a small sketch (Corollary 2.61);

(5) the categories axiomatizable by basic theories in first-order logic (Theorem 5.35).

The relationship between locally presentable and accessible categories is that (a) a category is locally presentable iff it is accessible and complete (Corollary 2.47) and (b) a category is accessible iff it is a full, cone-reflective subcategory of a locally presentable category, closed under λ-directed colimits (Theorem 2.53). Moreover, every accessible category has a full embedding into the category of graphs preserving λ-directed colimits (Theorem 2.65).

The natural choice of morphisms between λ-accessible categories is the λ-accessible functors, i.e., functors preserving λ-directed colimits. For example, each left or right adjoint between accessible categories is accessible (Proposition 2.23). In the last section we investigate properties of the 2-category of λ-accessible categories and λ-accessible functors. We show, for example, that the Eilenberg–Moore category of a λ-accessible monad is accessible (Theorem 2.78).

In Chapter 1 we proved that locally presentable categories are precisely those sketchable by a limit sketch, i.e., the categories of set-valued functors preserving specified limits (in a small category). Here we prove that accessible categories are precisely those sketchable by a sketch, i.e., the categories of set-valued functors preserving specified limits and colimits (in a small category); see Corollary 2.61. Our proof is simpler than any previously published proof and relies on the concept of a λ-pure subobject (which is, roughly speaking, a λ-directed colimit of split subobjects). The importance of λ-pure subobjects lies in the fact that every λ-accessible category has "enough" of them (Theorem 2.33). Consequently, a category is accessible iff it is equivalent to a full subcategory of a functor-category $\mathbf{Set}^{\mathcal{A}}$ closed under λ-directed colimits and λ-pure subobjects (Corollary 2.36).

2.A Accessible Categories

Definition and Examples

Whereas locally finitely presentable categories generalize algebraic lattices, finitely accessible categories are a direct generalization of Scott's domains. The concept of an accessible category has developed from model theory and category theory. We will explain the role that accessible and locally presentable categories play in model theory thoroughly in Chapter 5 below.

In the present chapter we treat the concept of a λ-accessible category as a natural generalization of a locally λ-presentable category. We just require, instead of cocompleteness, the existence of λ-directed colimits:

2.1 Definition. A category $\mathcal{K}$ is called *λ-accessible,* where λ is a regular cardinal, provided that

(1) $\mathcal{K}$ has λ-directed colimits,

(2) $\mathcal{K}$ has a set $\mathcal{A}$ of λ-presentable objects such that every object is a λ-directed colimit of objects from $\mathcal{A}$.

2.2 Remarks

(1) A category is called *accessible* if it is λ-accessible for some regular cardinal λ. (Then, as we show later, it is μ-accessible for arbitrarily large

regular cardinals μ.) For the case $\lambda = \omega$ we speak of *finitely accessible* categories.

(2) A λ-accessible category has λ-filtered colimits, see Remark 1.21.

(3) Every object of an accessible category is presentable (see Proposition 1.16).

(4) Every λ-accessible category has, up to isomorphism, only a set of λ-presentable objects. (The proof is analogous to Remark 1.9.) Therefore, we extend the notation of Remark 1.19 to accessible categories, and denote by $\mathbf{Pres}_\lambda \mathcal{K}$ a small, full subcategory of $\mathcal{K}$ formed by representatives of all λ-presentable objects of $\mathcal{K}$.

2.3 Examples

(1) Every locally λ-presentable category is λ-accessible. The converse holds for cocomplete categories, thus, e.g., **Top** is not accessible (see Example 1.18(5)).

(2) Finitely accessible posets are precisely Scott domains, i.e., posets with directed joins in which each element is a directed join of finite elements.

Scott domains which are not complete lattices are, obviously, not locally presentable.

(3) Every poset P of n elements is λ-accessible for any regular cardinal $\lambda > n$. (In fact, any λ-directed diagram in P has the greatest element.) However, no large ordered class is accessible.

(4) The following poset P is not finitely accessible, although it has directed colimits and a dense set of finitely presentable objects. (This contrasts with the characterization of locally finitely presentable categories in Theorem 1.11.) Let P be the set of all subsets of ω which are either infinite or singleton, ordered by inclusion. The singleton sets are finitely presentable, and they form a dense set in P. However, an infinite set is not a directed colimit of finitely presentable objects.

(5) The category **Fld** of fields and homomorphisms, which is far from being locally presentable (since morphisms are injective), is finitely accessible. It is evident that the category **Fld** has directed colimits, since it is closed under directed colimits in the category of commutative rings. Every field generated (in the usual algebraic sense) by finitely many el-

ements is, obviously, finitely presentable in **Fld**. Each field is a directed union (= directed colimit) of its finitely generated subfields. Finally, every finitely generated field is isomorphic to the field of quotients of some of the fields

$$\mathbb{Q}[x_1, \dots, x_n], \quad \mathbb{Z}_p[x_1, \dots, x_n] \qquad (n = 1, 2, 3, \dots)$$

where $\mathbb{Q}$ is the field of rationals, and $\mathbb{Z}_p = \mathbb{Z}/\text{mod}\, p$ is the field of integers modulo a prime p.

(6) The category of sets and one-to-one functions is finitely accessible (and is not locally presentable). Analogously with other locally presentable categories $\mathcal{K}$: the category of all $\mathcal{K}$-objects and all $\mathcal{K}$-monomorphisms is accessible, see Proposition 1.62 and Theorem 1.70.

(7) For each regular cardinal λ let $\mathbf{Pos}_\lambda$ denote the category whose objects are λ-directed posets and whose morphisms are order-embeddings (i.e., functions $f\colon (X, \leq) \to (Y, \leq)$ such that $x_1 \leq x_2$ iff $f(x_1) \leq f(x_2)$). $\mathbf{Pos}_\lambda$ has λ-directed colimits (calculated as in **Pos**). Every λ-directed poset is a λ-directed union (= colimit in $\mathbf{Pos}_\lambda$) of all its subposets of cardinality smaller than λ and with a largest element. Since the latter objects of $\mathbf{Pos}_\lambda$ are evidently λ-presentable, $\mathbf{Pos}_\lambda$ is λ-accessible.

(8) The category **Lin** of linearly ordered sets and strictly increasing maps f (i.e., $x < y \Rightarrow f(x) < f(y)$) is finitely accessible.

The full subcategory $\mathcal{W}$ consisting of well-ordered sets is ω_1-accessible but not finitely accessible (the obvious ω-chain $\{0\} \to \{0, -1\} \to \{0, -1, -2\} \to \cdots$ does not have a colimit in $\mathcal{W}$).

(9) The category **Hil** of Hilbert spaces is ω_1-accessible when considered as a full subcategory of the (locally ω_1-presentable) category **Ban** (see Example 1.48). If $d_i\colon B_i \to B$, $i \in I$, is an ω_1-directed colimit in **Ban**, then for any $x \in B$ there is $i \in I$ and $y \in B_i$ such that $d_i(y) = x$ and $\|y\| = \|x\|$. Since a Banach space is a Hilbert space iff it satisfies the parallelogram law

$$\|x + y\|^2 + \|x - y\|^2 = 2\|x\|^2 + 2\|y\|^2,$$

it immediately follows that **Hil** is closed in **Ban** under ω_1-directed colimits. Further, ω_1-presentable Hilbert spaces are either finite-dimensional or isomorphic to l_2. Any infinite-dimensional Hilbert space is an ω_1-directed colimit of copies of l_2.

The category **Hil** is self-dual (see Example 1.65(2)), which contrasts with the fact that by Theorem 1.64 the dual of a non-trivial locally presentable category is never locally presentable.

(10) Small categories (which, except for complete lattices, are not locally presentable) are often accessible, as we shall presently see.

2.4 Observation. Each accessible category has *split idempotents,* i.e., given $f: A \to A$ with $f \cdot f = f$ there exists a factorization $f = i \cdot p$ where $p \cdot i = id$.

In fact, each idempotent $f: A \to A$ defines a λ-filtered diagram of one object A and two morphisms id_A, f (for any λ). A colimit of this diagram consists of an object B and a morphism $p: A \to B$; since $f: A \to A$ forms a compatible cocone, we have a unique $i: B \to A$ with $f = i \cdot p$. The identity $p \cdot i = id$ is a consequence of $p \cdot i \cdot p = p$.

2.5 Remarks

(1) If a full, isomorphism-closed subcategory $\mathcal{K}$ of a category $\mathcal{L}$ has split idempotents, then $\mathcal{K}$ is closed in $\mathcal{L}$ under split subobjects.

In fact, if $i: L \to K$ is a split subobject of an object K from $\mathcal{K}$, we choose $p: K \to L$ with $p \cdot i = id_L$. Then $f = i \cdot p$ is an idempotent which we can split in $\mathcal{K}$ (by $i_0: K_0 \to K$ and $p_0: K \to K_0$). It is easy to see that $p \cdot i_0: K_0 \to L$ is an isomorphism (with inverse $p_0 \cdot i$).

(2) Every (small) category $\mathcal{K}$ has a (small) extension to a category $\widetilde{\mathcal{K}}$ with split idempotents which is universal. That is, there is a full embedding $E: \mathcal{K} \to \widetilde{\mathcal{K}}$ such that every functor from $\mathcal{K}$ to a category with split idempotents has an extension to $\widetilde{\mathcal{K}}$, unique up to natural isomorphism. See Exercise 2.b(2).

2.6 Proposition. *Each small category with split idempotents is accessible.*

PROOF. Let $\mathcal{K}$ be a small category with split idempotents, and let λ be a regular cardinal larger than the number of all $\mathcal{K}$-morphisms. We are going to prove that for the cardinal successor λ^+ of λ, the category $\mathcal{K}$ has λ^+-directed colimits preserved by the Yoneda embedding $Y: \mathcal{K} \to \mathbf{Set}^{\mathcal{K}^{op}}$. It follows that $\mathcal{K}$ is λ^+-accessible, because every object of $\mathcal{K}$ is clearly λ^+-presentable (since Y is a full embedding and each object YK is finitely presentable, hence, λ^+-presentable in $\mathbf{Set}^{\mathcal{K}^{op}}$, see Example 1.2(7)).

Thus, let $D\colon (I, \leq) \to \mathcal{K}$ be a λ^+-directed diagram in $\mathcal{K}$, and let $(YD_i \xrightarrow{c_i} C)_{i\in I}$ be a colimit of YD in $\mathbf{Set}^{\mathcal{K}^{op}}$. It is our task to show that C is a representable functor. To this end, we first observe that the comma-category $YD \downarrow C$ (of all arrows $YD_i \to C$ in $\mathbf{Set}^{\mathcal{K}^{op}}$) is λ^+-filtered: this follows from the fact that $(I, \leq)$ is λ^+-directed, and that each YD_i is finitely presentable in $\mathbf{Set}^{\mathcal{K}^{op}}$. Next, we prove that the number of morphisms of the category $YD \downarrow C$ is smaller than λ. Since Y is full and $\mathcal{K}$ has less than λ morphisms, it is sufficient to verify that the number of objects of $YD \downarrow C$ is smaller than λ. In fact, each object $YD_i \to C$ factorizes through some c_j, $j \in I$ (since YD_i is finitely presentable); thus, given λ distinct objects of $YD \downarrow C$, there exists $j_0 \in I$ such that they all factorize through c_{j_0} (in fact, $(I, \leq)$ is λ^+-directed). Since Y is full, this contradicts card $\bigcup_{i\in I} \hom(D_i, D_{j_0}) < \lambda$. Consequently, the λ^+-filtered category $YD \downarrow C$ contains a compatible cocone for itself. In other words, it contains an object $YD_{i_0} \xrightarrow{a} C$ such that there is a map which to each object $YD_i \xrightarrow{b} C$ assigns a morphism $b^*\colon YD_i \to YD_{i_0}$ with $b = a \cdot b^*$, compatibly with morphisms of $YD \downarrow C$, i.e., such that $b^* \cdot f = (b \cdot f)^*$ for each $f\colon YD_{i'} \to YD_i$. Consequently, we get a compatible cocone $\left(YD_i \xrightarrow{d_i^*} YD_{i_0}\right)_{i\in I}$ for YD. The unique factorization $\overline{a}\colon C \to YD_{i_0}$ with $\overline{a} \cdot d_i = d_i^*$ is an inverse to a; from

$$(a \cdot \overline{a}) \cdot d_i = a \cdot d_i^* = d_i \qquad (i \in I)$$

we conclude that $a \cdot \overline{a} = id$ and, since $a^* = id_{YD_{i_0}}$, from

$$a^* \cdot (\overline{a} \cdot a) = (a \cdot \overline{a} \cdot a)^* = a^*$$

we conclude that $\overline{a} \cdot a = id$. Thus, C is isomorphic to YD_{i_0}. □

2.7 Remark. It follows from the above proof that a category with split idempotents and with λ morphisms is $(\lambda^+)^+$-accessible. But it need not be $(\lambda)^+$-accessible: we will exhibit a countable category with split idempotents which is not ω_1-accessible, as follows.

Construct a countable infinite monoid C of self-maps of ω with the following properties:

(i) each $f \in C$ is one-to-one,

(ii) for each $f \in C - \{id\}$ the complement of $f(\omega)$ is infinite,

(iii) given $f, g \in C - \{id\}$ there exists $h \in C - \{id\}$ with $f = h \cdot g$.

Define an ω_1-chain $d_{i,j}$ $(i \leq j < \omega_1)$ in the one-object category C with $d_{i,j} \neq id$ for all $i < j$ by the following transfinite induction on j. The first step is $d_{0,0} = id$, and the isolated step is easy: given $d_{i,j}$, choose any element $d_{j,j+1}$ in $C - \{id\}$, and define $d_{i,j+1} = d_{j,j+1} \cdot d_{i,j}$. For the limit step, first express j as $j = \bigvee_{n \in \omega} i_n$ with $i_0 < i_1 < \cdots$, and define $d_{i_n,j}$ by induction on n: $d_{i_0,j}$ is any element of $C - \{id\}$; given $d_{i_n,j}$, there exists, by (iii) above, an element $d_{i_{n+1},j}$ in $C - \{id\}$ with $d_{i_{n+1},j} \cdot d_{i_n,i_{n+1}} = d_{i_n,j}$. We now define $d_{i,j} = d_{i_n,j} \cdot d_{i,i_n}$ for the smallest $n \in \omega$ with $i \le i_n$.

Suppose that C is an ω_1-accessible category. Then its unique object X is ω_1-presentable. Also, the above ω_1-chain, being ω_1-directed, has a colimit whose codomain is X, of course. It follows that id factorizes through one of the colimit maps of the considered colimit, but this is a contradiction, since id has no non-trivial factorization in C, due to (ii) above.

2.8. Recall the concept of a canonical functor $E\colon \mathcal{K} \to \mathbf{Set}^{\mathcal{A}^{\mathrm{op}}}$ from Notation 1.25.

Proposition. *If $\mathcal{K}$ is a λ-accessible category, then*

(i) $\mathbf{Pres}_\lambda\, \mathcal{K}$ *is dense in* $\mathcal{K}$,

(ii) *all canonical diagrams w.r.t.* $\mathbf{Pres}_\lambda\, \mathcal{K}$ *are λ-filtered,*

(iii) *the canonical functor* $E\colon \mathcal{K} \to \mathbf{Set}^{\mathcal{A}^{\mathrm{op}}}$, *where* $\mathcal{A} = \mathbf{Pres}_\lambda\, \mathcal{K}$, *is full, faithful and preserves λ-directed colimits.*

PROOF. (i) and (ii). Consider an arbitrary object K of $\mathcal{K}$ and express it as a colimit of a λ-directed diagram D with a colimit cocone $(D_i \xrightarrow{d_i} K)_{i \in I}$, where the D_i are λ-presentable. Then each morphism $f\colon A \to K$, with A λ-presentable, factorizes through some d_i, and given two such factorizations, $f = d_i \cdot p = d_{i'} \cdot p'$, there exists an upper bound j of i and i' such that $D(i \to j) \cdot p = D(i' \to j) \cdot p'$. Consequently, K is a canonical colimit of λ-presentable objects (see Exercise 1.o), and the canonical diagram is λ-filtered.

(iii) See Proposition 1.26. □

Raising the Index of Accessibility

We know that each locally λ-presentable category is locally μ-presentable for all regular cardinals $\mu \ge \lambda$ (see Remark 1.20). The corresponding result for λ-accessible categories is not true, as we presently demonstrate. However, each λ-accessible category will be proved to be μ-accessible for arbitrarily large regular cardinals μ.

2.9 Notation. For two cardinals β and λ we denote

$$\beta^{<\lambda} = \sum_{\alpha<\lambda} \beta^\alpha$$

(where β^α is the cardinality of the set of all functions from α to β). Observe that for each set X of cardinality $\geq \lambda$ we have

$$\operatorname{card} P_\lambda(X) = (\operatorname{card} X)^{<\lambda}$$

where $P_\lambda(X)$ denotes the collection of all subsets of X of cardinality smaller than λ.

2.10 Lemma. *If λ is a regular cardinal, then $\left(\beta^{<\lambda}\right)^{<\lambda} = \beta^{<\lambda}$ for each cardinal $\beta \geq \lambda$.*

Remark. Equivalently: $\left(\beta^{<\lambda}\right)^\alpha = \beta^{<\lambda}$ for each cardinal $\alpha < \lambda$.

PROOF. We shall prove the remark. We consider each cardinal as the set of all smaller ordinals. Let F be the set of all maps from α to λ. Since λ is regular and $\alpha < \lambda \leq \beta$, we have

$$\operatorname{card} F = \lambda^\alpha \leq \beta^\alpha \leq \beta^{<\lambda}$$

and for each $f \in F$

$$\beta^{\sum_{\gamma<\alpha} \operatorname{card} f(\gamma)} \leq \beta^{<\lambda}.$$

Therefore

$$\begin{aligned}
\left(\beta^{<\lambda}\right)^\alpha &= \left(\sum_{\delta<\lambda} \beta^\delta\right)^\alpha \\
&= \sum_{f\in F} \prod_{\gamma<\alpha} \beta^{\operatorname{card} f(\gamma)} \\
&= \sum_{f\in F} \beta^{\sum_{\gamma<\alpha} \operatorname{card} f(\gamma)} \\
&\leq \sum_{f\in F} \beta^{<\lambda} \\
&\leq \beta^{<\lambda}.
\end{aligned}$$

□

2.11 Theorem. *For regular cardinals $\lambda < \mu$ the following conditions are equivalent:*

(i) *Each λ-accessible category is μ-accessible.*

(ii) *The (λ-accessible) category* $\mathbf{Pos}_\lambda$ *of Example 2.3(7) is* μ*-accessible.*

(iii) *For each set* X *of less than* μ *elements the poset* $P_\lambda(X)$ *has a cofinal set of cardinality less than* μ.

(iv) *In each* λ*-directed poset every subset of less than* μ *elements is contained in a* λ*-directed subset of less than* μ *elements.*

PROOF. (i) $\Rightarrow$ (ii) is clear.

(ii) $\Rightarrow$ (iii). We will first show that an object A in $\mathbf{Pos}_\lambda$ is μ-presentable iff $\operatorname{card} A < \mu$. Sufficiency is clear. To show necessity, let A in $\mathbf{Pos}_\lambda$ be μ-presentable, and let $A' = A \cup \{T\}$ be the extension of A by a greatest element T. For every subposet $B \subseteq A$ with $\operatorname{card} B < \mu$ put $B' = B \cup \{T\}$. Since A' is a μ-directed union of all those subposets B' ($B \subseteq A$, $\operatorname{card} B < \mu$) and each B' lies in $\mathbf{Pos}_\lambda$, the inclusion $A \to A'$ factorizes through one of those subposets. Therefore $\operatorname{card} A < \mu$.

The poset $P_\lambda(X)$ is an object of $\mathbf{Pos}_\lambda$. Thus, it is a μ-directed colimit of μ-presentable objects. The images of the colimit maps form a collection Y_i ($i \in I$) of λ-directed subsets of $P_\lambda(X)$ of cardinalities less than μ; moreover, the collection is μ-directed. Each singleton subset of X lies in some Y_i, and since X has less than μ elements, there exists $i_0 \in I$ such that Y_{i_0} contains all singleton sets. Since, moreover, Y_{i_0} is λ-directed, it is a cofinal set in $P_\lambda(X)$ of cardinality less than μ.

(iii) $\Rightarrow$ (iv). Let $(Z, \leq)$ be a λ-directed poset, and let $M \subseteq Z$ be a set of cardinality less than μ. Then we define a chain $M = M_0 \subseteq M_1 \subseteq \cdots \subseteq M_i \subseteq \dots$ ($i < \lambda$) of subsets of cardinalities less than μ such that $\bigcup_{i<\lambda} M_i$ is the required λ-directed subset. The limit steps are given by the union $M_i = \bigcup_{j<i} M_j$ (which satisfies the cardinality condition since $\operatorname{card} i < \mu$ and μ is regular). For the isolated steps, given M_i, choose a cofinal set Y in $P_\lambda(M_i)$ of cardinality less than μ. For any $A \in Y$ we can choose an upper bound z_A in Z. Put

$$M_{i+1} = \bigcup_{A \in Y} A \cup \{z_A\}.$$

It is clear that $\bigcup_{i<\lambda} M_i$ is λ-directed and has a cardinality less than μ.

(iv) $\Rightarrow$ (i) Each λ-accessible category $\mathcal{K}$ has μ-directed colimits. Let $\mathcal{A}$ be the collection of all objects which are λ-directed colimits of less than μ objects in $\mathbf{Pres}_\lambda \mathcal{K}$; since $\mathbf{Pres}_\lambda \mathcal{K}$ is small, so is $\mathcal{A}$. Each $\mathcal{A}$-object is μ-presentable by Proposition 1.16. It is now sufficient to prove that each object K in $\mathcal{K}$ is a μ-directed colimit of $\mathcal{A}$-objects. Let D be a λ-directed diagram in $\mathbf{Pres}_\lambda \mathcal{K}$ with a colimit $(A_i \xrightarrow{a_i} K)_{i \in I}$. Let $\hat{I}$ be the poset of all λ-directed subsets of I of cardinalities less than μ (ordered by inclusion). By (iv), $\hat{I}$ is μ-directed. For each $M \in \hat{I}$ let $B_M \in \mathcal{A}$ be a colimit of the subdiagram of D indexed by M. Then for $M \subseteq M'$ we have a canonical morphism $b_{M,M'}\colon B_M \to B_{M'}$ as well as a canonical morphism $b_M\colon B_M \to K = \operatorname{colim} D$. The diagram $\hat{D}$ of all $b_{M,M'}\colon B_M \to B_{M'}$ ($M \subseteq M'$ in $\hat{I}$) is μ-directed, and it has a colimit $(B_M \xrightarrow{b_M} K)_{M\in\hat{I}}$. Consequently, $\mathcal{K}$ is μ-accessible. □

2.12 Definition. A regular cardinal λ is said to be *sharply smaller* than a regular cardinal μ provided that the equivalent conditions of the preceding theorem are satisfied. Notation:

$$\lambda \lhd \mu$$

2.13 Examples

(1) $\omega \lhd \lambda$ for each uncountable regular cardinal λ. (See Theorem 2.11(iii).)

(2) For each regular cardinal λ we have

$$\lambda \lhd \lambda^+$$

(where λ^+ denotes the cardinal successor of λ). In fact, if X has cardinality less than λ^+, i.e., $\operatorname{card} X \leq \lambda$, then we can write $X = \{x_i \mid i < \lambda\}$ and the desired cofinal subset of $P_\lambda(X)$ consists of $\{x_i \mid i < \alpha\}$ for all $\alpha < \lambda$.

(3) For arbitrary regular cardinals λ, μ we have

$$\lambda \leq \mu \quad \Rightarrow \quad \lambda \lhd (2^\mu)^+.$$

In fact, given cardinals $\alpha < \lambda \leq \mu$ and $\beta < \left(2^\mu\right)^+$, then

$$\beta^\alpha \leq 2^{\mu\cdot\alpha} = 2^\mu < \left(2^\mu\right)^+$$

and the implication is a consequence of the following fact:

(4) Given regular cardinals $\lambda < \mu$ such that for all cardinals $\alpha < \lambda$ and $\beta < \mu$ we have $\beta^\alpha < \mu$, then $\lambda \lhd \mu$. Indeed, let $\operatorname{card} X = \beta < \mu$. Then $\operatorname{card} P_\lambda(X) = \sum_{\alpha<\lambda} \beta^\alpha < \mu$ and we can take $P_\lambda(X)$ as a cofinal subset of itself.

(5) For any regular cardinal μ, we have

$$\mu \lhd \left(\mu^{<\mu}\right)^+.$$

This follows from (4) because for $\alpha < \mu$ and $\beta \leq \mu^{<\mu}$ we have $\beta^\alpha \leq \left(\mu^{<\mu}\right)^\alpha = \mu^{<\mu}$ (see Remark 2.10).

(6) For each set L of regular cardinals there exist arbitrarily large regular cardinals μ with $\lambda \lhd \mu$ for each $\lambda \in L$. This follows from (3).

(7) The relation $\lhd$ is transitive (see Theorem 2.11(i)).

(8) $\aleph_1 \lhd \aleph_{\omega+1}$ does not hold. In fact, consider a set $X = \bigcup_{n<\omega} X_n$ with $\operatorname{card} X_n = \aleph_n$. Then $\operatorname{card} X = \aleph_\omega$. No subset $\mathcal{A}$ of $P_{\aleph_1}(X)$ of cardinality less than $\aleph_{\omega+1}$ is cofinal in $P_{\aleph_1}(X)$. To show this, choose a surjective function $f\colon X \to \mathcal{A}$. For each $n < \omega$ choose $z_n \in X_{n+1} - \bigcup_{x \in X_n} f(x)$; this is possible since $\operatorname{card} \bigcup_{x\in X_n} f(x) = \aleph_n$. We will show that $\mathcal{A}$ is not cofinal in $P_{\aleph_1}(X)$ since the set $Z = \{z_n \mid n < \omega\}$ is not contained in any element of $\mathcal{A}$. In fact, if $Z \subseteq A \in \mathcal{A}$, then for $x \in X$ with $A = f(x)$ we can find n with $x \in X_n$. It follows that $z_n \notin A$, in contradiction to $Z \subseteq A$.

2.14 Corollary. *For each accessible category $\mathcal{K}$ there exist arbitrarily large regular cardinals λ such that $\mathcal{K}$ is λ-accessible.*

Remark. In fact, for any set $\mathcal{K}_i$, $i \in I$, of accessible categories there exist arbitrarily large cardinals λ such that $\mathcal{K}_i$ is λ-accessible for each $i \in I$.

PROOF. Combine (i) of Theorem 2.11 and Example 2.13(6). □

2.15 Remark. Let $\mathcal{K}$ be a λ-accessible category. Given $\mu \rhd \lambda$, an object of $\mathcal{K}$ is μ-presentable iff it is a split subobject of a μ-small λ-directed colimit of λ-presentable objects. In fact:

(1) Each colimit of a μ-small λ-directed diagram in $\mathbf{Pres}_\lambda\, \mathcal{K}$ is μ-presentable, see Proposition 1.16, and each split subobject of a μ-presentable object is μ-presentable (see Remark 1.16).

(2) Conversely, given a μ-presentable object K and given a λ-directed diagram D of λ-presentable objects with a colimit $(A_i \xrightarrow{a_i} K)_{i\in I}$, we form a new diagram $\widehat{D}$ as follows. Let $\hat{I}$ be the set of all λ-directed subsets of I of cardinality smaller than μ. Since $\lambda \lhd \mu$, it follows from Theorem 2.11(iv) that $\hat{I}$, ordered by inclusion, is μ-directed. For each $M \in \hat{I}$ let B_M be a colimit object of the restriction of D to M. This defines a μ-directed diagram $\widehat{D}$ with canonical morphisms $B_M \to B_{M'}$ for $M \subseteq M'$ in $\hat{I}$. A colimit of $\widehat{D}$ is formed by the canonical morphisms $b_M\colon B_M \to K = \operatorname{colim} D$. Since K is μ-presentable, id_K factorizes through some b_M. Thus, K is a split subobject of B_M.

Let us add that a stronger result holds: μ-presentable objects are μ-small λ-directed colimits of λ-presentable objects. The proof (substantially more complicated then the one above) can be found in [Makkai, Paré 1989] (Proposition 2.3.11).

2.B Accessible Functors

2.16 Definition. A functor $F\colon \mathcal{K} \to \mathcal{L}$ is called *λ-accessible* (where λ is a regular cardinal) if $\mathcal{K}$ and $\mathcal{L}$ are λ-accessible categories, and F preserves λ-directed colimits.

A functor is called *accessible* if it is λ-accessible for some regular cardinal λ.

2.17 Examples

(1) For each accessible category $\mathcal{K}$ all hom-functors $\hom(K, -)\colon \mathcal{K} \to \mathbf{Set}$ are accessible. In fact, if K is λ-presentable, then $\hom(K, -)$ preserves λ-directed colimits.

(2) Let $\mathcal{K}$ be an accessible category. A functor $F\colon \mathcal{K} \to \mathbf{Set}$ is accessible iff it is a colimit of a small diagram of hom-functors.

In fact, each hom-functor $\hom(A, -)$, where A is λ-presentable in $\mathcal{K}$, preserves λ-directed colimits. Thus, for each small diagram of hom-functors there exists a regular cardinal λ such that (i) $\mathcal{K}$ is λ-accessible, (ii) the diagram is λ-small, and (iii) each of the hom-functors preserves λ-directed colimits (see Corollary 2.14). It follows that the colimit of that diagram is a λ-accessible functor.

Conversely, let $F\colon \mathcal{K} \to \mathbf{Set}$ be λ-accessible. Since $\mathbf{Pres}_\lambda\, \mathcal{K}$ is small, the restriction F_0 of F to $\mathbf{Pres}_\lambda\, \mathcal{K}$ is a colimit of a (small) diagram of hom-functors $\hom(A_i, -)\colon \mathbf{Pres}_\lambda\, \mathcal{K} \to \mathbf{Set}$ $(i \in I)$. It is evident that F is a colimit of the corresponding diagram of hom-functors $\hom(A_i, -)\colon \mathcal{K} \to \mathbf{Set}$ $(i \in I)$.

(3) Let $\mathcal{K}$ be a λ-accessible category. For a full subcategory $\mathcal{A}$ the inclusion functor $\mathcal{A} \hookrightarrow \mathcal{K}$ is λ-accessible iff $\mathcal{A}$ is λ-accessible and closed under λ-directed colimits. Whether or not *each* inclusion of a full accessible subcategory into an accessible category is an accessible functor is undecidable—see Theorem 6.9 and Example 6.13.

(4) An example of a functor $\mathbf{Set} \to \mathbf{Set}$ which is not accessible is the power-set functor.

2.18 Remarks

(1) A λ-accessible functor from $\mathcal{K}$ to $\mathcal{L}$ is determined, up to natural isomorphism, by its domain restriction to $\mathbf{Pres}_\lambda\, \mathcal{K}$.

(2) Every λ-accessible functor is λ'-accessible for all regular cardinal cardinals $\lambda' \rhd \lambda$. In particular, a composite of two accessible functors is accessible.

In fact, if $F: \mathcal{K} \to \mathcal{L}$ is λ-accessible and $\lambda' \rhd \lambda$ then both $\mathcal{K}$ and $\mathcal{L}$ are λ'-accessible categories. Since F preserves λ-directed colimits, it preserves λ'-directed colimits, of course.

(3) Thus, each accessible functor is λ-accessible for arbitrarily large regular cardinals λ. More can be said:

2.19 Uniformization Theorem. *For each accessible functor $F: \mathcal{K} \to \mathcal{L}$ there exist arbitrarily large regular cardinals λ such that F is λ-accessible and preserves λ-presentable objects (i.e., if K is a λ-presentable object of $\mathcal{K}$, then FK is a λ-presentable object of $\mathcal{L}$).*

Remark. In fact, a more general result can be proved: for any small collection F_i $(i \in I)$ of accessible functors there exist arbitrarily large regular cardinals λ such that each F_i is λ-accessible and preserves λ-presentability.

PROOF. Let $F_i: \mathcal{K} \to \mathcal{L}$ be λ_0-accessible $(i \in I)$, and let λ_1 be a regular cardinal such that $\lambda_0 \lhd \lambda_1$ and for each λ_0-presentable object K of $\mathcal{K}$ and each $i \in I$ the object F_iK is λ_1-presentable in $\mathcal{L}$. For each regular cardinal $\lambda \rhd \lambda_1$ we know that the F_i are λ-accessible. Let us prove that whenever K is a λ-presentable object of $\mathcal{K}$ and $i \in I$, then F_iK is λ-presentable in $\mathcal{L}$. By Remark 2.15, K is a split subobject of a colimit of a λ_0-directed, λ-small diagram in $\mathbf{Pres}_{\lambda_0} \mathcal{K}$. Since F_i preserves that colimit, we see that F_iK is a split subobject of a colimit of a λ_0-directed, λ-small diagram in $F_i(\mathbf{Pres}_{\lambda_0} \mathcal{K}) \subseteq \mathbf{Pres}_{\lambda_1} \mathcal{L}$. It follows (by Proposition 1.16) that F_iK is λ-presentable. □

2.20 Remark. If a λ-accessible functor F preserves λ-presentable objects and if $\lambda \lhd \mu$, then F preserves μ-presentable objects as well. This follows from Remark 2.15.

2.21 Proposition. *Let $\mathcal{K}$ be a λ-accessible category. Then all hom-functors of λ-presentable objects jointly reflect*

(i) *isomorphisms (i.e., a morphism f is an isomorphism in $\mathcal{K}$ if and only if* $\hom(K, f)$ *is an isomorphism in* **Set** *for each λ-presentable object K),*

and

(ii) *λ-directed colimits (i.e., a compatible cocone of a λ-directed diagram D in $\mathcal{K}$ is a colimit of D iff its image under* $\hom(K, -)$ *is a colimit of* $\hom(K, -) \cdot D$ *for each λ-presentable object K).*

Proof. (i). Suppose that for $f\colon A \to B$, all $\hom(K,f)\colon \hom(K,A) \to \hom(K,B)$ with K λ-presentable are isomorphisms. Let D be a λ-directed diagram of λ-presentable objects with a colimit $(Dd \xrightarrow{b_d} B)_{d\in\mathcal{D}^{\mathrm{obj}}}$. For each d there exists a unique $b_d^*\colon Dd \to A$ with $b_d = f \cdot b_d^*$ (since $\hom(Dd,f)$ is a bijection). The cocone of all b_d^*'s is compatible with D (due to uniqueness). The unique morphism $f^*\colon B \to A$ with $f^* \cdot b_d = b_d^*$ for all $d \in \mathcal{D}^{\mathrm{obj}}$ satisfies $f \cdot f^* \cdot b_d = b_d$ for each d, thus, $f \cdot f^* = id$. Besides, f is a monomorphism because all λ-presentable objects form a generator of $\mathcal{K}$, and for each λ-presentable object K, $\hom(K,f)$ is a monomorphism. Thus, $f^* = f^{-1}$.

(ii). Let $(Dd \xrightarrow{c_d} C)$ be an arbitrary cocone with the above property, and let $(Dd \xrightarrow{c_d^*} C^*)$ be a colimit of D in $\mathcal{K}$. The factorization map $u\colon C^* \to C$ (defined by $c_d = u \cdot c_d^*$) is an isomorphism by (i) above: for each λ-presentable object K in $\mathcal{K}$, $\hom(K,u)$ is, obviously, an isomorphism in **Set**. □

2.22 Corollary. *(Hom-functors are generic accessible functors.) Let $\mathcal{K}$ and $\mathcal{L}$ be accessible categories. A functor $F\colon \mathcal{K} \to \mathcal{L}$ is accessible iff for each hom-functor* $\hom(L,-)\colon \mathcal{L} \to$ **Set** *the composite* $\hom(L,-)\cdot F\colon \mathcal{K} \to$ **Set** *is accessible.*

Proof. If F is accessible, then its composite with each of the (accessible, cf. Example 2.17(1)) hom-functors is accessible. Conversely, let $F\colon \mathcal{K} \to \mathcal{L}$ have accessible composites with hom-functors. There exists a regular cardinal λ such that $\mathcal{K}$ and $\mathcal{L}$ are λ-accessible. We can find a regular cardinal $\lambda' \rhd \lambda$ such that $\hom(L,-)\cdot F$ is λ'-accessible for each λ-presentable object L of $\mathcal{L}$. Since $\mathcal{L}$ is λ-accessible, the hom-functors of λ-presentable objects reflect λ-directed (and thus, λ'-directed) colimits (by Proposition 2.21(ii)), hence, F preserves λ'-directed colimits. □

2.23 Proposition. *Every left or right adjoint between accessible categories is accessible.*

Proof. Let $\mathcal{K}$ and $\mathcal{L}$ be λ-accessible categories. Every left adjoint

$$G\colon \mathcal{K} \to \mathcal{L}$$

preserves (λ-directed) colimits, and hence, is λ-accessible. For every right adjoint $F\colon \mathcal{K} \to \mathcal{L}$ we choose a left adjoint $G \dashv F$; due to Theorem 2.19, we can assume that G preserves λ-presentable objects. We shall prove that F

then preserves λ-directed colimits. Let $D: I \to \mathcal{K}$ be a λ-directed diagram in $\mathcal{K}$, and let

$$u: \operatorname{colim} FD_i \to F(\operatorname{colim} D_i)$$

be the induced morphism. To prove that u is an isomorphism, it is sufficient, by Proposition 2.21, to show that each $\hom(L,-)$ with L λ-presentable maps u to an isomorphism

$$\hom(L,u): \hom(L, \operatorname{colim} FD_i) \to \hom\big(L, F(\operatorname{colim} D_i)\big).$$

In fact, $\hom(L,u)$ is the composite of the following isomorphisms:

$$\begin{aligned}\hom(L, \operatorname{colim} FD_i) &\cong \operatorname{colim} \hom(L, FD_i) && (L \text{ is } \lambda\text{-presentable})\\ &\cong \operatorname{colim} \hom(GL, D_i) && (G \dashv F)\\ &\cong \hom(GL, \operatorname{colim} D_i) && (GL \text{ is } \lambda\text{-presentable})\\ &\cong \hom(L, F(\operatorname{colim} D_i)) && (G \dashv F).\end{aligned}$$

□

2.C Accessible Categories as Free Cocompletions

In the present section we prove a representation theorem for λ-accessible categories: they are just the free completions of small categories w.r.t. λ-directed colimits. (We will later prove other equivalent conditions.) Recall from Theorem 1.46 that a category is locally λ-presentable iff it is equivalent to a λ-free cocompletion of a small category.

2.24. Recall from Remark 1.45 the concept of a Kan extension

$$F^*: \mathbf{Set}^{\mathcal{A}^{op}} \to \mathbf{Set}$$

of a functor $F: \mathcal{A} \to \mathbf{Set}$.

Lemma. *For each functor $F: \mathcal{A} \to \mathbf{Set}$, $\mathcal{A}$ small, the following conditions are equivalent:*

(i) *F is a λ-directed colimit of hom-functors in $\mathbf{Set}^{\mathcal{A}}$;*

(ii) *F^* preserves λ-small limits;*

(iii) *F^* preserves λ-small limits of hom-functors (i.e., limits of λ-small diagrams in $\mathbf{Set}^{\mathcal{A}^{op}}$ factorizing through $Y: \mathcal{A} \to \mathbf{Set}^{\mathcal{A}^{op}}$).*

PROOF. (i) ⇒ (ii). Observe that hom-functors have the property that a Kan extension can always be chosen as a hom-functor:

$$\hom(A,-)^* = \hom(\hom(-,A),-)$$

This follows from the fact that for each functor $S\colon \mathbf{Set}^{\mathcal{A}^{\mathrm{op}}} \to \mathbf{Set}$ we have, by the Yoneda lemma, the following bijective correspondence:

$$\frac{\hom(A,-) \to SY}{\hom(\hom(-,A),-) \to S}.$$

Let $F = \operatorname{colim} H_i$ be a λ-directed colimit of hom-functors. Then $F^* = \operatorname{colim} H_i^*$ (see Exercise 1.p(3)) is a λ-directed colimit of hom-functors, and we conclude that F^* preserves λ-small limits (since, by Proposition 1.59, they commute in **Set** with λ-directed colimits).

(ii) ⇒ (iii) is clear.

(iii) ⇒ (i). Let F^* preserve λ-small limits of hom-functors. We will prove that the canonical diagram

$$D\colon Y'(\mathcal{A}^{\mathrm{op}}) \downarrow F \to \mathbf{Set}^{\mathcal{A}}$$

of F is λ-filtered (where $Y'\colon \mathcal{A}^{\mathrm{op}} \to \mathbf{Set}^{\mathcal{A}}$ is the dual Yoneda embedding). It follows that F is a λ-filtered (thus, also a λ-directed) colimit of hom-functors. Let $\mathcal{D}_0$ be a subcategory of $Y'(\mathcal{A}^{\mathrm{op}}) \downarrow F$ of less than λ morphisms, let $\hom(A_i,-) \to F$ $(i \in I)$ be the objects of $\mathcal{D}_0$, and let $a_i \in FA_i$ be the corresponding points. Let $D_0\colon \mathcal{D}_0 \to \mathcal{A}^{\mathrm{op}}$ be the composition of the domain restriction of D with the canonical isomorphism of $Y'(\mathcal{A}^{\mathrm{op}})$ with $\mathcal{A}^{\mathrm{op}}$. We form a limit $\big(H \xrightarrow{f_i} \hom(-,A_i)\big)_{i\in I}$ of $YD_0^{\mathrm{op}}\colon \mathcal{D}_0^{\mathrm{op}} \to \mathbf{Set}^{\mathcal{A}^{\mathrm{op}}}$, and by assumption we have a limit

$$\big(F^*(H) \xrightarrow{F^*f_i} F^*(\hom(-,A_i)) = FA_i\big)_{i\in I}$$

of $F^*YD_0^{\mathrm{op}}$ in **Set**. Consequently, for the (compatible) collection of points $a_i \in FA_i$ there exists a unique point $b \in F^*(H)$ with $a_i = (F^*f_i)(b)$ for all $i \in I$. Furthermore, F^* preserves the canonical colimit of H w.r.t. hom-functors; thus, for $b \in F^*(H)$ there exists a morphism $f\colon \hom(-,A) \to H$ and a point $a \in FA$ with $b = (F^*f)(a)$. Consequently,

$$a_i = F^*(f_i \cdot f)(a) \qquad \text{for all } i \in I.$$

Therefore, the object

$$\hom(A,-) \to F \text{ of } Y(\mathcal{A}^{\mathrm{op}}) \downarrow F$$

corresponding to $a \in FA$ is a codomain of a compatible cocone of the diagram $\mathcal{D}_0$ in $Y(\mathcal{A}^{\mathrm{op}}) \downarrow F$. □

2.25 Definition. Let $\mathcal{A}$ be a small category, and λ a regular cardinal. A full embedding $E\colon \mathcal{A} \to \mathcal{A}^*$ is called a *free cocompletion of $\mathcal{A}$ with respect to λ-directed colimits* provided that $\mathcal{A}^*$ has λ-directed colimits and that for each functor $F\colon \mathcal{A} \to \mathcal{B}$ such that $\mathcal{B}$ has λ-directed colimits there exists an extension $F^*\colon \mathcal{A}^* \to \mathcal{B}$ preserving λ-directed colimits, unique up to a natural isomorphism.

2.26 Representation Theorem. *For each regular cardinal λ and each category $\mathcal{K}$ the following conditions are equivalent:*

(i) *$\mathcal{K}$ is λ-accessible,*

(ii) *$\mathcal{K}$ is a free cocompletion of a small category $\mathcal{A}$ with respect to λ-directed colimits,*

(iii) *$\mathcal{K}$ is equivalent to the full subcategory of $\mathbf{Set}^{\mathcal{A}^{op}}$ formed by all λ-directed colimits of hom-functors for some small category $\mathcal{A}$,*

(iv) *$\mathcal{K}$ is equivalent to the category of all functors from $\mathbf{Set}^{\mathcal{A}}$ to $\mathbf{Set}$ preserving colimits and λ-small limits (and of all natural transformations) for some small category $\mathcal{A}$.*

Remarks

(1) In conditions (ii)–(iv) we can choose as $\mathcal{A}$ the subcategory $\mathbf{Pres}_\lambda\,\mathcal{K}$. This generalizes the well-known fact that each Scott domain is a free completion of its finite elements.

(2) The condition (iv) represents $\mathcal{K}$ as a full subcategory of $\mathbf{Set}^{\mathbf{Set}^{\mathcal{A}}}$, which is not a legitimate category. However, we will see in Theorem 2.58 that a natural restriction can be made as follows: there is a small subcategory $\mathcal{L}$ of $\mathbf{Set}^{\mathcal{A}}$ and a small collection of colimits in $\mathbf{Set}^{\mathcal{L}}$ such that $\mathcal{K}$ is equivalent to the category of all functors in $\mathbf{Set}^{\mathcal{L}}$ preserving the specified colimits and λ-small limits.

PROOF. (iv) $\Leftrightarrow$ (iii) by Lemma 2.24 and Proposition 1.45(i) applied to $\mathcal{A}^{op}$.

(iii) $\Rightarrow$ (i). Since $\mathcal{K}$ is closed under λ-directed colimits in $\mathbf{Set}^{\mathcal{A}^{op}}$, all hom-functors are λ-presentable in $\mathcal{K}$ (since they are finitely presentable in $\mathbf{Set}^{\mathcal{A}^{op}}$, see Example 1.2(7)).

(i) $\Rightarrow$ (ii). Let $F\colon \mathbf{Pres}_\lambda\,\mathcal{K} \to \mathcal{B}$ be a functor, and let $\mathcal{B}$ have λ-directed colimits. For each object K of $\mathcal{K}$ we choose a colimit of FD_K in $\mathcal{B}$, where

$D_K\colon \mathbf{Pres}_\lambda \mathcal{K} \downarrow K \to \mathbf{Pres}_\lambda \mathcal{K}$ is the canonical (λ-filtered, see Proposition 2.8) diagram of K, with a colimit cocone

$$\left(FA \xrightarrow{a^*} F^*K\right)_{a\colon A\to K}$$

The only condition put upon our choice is that if K lies in $\mathbf{Pres}_\lambda \mathcal{K}$, we choose $F^*K = FK$ and $a^* = Fa$. For each morphism $f\colon K_1 \to K_2$ in $\mathcal{K}$ we have a unique morphism

$$F^*f\colon F^*K_1 \to F^*K_2 \qquad \text{with} \quad F^*f \cdot a^* = (f \cdot a)^*.$$

It is easy to verify that $F^*\colon \mathcal{K} \to \mathcal{B}$ is a well-defined functor extending F. Since every object of $\mathbf{Pres}_\lambda \mathcal{K}$ is λ-presentable, F^* preserves λ-directed colimits: whenever $(K_i \xrightarrow{k_i} K)_{i\in I}$ is a λ-directed colimit of D in $\mathcal{K}$, then each morphism $a \in \mathbf{Pres}_\lambda \mathcal{K} \downarrow K$ factorizes (essentially uniquely) through some morphism in $\mathbf{Pres}_\lambda \mathcal{K} \downarrow K_i$, thus, $(F^*K_i \xrightarrow{F^*k_i} F^*K)_{i\in I}$ is a colimit of F^*D in $\mathcal{B}$. It is obvious that F^* is unique up-to natural isomorphism.

(ii) $\Rightarrow$ (iii). It is our task to show that for each small category $\mathcal{A}$, if $\mathcal{K}$ denotes the full subcategory of $\mathbf{Set}^{\mathcal{A}^{\mathrm{op}}}$ formed by all λ-directed colimits of hom-functors, then the Yoneda embedding $Y\colon \mathcal{A} \to \mathcal{K}$ is a free cocompletion w.r.t. λ-directed colimits. We know by (iii) $\Rightarrow$ (i) that $\mathcal{K}$ is λ-accessible.

(a) Suppose that $\mathcal{A}$ has split idempotents (see Observation 2.4). It follows from Remark 2.5(1) that in $\mathbf{Set}^{\mathcal{A}^{\mathrm{op}}}$ any split subobject of a hom-functor is naturally isomorphic to a hom-functor. Therefore λ-presentable objects of $\mathcal{K}$ are, up to natural isomorphism, precisely the hom-functors, thus, $\mathbf{Pres}_\lambda \mathcal{K}$ is equivalent to $\mathcal{A}$. By the already proved implication (i) $\Rightarrow$ (ii) we conclude that $\mathcal{K}$ is a free cocompletion of $\mathcal{A}$ w.r.t. λ-directed colimits.

(b) If $\mathcal{A}$ is an arbitrary small category, we can find a small extension $E\colon \mathcal{A} \to \widetilde{\mathcal{A}}$ with split idempotents which is universal in the sense of Remark 2.5(2). Then $\mathcal{A}$ and $\widetilde{\mathcal{A}}$ have the same free cocompletion w.r.t. λ-directed colimits (more precisely, if $E_0\colon \widetilde{\mathcal{A}} \to \mathcal{B}$ is such a cocompletion for $\widetilde{\mathcal{A}}$, then $E_0 \cdot E\colon \mathcal{A} \to \mathcal{B}$ is such a cocompletion for $\mathcal{A}$, and vice versa). Thus, (ii) for $\mathcal{A}$ implies (ii) for $\widetilde{\mathcal{A}}$, from which we know that (iii) follows. □

2.D Pure Subobjects

Pure subobjects are, roughly speaking, directed colimits of split subobjects. This is one of the central concepts of the theory of accessible categories. We will see, for example, that if a full subcategory of an accessible category is closed under λ-directed colimits, then it is accessible iff it is closed under

λ'-pure subobjects for some regular cardinal λ'. Although the concept of a pure subobject is rather abstract (subobjects are seldom pure), we are going to prove that every accessible category has "enough" pure subobjects.

Basic Properties of Pure Subobjects

2.27 Definition. A morphism $f: A \to B$ is said to be λ-*pure* (λ a regular cardinal) provided that in each commutative square

$$\begin{array}{ccc} A' & \xrightarrow{f'} & B' \\ {\scriptstyle u}\downarrow & & \downarrow{\scriptstyle v} \\ A & \xrightarrow[f]{} & B \end{array}$$

with A' and B' λ-presentable, u factorizes through f' (i.e., $u = \overline{u} \cdot f'$ for some $\overline{u}: B' \to A$).

2.28 Examples

(1) Split monomorphisms are λ-pure for any λ.

(2) In **Set**, λ-pure morphisms are precisely the split monomorphisms (for any λ).

(3) In **Pos** the embedding $f: \omega \to \omega \cup \{\infty\}$, where ∞ is the largest element of $\omega \cup \{\infty\}$, is ω-pure. In fact, given a commutative square as above with A' and B' finite, let $n \in \omega$ be an upper bound of the image of u, then the map

$$\overline{u}: B' \to \omega, \qquad \overline{u}(b) = \begin{cases} v(b) & \text{if } v(b) \neq \infty \\ n & \text{if } v(b) = \infty \end{cases}$$

is order-preserving, and $u = \overline{u} \cdot f'$.

(4) In the category of complete semilattices, all non-constant morphisms are λ-pure for any λ. (Recall that non-trivial complete semilattices are not presentable, see Example 1.14(4).)

Remarks

(1) A composite of λ-pure morphisms is λ-pure.

(2) If $f \cdot g$ is λ-pure, then g is λ-pure.

(3) Every λ-pure morphism is λ'-pure for all $\lambda' \geq \lambda$.

2.29 Proposition. *Every λ-pure morphism in a λ-accessible category is a monomorphism.*

PROOF. Let $f: A \to B$ be a λ-pure morphism in a λ-accessible category $\mathcal{K}$. Since λ-presentable objects form a generator $\mathbf{Pres}_\lambda\,\mathcal{K}$ of $\mathcal{K}$, it is sufficient to prove that, given morphisms $p, q: K \to A$ with K λ-presentable, then $f \cdot p = f \cdot q$ implies $p = q$. Since A is a filtered colimit of its canonical diagram w.r.t. $\mathbf{Pres}_\lambda\,\mathcal{K}$, there exists a morphism $u: A' \to A$ with A' λ-presentable such that both p and q factorize through u:

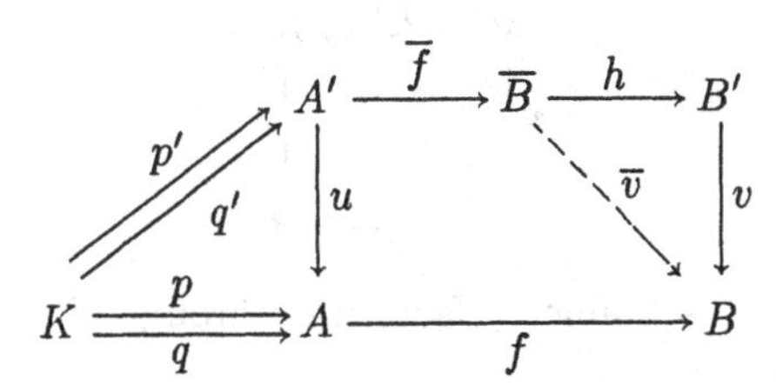

Since B is also a filtered colimit of its canonical diagram w.r.t. $\mathbf{Pres}_\lambda\,\mathcal{K}$, there exists a morphism $\overline{v}: \overline{B} \to B$ with $\overline{B}$ λ-presentable such that $f \cdot u$ factorizes through $\overline{v}$ (say, $f \cdot u = \overline{v} \cdot \overline{f}$), and then the equality

$$\overline{v} \cdot (\overline{f} \cdot p') = f \cdot p = f \cdot q = \overline{v} \cdot (\overline{f} \cdot q')$$

implies that there is a morphism $h: (\overline{B} \xrightarrow{\overline{v}} B) \to (B' \xrightarrow{v} B)$ in the canonical diagram of B with $h \cdot (\overline{f} \cdot p') = h \cdot (\overline{f} \cdot q')$. Since f is λ-pure, u factorizes through $f' = h \cdot \overline{f}$, say, $u = \overline{u} \cdot f'$. Then

$$p = u \cdot p' = \overline{u} \cdot h \cdot \overline{f} \cdot p' = \overline{u} \cdot h \cdot \overline{f} \cdot q' = u \cdot q' = q. \qquad \square$$

Remark. We thus speak about λ-pure subobjects rather than λ-pure morphisms. We are now going to show that λ-pure subobjects in a locally λ-presentable category $\mathcal{K}$ are just the λ-directed colimits of split subobjects. The colimit relates here to the category $\mathcal{K}^{\mathbf{2}}$ of $\mathcal{K}$-morphisms (cf. Example 1.55):

2.30 Proposition. (λ-pure subobjects are exactly the λ-directed colimits of split subobjects.)

(i) *If $\mathcal{K}$ is a λ-accessible category, then the full subcategory of $\mathcal{K}^{\mathbf{2}}$ consisting of λ-pure morphisms is closed under λ-directed colimits and contains all split subobjects.*

(ii) *If $\mathcal{K}$ is a locally λ-presentable category, then every λ-pure morphism in $\mathcal{K}$ is a λ-directed colimit (in $\mathcal{K}^{\mathbf{2}}$) of split monomorphisms.*

Proof. (i). Given a λ-directed diagram in $\mathcal{K}^2$ with a colimit

$$\left(f_i \xrightarrow{(a_i,b_i)} f\right)_{i\in I}$$

if each f_i is λ-pure, then so is f.

In fact, for each square as in Definition 2.27 there exists $i \in I$ with the following factorization:

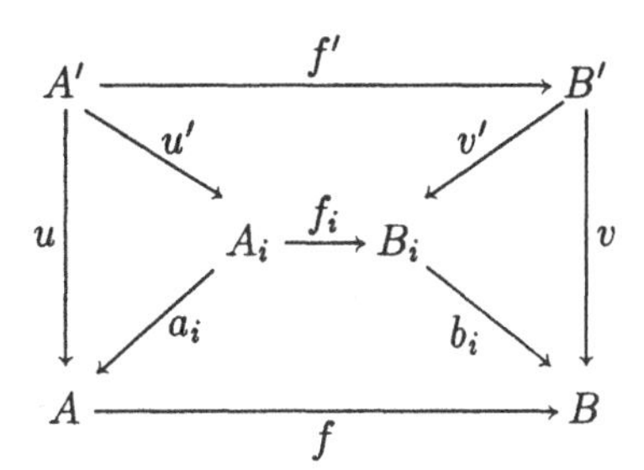

since, obviously, both $(A_i \xrightarrow{a_i} A)_{i\in I}$ and $(B_i \xrightarrow{b_i} B)_{i\in I}$ are λ-directed colimits in $\mathcal{K}$, and A' and B' are λ-presentable. Then u' factorizes through f', thus, u also factorizes through f'.

(ii) Let $f\colon A \to B$ be a λ-pure morphism in a locally λ-presentable category $\mathcal{K}$. Following Example 1.55(2), we can express f as a λ-directed colimit $\left(f_i \xrightarrow{(u_i,v_i)} f\right)_{i\in I}$ with $f_i\colon A_i \to B_i$ such that A_i, B_i are both λ-presentable $(i \in I)$:

$$\begin{array}{ccc} A_i & \xrightarrow{f_i} & B_i \\ {\scriptstyle u_i}\downarrow & & \downarrow{\scriptstyle v_i} \\ A & \xrightarrow[f]{} & B \end{array}$$

Since f is λ-pure, u_i factorizes through f_i for each $i \in I$; consequently, in the following pushout

$$\begin{array}{ccc} A_i & \xrightarrow{f_i} & B_i \\ {\scriptstyle u_i}\downarrow & & \downarrow{\scriptstyle \overline{u}_i} \\ A & \xrightarrow[\overline{f}_i]{} & \overline{B}_i \end{array}$$

$\overline{f}_i$ is a split monomorphism $(i \in I)$. The induced morphisms $\overline{b}_{i,j} : \overline{B}_i \to \overline{B}_j$, $i \le j$, constitute the λ-directed diagram $D = \left(\overline{f}_i \xrightarrow{(id_A,\overline{b}_{i,j})} \overline{f}_j\right)$ in $\mathcal{K}^2$. For the unique $\overline{v}_i \colon \overline{B}_i \to B$ with $f = \overline{v}_i \cdot \overline{f}_i$ and $v_i = \overline{v}_i \cdot \overline{u}_i$ we have a λ-directed colimit

$$(\overline{f}_i \xrightarrow{(id_A, \overline{v}_i)} f)_{i \in I}$$

of D in $\mathcal{K}^2$. □

Corollary. *In a locally λ-presentable category, λ-pure morphisms are precisely the λ-directed colimits of split monomorphisms.*

Remark. 2.30(ii) holds in every λ-accessible category with pushouts (use Exercise 2.c).

2.31. We have seen above that λ-pure morphisms in an accessible category are monomorphisms. It is an open problem whether they must be regular monomorphisms. This is true in any locally presentable category:

Proposition. *Every λ-pure morphism $f \colon A \to B$ in a locally λ-presentable category is a regular monomorphism.*

Remark. We will prove that f is an equalizer of a pair $B \rightrightarrows B^*$ such that B^* is a λ-directed colimit of powers of B.

PROOF. In Proposition 2.30 we have found a λ-directed collection of split monomorphisms $f_i \colon A \to B_i$ $(i \in I)$ whose colimit is f; more precisely, we have found a colimit $(B_i \xrightarrow{b_i} B)_{i \in I}$ of a λ-directed diagram D (of objects B_i and morphisms $b_{i,j}$ for $i \leq j$) such that $f = b_i \cdot f_i$ for each $i \in I$ and $b_{i,j} \cdot f_i = f_j$ for all $i \leq j$. Choose $\overline{f}_i \colon B_i \to A$ with $\overline{f}_i \cdot f_i = id$.

Let D^* be the following λ-directed diagram: D_i^* is the power of B to the set $\uparrow i = \{j \in I \mid i \leq j\}$, for $i \in I$, and $d_{i,j}^* \colon D_i^* \to D_j^*$ is the canonical projection for each $i \leq j$. Let $(d_i^* \colon D_i^* \to B^*)_{i \in I}$ be a colimit of D^*. The morphisms $p_i \colon B_i \to D_i^*$ whose jth components are

$$b_{i,j} \colon B_i \to B_j \qquad\qquad (i \leq j)$$

form a natural transformation from D to D^*, which defines a unique factorization $p \colon B \to B^*$. Analogously, the morphisms $q_i \colon B_i \to D_i^*$ whose jth components are $f_j \cdot \overline{f}_j \cdot b_{i,j}$ yield a morphism $q \colon B \to B^*$. We claim that f is an equalizer of p and q.

(1) $p \cdot f = q \cdot f$ because for any $i \in I$ we see that $p_i \cdot f_i = q_i \cdot f_i$ (since the components of the right-hand side are $f_j \cdot \overline{f}_j \cdot b_{i,j} \cdot f_i = f_j \cdot \overline{f}_j \cdot f_j =$

$f_j = b_{i,j} \cdot f_i$). Then

$$\begin{aligned} p \cdot f &= p \cdot b_i \cdot f_i \\ &= d_i^* \cdot p_i \cdot f_i \\ &= d_i^* \cdot q_i \cdot f_i \\ &= q \cdot b_i \cdot f_i \\ &= q \cdot f. \end{aligned}$$

(2) To prove the universal property of f, it is sufficient to show that, for each λ-presentable object H, each morphism $h \colon H \to B$ merging p and q factorizes uniquely through f. We know that h factorizes through some b_i, say, $h = b_i \cdot h'$. Then

$$\begin{aligned} d_i^* \cdot (p_i \cdot h') &= p \cdot b_i \cdot h' \\ &= p \cdot h \\ &= q \cdot h \\ &= q \cdot b_i \cdot h' \\ &= d_i^* \cdot (q_i \cdot h') \end{aligned}$$

which implies, since H is λ-presentable, that $d_{i,j}^* \cdot p_i \cdot h' = d_{i,j}^* \cdot q_i \cdot h'$ for some $j \geq i$. Thus, for $h'' = b_{i,j} \cdot h'$ we have $p_j \cdot h'' = q_j \cdot h'' \colon H \to D_j^* = \prod_{j \leq k} B$. The jth components of these two morphisms are h'' and $f_j \cdot \overline{f}_j \cdot h''$, thus, $h'' = f_j \cdot \overline{f}_j \cdot h''$. Therefore, h factorizes through f:

$$\begin{aligned} h &= b_j \cdot b_{i,j} \cdot h' \\ &= b_j \cdot h'' \\ &= b_j \cdot f_j \cdot \overline{f}_j \cdot h'' \\ &= f \cdot \overline{f}_j \cdot h''. \end{aligned}$$

Unicity follows from Proposition 2.29. □

Accessible Categories Have Enough Pure Subobjects

2.32 Proposition. *Every accessible category is equivalent to a full subcategory of* $\mathbf{Set}^{\mathcal{A}}$ *closed under* λ*-directed colimits and* λ*-pure subobjects (for some small category* $\mathcal{A}$ *and some regular cardinal* λ*).*

PROOF. By Proposition 2.8 we know that every accessible category is equivalent to a full subcategory of $\mathbf{Set}^{\mathcal{A}}$ closed under λ-directed colimits for some

$\mathcal{A}$ and λ. Thus, it is sufficient to show that for each full accessible subcategory $\mathcal{K}$ of $\mathbf{Set}^{\mathcal{A}}$ closed under λ-directed colimits there exists a regular cardinal $\lambda^* \geq \lambda$ such that $\mathcal{K}$ is closed under λ^*-pure subobjects. This cardinal is obtained from the uniformization theorem 2.19: suppose that $\mathcal{K}$ is λ^*-accessible and that every λ^*-presentable object of $\mathcal{K}$ is λ^*-presentable in $\mathbf{Set}^{\mathcal{A}}$. Then we shall prove that for each λ^*-pure morphism $f\colon L \to K$ with K in $\mathcal{K}$ we have $L \in \mathcal{K}^{\mathrm{obj}}$. Let us express L as a canonical colimit of λ^*-presentable objects $(L_i \xrightarrow{u_i} L)_{i \in I}$ in $\mathbf{Set}^{\mathcal{A}}$. Since $\mathcal{K}$ is λ^*-accessible, K is a λ^*-directed colimit of objects which are λ^*-presentable in $\mathcal{K}$. Thus, for each $i \in I$ there exists a λ^*-presentable object K_i in $\mathcal{K}$ and a commutative square

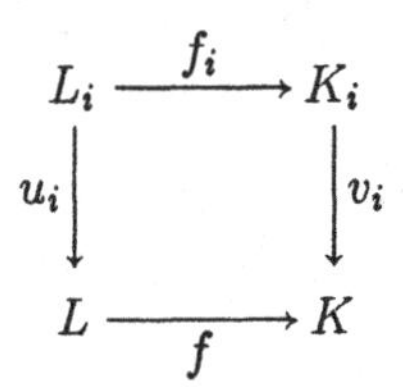

Since f is λ^*-pure in $\mathbf{Set}^{\mathcal{A}}$, and since K_i is λ^*-presentable in $\mathbf{Set}^{\mathcal{A}}$, we have $\overline{u}_i\colon K_i \to L$ with $u_i = \overline{u}_i \cdot f_i$. Now $\overline{u}_i\colon K_i \to L$ is one of the colimit morphisms of the canonical colimit for L, thus, the canonical diagram has a cofinal subdiagram of objects in $\mathcal{K}$. That subdiagram is λ^*-directed, see Exercise 1.o(3). Since L is a colimit of that subdiagram and $\mathcal{K}$ is closed under λ^*-directed colimits, this proves that $L \in \mathcal{K}^{\mathrm{obj}}$. □

2.33 Theorem. (Every λ-accessible category has enough λ-pure subobjects.) *For each λ-accessible category $\mathcal{K}$ there exist arbitrary large regular cardinals $\gamma \rhd \lambda$ such that every subobject $A \hookrightarrow B$ in $\mathcal{K}$ with A γ-presentable is contained in a λ-pure subobject $\overline{A} \hookrightarrow B$ with $\overline{A}$ also γ-presentable.*

Remark. We will prove a stronger result: every morphism $f\colon A \to B$ with A γ-presentable factorizes through a λ-pure morphism $\overline{f}\colon \overline{A} \to B$ with $\overline{A}$ γ-presentable.

PROOF. We know from Proposition 2.32 that $\mathcal{K}$ is equivalent to a full subcategory of $\mathbf{Set}^{\mathcal{A}}$ closed under μ-directed colimits and μ-pure subobjects (for some small category $\mathcal{A}$ and some regular cardinal μ). We can assume that $\mathcal{K}$ actually *is* such a subcategory of $\mathbf{Set}^{\mathcal{A}}$. Moreover, by uniformization theorem 2.19 (and Remark 2.28(3)), we can assume that every μ-presentable object of $\mathcal{K}$ is μ-presentable in $\mathbf{Set}^{\mathcal{A}}$). Finally, we can assume that μ is larger than the number of morphisms of $\mathcal{A}$. (In fact, μ can be chosen to be arbitrarily large in all these considerations.)

I. For each morphism $f\colon A \to B$ in $\mathbf{Set}^{\mathcal{A}}$ we construct a factorization through a μ-pure morphism $\overline{f}\colon \overline{A} \to B$ in $\mathbf{Set}^{\mathcal{A}}$. To this end, we define a chain $f_i\colon A_i \to B$, $i \leq \mu$, of subobjects of B by the following transfinite induction:

First step: $A_0 = f(A)$ is the image of f and $f_0\colon A_0 \to B$ is the inclusion.

Isolated step: let $f_i\colon A_i \to B$ be given. Consider a set of representatives of all spans $\big(A_i \xleftarrow{u} A' \xrightarrow{f'} B'\big)$ with A', B' μ-presentable, such that $f_i \cdot u$ factorizes through f'. For any such span let us choose $v\colon B' \to B$ with $f_i \cdot u = v \cdot f'$. Denote by $f_{i+1}\colon A_{i+1} \to B$ the subobject which is the union of f_i with the images $v(B') \subseteq B$ of all the chosen v's. This subobject contains f_i (say $f_i = f_{i+1} \cdot a_{i,i+1}$) and has the following property: in each commutative square

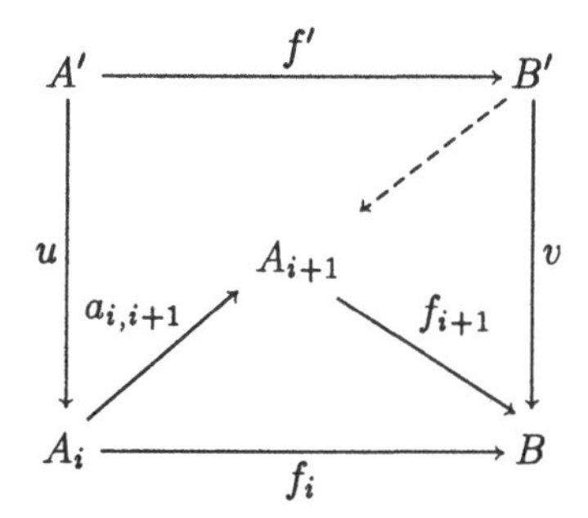

with A', B' μ-presentable, the morphism $a_{i,i+1} \cdot u$ factorizes through f'. (In fact, v factorizes through f_{i+1}, and f_{i+1} is a monomorphism.)

Limit step: $f_i = \bigcup_{j<i} f_j$ for each limit ordinal $i \leq \mu$.

The morphism $\overline{f} = f_\mu\colon A_\mu \to B$ is μ-pure. In fact, for each commutative square

$$\begin{array}{ccc} A' & \xrightarrow{f'} & B' \\ {\scriptstyle u}\downarrow & & \downarrow{\scriptstyle v} \\ A_\mu & \xrightarrow[\overline{f}]{} & B \end{array}$$

with A', B' μ-presentable, the morphism u factorizes through some $a_{i,\mu}$, say, $u = a_{i,\mu} \cdot u'$, since $(A_i \xrightarrow{a_{i,\mu}} A_\mu)$ is a μ-directed colimit in $\mathbf{Set}^{\mathcal{A}}$, see Corollary 1.63. Then $\big(A_i \xleftarrow{u'} A' \xrightarrow{f'} B'\big)$ is one of the above spans; thus, $a_{i,i+1} \cdot u'$ factorizes through f'. Consequently, $u = a_{i+1,\mu} \cdot (a_{i,i+1} \cdot u')$ also factorizes through f'. This proves that $\overline{f}$ is μ-pure.

II. Let $\gamma = (\mu^{<\mu})^+$ be the cardinal successor of $\mu^{<\mu}$ (see Notation 2.9). We will prove that if A is a γ-presentable object in the construction I, then $\overline{A}$ is also γ-presentable. This will conclude the proof of the theorem: we have $\mu < \gamma$, and we can choose μ arbitrarily large. Since $\mu \lhd \gamma$ (see Example 2.13(5)) and the forgetful functor from $\mathcal{K}$ to $\mathbf{Set}^S$, where $S = \mathcal{A}^{\mathrm{obj}}$, preserves μ-presentable objects (by our assumption at the beginning and Examples 1.31, 1.14(1)), the functor also preserves γ-presentable objects, see Remark 2.20. Consequently, A is a γ-presentable object of $\mathcal{K}$ iff $|A| < \gamma$, where $|A|$ is the sum of cardinalities of the sets in the image of $A: \mathcal{A} \to \mathbf{Set}$.

We will prove by transfinite induction that $|A| < \gamma$ implies $|A_i| < \gamma$ for all $i \leq \mu$. Clearly, $|A_0| < \gamma$. For the isolated step, let us assume $|A_i| < \gamma$, i.e., $|A_i| \leq \mu^{<\mu}$. The number of representative spans $(A_i \xleftarrow{u} A' \xrightarrow{f} B')$, where A', B' are μ-presentable (thus, $|A'| < \mu$ and $|B'| < \mu$) is at most

$$|A_i|^{<\mu} \leq (\mu^{<\mu})^{<\mu} = \mu^{<\mu}$$

(see Lemma 2.10), thus

$$|A_{i+1}| \leq |A_i| + \sum_{(u,f')} \mu \leq \mu^{<\mu} + \mu \cdot \mu^{<\mu} = \mu^{<\mu} < \gamma.$$

The limit steps are obvious since $f_i = \bigcup_{j<i} f_j$ implies $|A_i| = \sup_{j<i} |A_j|$. □

2.34 Theorem. *For any λ-accessible category $\mathcal{K}$, the subcategory $\mathbf{Pure}_\lambda\,\mathcal{K}$ of $\mathcal{K}$ consisting of all $\mathcal{K}$-objects and all λ-pure morphisms is accessible.*

Remark. We will also prove that $\mathbf{Pure}_\lambda\,\mathcal{K}$ is closed under λ-directed colimits in $\mathcal{K}$.

PROOF. I. We first prove the remark. Let $D: (I, \leq) \to \mathbf{Pure}_\lambda\,\mathcal{K}$ be a λ-directed diagram in $\mathbf{Pure}_\lambda\,\mathcal{K}$, and $(D_i \xrightarrow{k_i} K)_{i \in I}$ be a colimit of D in $\mathcal{K}$. It follows from Proposition 2.30 that the morphisms k_i are λ-pure (apply 2.30(i) to $f_j = d_{i,j}: D_i \to D_j,\ j \in I$). Let $(\overline{k}_i: D_i \to \overline{K})_{i \in I}$ be a compatible cocone in $\mathbf{Pure}_\lambda\,\mathcal{K}$. We have to prove that the induced morphism $h: K \to \overline{K}$ is λ-pure. Consider a commutative square

$$\begin{array}{ccc} A & \xrightarrow{f} & B \\ {\scriptstyle u}\downarrow & & \downarrow{\scriptstyle v} \\ K & \xrightarrow[h]{} & \overline{K} \end{array}$$

in $\mathcal{K}$ with A, B λ-presentable. There exists $i \in I$ and $u' : A \to D_i$ with $k_i \cdot u' = u$. Since $\overline{k}_i$ is λ-pure, there exists $g : B \to D_i$ with $g \cdot f = u'$. Then $k_i \cdot g$ yields a factorization of u through f, which proves that h is λ-pure:

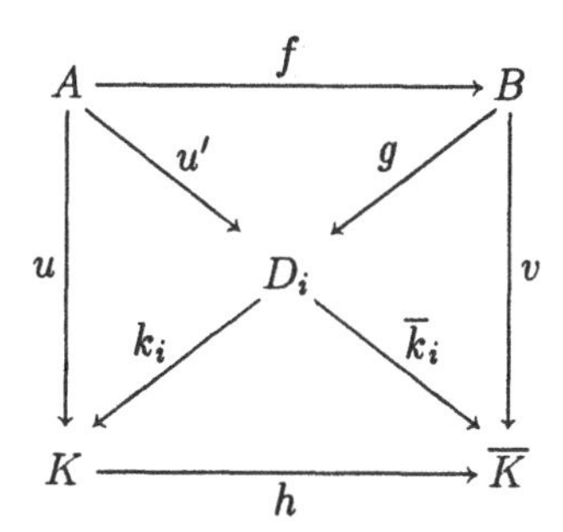

II. Let $\gamma \rhd \lambda$ be a cardinal from Theorem 2.33. We will prove that $\mathbf{Pure}_\lambda \mathcal{K}$ is γ-accessible. We know that $\mathbf{Pure}_\lambda \mathcal{K}$ has γ-directed colimits and that, for every B in $\mathcal{K}$, any $\mathcal{K}$-morphism $f : A \to B$ with A γ-presentable has a factorization

$$(*) \qquad \begin{array}{ccc} A & \xrightarrow{\;f\;} & B \\ \downarrow & \nearrow_{\overline{f}} & \\ \overline{A} & & \end{array}$$

with $\overline{A}$ γ-presentable and $\overline{f}$ λ-pure. Since $\mathcal{K}$ is γ-accessible, the comma-category $\mathbf{Pres}_\gamma \mathcal{K} \downarrow B$ (in $\mathcal{K}$) is γ-filtered. Let $\mathcal{P}$ be the full subcategory of $\mathbf{Pres}_\gamma \mathcal{K} \downarrow B$ over all λ-pure morphisms $f : A \to B$ with A γ-presentable. If h is a morphism in $\mathcal{P}$ from $f : A \to B$ to $f' : A' \to B$, then $h : A \to A'$ is λ-pure in $\mathcal{K}$ by Remark 2.28(2). The above factorizations $(*)$ show that $\mathcal{P}$ is cofinal in $\mathbf{Pres}_\gamma \mathcal{K} \downarrow B$. Therefore B is a γ-directed colimit in $\mathbf{Pure}_\lambda \mathcal{K}$ of objects from $\mathbf{Pres}_\gamma \mathcal{K}$ (see Exercise 1.o(3)). Hence $\mathbf{Pure}_\lambda \mathcal{K}$ is γ-accessible. $\square$

2.35 Definition. A subcategory $\mathcal{A}$ of a category $\mathcal{K}$ is called *accessibly embedded* if it is full and there is a regular cardinal λ such that $\mathcal{A}$ is closed under λ-directed colimits in $\mathcal{K}$.

Remark. Let $\mathcal{K}$ be an accessible category. Every accessibly embedded subcategory of $\mathcal{K}$ is

(1) isomorphism-closed

and

(2) closed under split subobjects (see Observation 2.4 and Remark 2.5(1).

2.36 Corollary. *Let $\mathcal{K}$ be an accessible category, and let $\mathcal{A}$ be an accessibly embedded subcategory of $\mathcal{K}$. Then $\mathcal{A}$ is accessible iff it is closed in $\mathcal{K}$ under λ-pure subobjects for some regular cardinal λ.*

In particular, a category is accessible iff it is equivalent to a full subcategory of $\mathbf{Set}^{\mathcal{A}}$, $\mathcal{A}$ small, closed under λ-directed colimits and λ-pure subobjects for some regular cardinal λ.

PROOF. The latter statement follows from the former one by Proposition 2.32. To prove the former statement, let $\mathcal{A}$ be closed under λ-pure subobjects. We can assume that $\mathcal{K}$ is λ-accessible and $\mathcal{A}$ is closed in $\mathcal{K}$ under λ-directed colimits. Since $\mathbf{Pure}_\lambda\, \mathcal{K}$ is μ-accessible for some $\mu \rhd \lambda$, any object of $\mathcal{A}$ is a μ-directed colimit of a diagram in $\mathcal{K}$ consisting of μ-presentable objects and λ-pure morphisms. Since $\mathcal{A}$ is closed under λ-pure subobjects, this diagram lies in $\mathcal{A}$, and it clearly consists of objects μ-presentable in $\mathcal{A}$. Therefore, $\mathcal{A}$ is μ-accessible.

The converse follows from the proof of Proposition 2.32. □

Remark. In particular, any accessibly embedded subcategory of an accessible category closed under subobjects is accessible.

2.37 Corollary. *Let $\mathcal{K}$ be an accessible category. An intersection of a set of accessible and accessibly embedded subcategories of $\mathcal{K}$ is an accessible, accessibly embedded subcategory of $\mathcal{K}$.*

2.38 Proposition. *Each λ-accessible functor preserving λ-presentable objects preserves λ-pure morphisms.*

PROOF. Let $F\colon \mathcal{K} \to \mathcal{L}$ be λ-accessible and let it preserve λ-presentable objects. Consider a commutative square

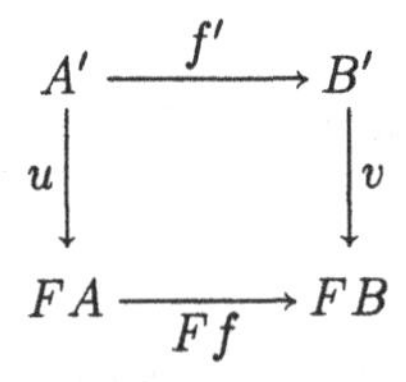

in $\mathcal{L}$ with A' and B' λ-presentable in $\mathcal{L}$ and $f\colon A \to B$ λ-pure in $\mathcal{K}$. Since $\mathcal{K}$ is a λ-accessible category, we can express both A and B as λ-directed colimits $(A_i \xrightarrow{a_i} A)_{i\in I}$ and $(B_j \xrightarrow{b_j} B)_{j\in J}$ of λ-presentable objects. Since F preserves those colimits and since A' and B' are λ-presentable, there exists $i \in I$ such that u factorizes through Fa_i, and $j_0 \in J$ such that v factorizes through Fb_{j_0}. Moreover, $f \cdot a_i \colon A_i \to B$ factorizes through some b_{j_1} (since A_i is λ-presentable); without loss of generality assume $j_0 \leq j_1$. Thus, we have $u = Fa_i \cdot u^*$, $v = Fb_{j_1} \cdot v^*$, and $f \cdot a_i = b_{j_1} \cdot f^*$:

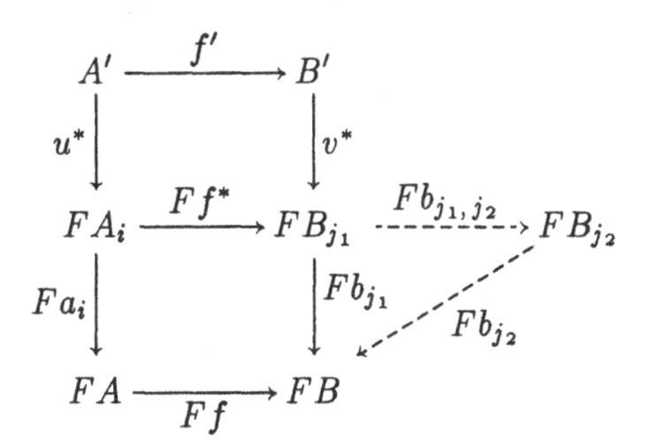

Since A' is λ-presentable and F preserves the colimit expressing B above, it follows from the equation

$$Fb_{j_1} \cdot (Ff^* \cdot u^*) = Fb_{j_1} \cdot (v^* \cdot f')$$

that there exists $j_2 \geq j_1$ such that

$$Fb_{j_1,j_2} \cdot (Ff^* \cdot u^*) = Fb_{j_1,j_2} \cdot (v^* \cdot f').$$

The morphism f is λ-pure, and A_i and B_{j_2} are λ-presentable objects. Thus, from $f \cdot a_i = b_{j_2} \cdot (b_{j_1,j_2} \cdot f^*)$ it follows that there exists $h \colon B_{j_2} \to A$ with

$$a_i = h \cdot b_{j_1,j_2} \cdot f^*.$$

The morphism $h^* = F(h \cdot b_{j_1,j_2}) \cdot v^*$ then fulfils

$$\begin{aligned} u &= Fa_i \cdot u^* \\ &= F(h \cdot b_{j_1,j_2}) \cdot Ff^* \cdot u^* \\ &= F(h \cdot b_{j_1,j_2}) \cdot v^* \cdot f' \\ &= h^* \cdot f', \end{aligned}$$

which proves that Ff is a λ-pure morphism. □

2.E Properties of Accessible Categories

In the present section we will prove that if $\mathcal{K}$ is an accessible category, then all the functor categories $\mathcal{K}^{\mathcal{A}}$ and various comma-categories are accessible.

Furthermore, (co-)complete and accessible categories are precisely the locally presentable categories. We will also prove that a full subcategory of an accessible category is accessible iff its embedding satisfies the solution-set condition. We postpone to Section 2.H the limit theorem which asserts that a lax limit of accessible categories is accessible, and some other properties of accessible categories (e.g., that algebras of an accessible monad form an accessible category).

2.39 Theorem. *For each accessible category $\mathcal{K}$, all functor-categories $\mathcal{K}^{\mathcal{A}}$ ($\mathcal{A}$ small) are accessible.*

PROOF. By Proposition 2.32 we can assume that $\mathcal{K}$ is a full subcategory of some $\mathbf{Set}^{\mathcal{B}}$ ($\mathcal{B}$ small) closed under λ-directed colimits and λ-pure subobjects for some regular cardinal λ. Then $\mathcal{K}^{\mathcal{A}}$ is a full subcategory of $\left(\mathbf{Set}^{\mathcal{B}}\right)^{\mathcal{A}}$ closed under λ-directed colimits (since these are computed component-wise). For each object A of $\mathcal{A}$ the functor $\Phi_A\colon \left(\mathbf{Set}^{\mathcal{B}}\right)^{\mathcal{A}} \to \mathbf{Set}^{\mathcal{B}}$ given by

$$\Phi_A(F) = F(A) \quad \text{and} \quad \Phi_A(\varphi) = \varphi_A$$

is ω-accessible. By Remark 2.19, there exists a regular cardinal $\mu > \lambda$ such that each Φ_A preserves μ-presentable objects. It follows from Proposition 2.38 that Φ_A preserve μ-pure morphisms, thus, $\mathcal{K}^{\mathcal{A}}$ is closed in $\left(\mathbf{Set}^{\mathcal{B}}\right)^{\mathcal{A}}$ under μ-pure subobjects. Hence $\mathcal{K}^{\mathcal{A}}$ is accessible by Corollary 2.36. □

2.40 Remark. We know from Corollary 1.54 that $\mathcal{K}^{\mathcal{A}}$ is locally λ-presentable whenever $\mathcal{K}$ is locally λ-presentable. The situation for accessible categories is different:

2.41 Example. The category $\mathbf{Set}_\infty$ of infinite sets and mappings is ω_1-accessible. However, the uncountable product $\mathbf{Set}_\infty^{\omega_1} = \prod_{\omega_1} \mathbf{Set}_\infty$ is not ω_1-accessible: it has no ω_1-presentable object. To see this, consider, for each object $K = (X_k)_{k\in\omega_1}$ of $\mathbf{Set}_\infty^{\omega_1}$, the following ω_1-chain $K_i \to K_j$ ($i \le j < \omega_1$) in $\mathbf{Set}_\infty^{\omega_1}$: for each $k < \omega_1$ choose an infinite set $Y_k \subsetneqq X_k$, and put

$$K_i = \left(X_{i,k}\right)_{k<\omega_1} \quad \text{with} \quad X_{i,k} = \begin{cases} Y_k & \text{for } k \ge i \\ X_k & \text{for } k < i \end{cases}$$

and let $K_i \to K_j$ be the morphism whose components are the inclusion maps. A colimit of this ω_1-chain is K together with component-wise inclusion maps $k_i\colon K_i \to K$ ($i < \omega_1$). Since id_K does not factorize through any k_i, it follows that K is not ω_1-presentable.

2.42 Notation. Given functors $F_1 : \mathcal{K}_1 \to \mathcal{L}$ and $F_2 : \mathcal{K}_2 \to \mathcal{L}$, we denote by

$$F_1 \downarrow F_2$$

the comma-category of all arrows $F_1K_1 \xrightarrow{f} F_2K_2$ in $\mathcal{L}$ (with $K_i \in \mathcal{K}_i^{\mathrm{obj}}$) whose morphisms from $F_1K_1 \xrightarrow{f} F_2K_2$ into $F_1K_1' \xrightarrow{f'} F_2K_2'$ are those morphisms (k_1, k_2) of $\mathcal{K}_1 \times \mathcal{K}_2$ for which the square

$$\begin{array}{ccc} F_1K_1 & \xrightarrow{f} & F_2K_2 \\ {\scriptstyle F_1k_1}\downarrow & & \downarrow{\scriptstyle F_2k_2} \\ F_1K_1' & \xrightarrow[f']{} & F_2K_2' \end{array}$$

commutes. Composition and identity morphisms are defined as in $\mathcal{K}_1 \times \mathcal{K}_2$.

2.43 Theorem. *The category $F_1 \downarrow F_2$ is accessible for arbitrary accessible functors $F_i : \mathcal{K}_i \to \mathcal{L}$ $(i = 1, 2)$.*

Remark. It will also be clear from the proof that the natural forgetful functors $P_i : F_1 \downarrow F_2 \to \mathcal{K}_i$ are accessible for $i = 1, 2$.

PROOF. By Remark 2.19 there exists a regular cardinal λ such that

(1) the categories $\mathcal{K}_1$, $\mathcal{K}_2$, and $\mathcal{L}$ are λ-accessible,

and

(2) the functors F_1 and F_2 are λ-accessible and preserve λ-presentable objects.

We will prove that $F_1 \downarrow F_2$ is then a λ-accessible category. Since F_1, F_2 preserve λ-directed colimits, it is obvious that the category $F_1 \downarrow F_2$ has λ-directed colimits computed on the level of $\mathcal{K}_1 \times \mathcal{K}_2$. Next we prove that each object $F_1K_1 \xrightarrow{f} F_2K_2$ of $F_1 \downarrow F_2$ is a λ-directed colimit of objects $F_1A_1 \xrightarrow{a} F_2A_2$ where A_j is λ-presentable in $\mathcal{K}_j$ for $j = 1, 2$. As this implies that F_jA_j is λ-presentable in $\mathcal{L}$, it is clear that these objects $F_1A_1 \xrightarrow{a} F_2A_2$ are λ-presentable in $F_1 \downarrow F_2$. Since they form, up to isomorphism, a set, this will conclude the proof.

There exist a λ-directed diagram D of λ-presentable objects D_i (and morphisms $d_{i,i'}$) in $\mathcal{K}_1$ with a colimit $(D_i \xrightarrow{k_i} K_1)_{i \in I}$, and a λ-directed diagram D^* of λ-presentable objects D_j^* (and morphisms $d_{j,j'}^*$) in $\mathcal{K}_2$ with a colimit $\left(D_j^* \xrightarrow{k_j^*} K_2\right)_{j \in J}$. We define a λ-directed diagram $\widehat{D}$ in $F_1 \downarrow F_2$ as follows. The underlying poset of $\widehat{D}$ is the set T of all arrows $F_1D_i \xrightarrow{\overline{f}} F_2D_j^*$ in $\mathcal{L}$ ($i \in I$, $j \in J$) such that $F_2k_j^* \cdot \overline{f} = f \cdot F_1k_i$. The ordering is given via the following square:

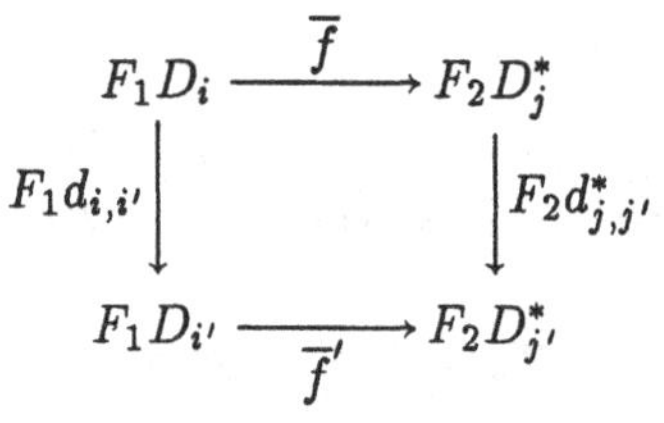

$$\begin{array}{ccc} F_1D_i & \xrightarrow{\quad \overline{f} \quad} & F_2D_j^* \\ {\scriptstyle F_1d_{i,i'}}\downarrow & & \downarrow{\scriptstyle F_2d^*_{j,j'}} \\ F_1D_{i'} & \xrightarrow[\quad \overline{f}' \quad]{} & F_2D_{j'}^* \end{array}$$

That is, $\overline{f}$ is below $\overline{f}'$ in T iff $i \leq i'$ in I, $j \leq j'$ in J, and the above square commutes. Let us prove that T is λ-directed. Given a set $T_0 \subseteq T$ of less than λ elements, there exists an upper bound i_0 of all the corresponding i's in I, and an upper bound j_0 of all the corresponding j's in J. Besides, since $F_1D_{i_0}$ is a λ-presentable object of $\mathcal{L}$ and F_2 preserves the colimit of D^*, the morphism $f \cdot F_1k_{i_0}$ factorizes through $F_2k_j^*$ for some $j \in J$—without loss of generality, we can assume that this holds for j_0. Thus, we have a morphism $F_1D_{i_0} \xrightarrow{f_0} F_2D_{j_0}^*$ in L:

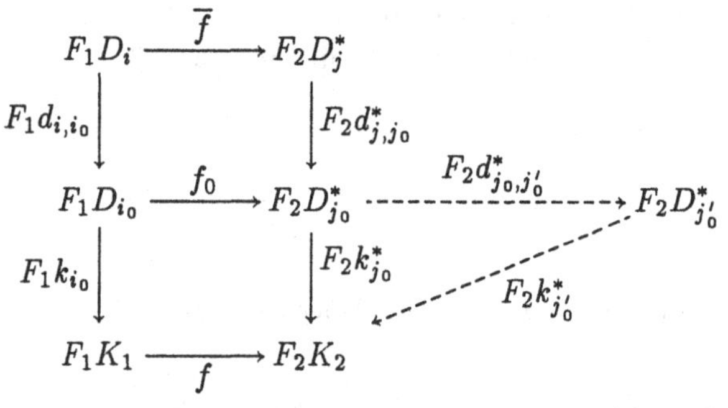

$$\begin{array}{ccccc} F_1D_i & \xrightarrow{\quad \overline{f} \quad} & F_2D_j^* & & \\ {\scriptstyle F_1d_{i,i_0}}\downarrow & & \downarrow{\scriptstyle F_2d^*_{j,j_0}} & & \\ F_1D_{i_0} & \xrightarrow{\quad f_0 \quad} & F_2D_{j_0}^* & \overset{F_2d^*_{j_0,j_0'}}{\dashrightarrow} & F_2D^*_{j_0'} \\ {\scriptstyle F_1k_{i_0}}\downarrow & & \downarrow{\scriptstyle F_2k^*_{j_0}} & \swarrow{\scriptstyle F_2k^*_{j_0'}} & \\ F_1K_1 & \xrightarrow[\quad f \quad]{} & F_2K_2 & & \end{array}$$

For each object $F_1D_i \xrightarrow{\overline{f}} F_2D_j^*$ of T_0 we know that F_1D_i is a λ-presentable object of $\mathcal{L}$ such that

$$F_2k^*_{j_0} \cdot (f_0 \cdot F_1d_{i,i_0}) = F_2k^*_{j_0} \cdot (F_2d^*_{j,j_0} \cdot \overline{f}).$$

Since F_2 preserves the colimit of D^*, there exists $j_0' \geq j_0$ in J such that

$$F_2d^*_{j,j_0'} \cdot (f_0 \cdot F_1d_{i,i_0}) = F_2D^*(j_0 \to j_0') \cdot (F_2d^*_{j,j_0} \cdot \overline{f}).$$

Moreover, since J is λ-directed and T_0 has less than λ elements, we can choose such a j_0' independent of the concrete object of T_0. It follows that the object

$$F_1D_{i_0} \xrightarrow{F_2d^*_{j_0,j_0'} \cdot f_0} F_2D^*_{j_0'}$$

is an upper bound of T_0.

The diagram $\widehat{D}$ assigns to each object $F_1 D_i \xrightarrow{\bar{f}} F_2 D_j^*$ the object itself and, to each pair of objects for which the square above commutes, it assigns the morphism $(d_{i,i'}, d_{j,j'}^*)$. The cocone of all

$$(k_i, k_j^*) \colon (F_1 D_i \xrightarrow{\bar{f}} F_2 D_j^*) \to (F_1 K_1 \xrightarrow{f} F_2 K_2)$$

is a colimit of $\widehat{D}$—this is obvious since F_1 preserves the colimit of D, and F_2 preserves the colimit of D^*. □

2.44 Corollary. *For each accessible category $\mathcal{K}$, every comma-category $K \downarrow \mathcal{K}$ and $\mathcal{K} \downarrow K$ is accessible.*

In fact, $K \downarrow \mathcal{K} = C_K \downarrow Id_{\mathcal{K}}$ and $\mathcal{K} \downarrow K = Id_{\mathcal{K}} \downarrow C_K$, where C_K is the constant functor of value K. □

2.45 Corollary. *Every accessible functor satisfies the solution-set condition.*

In fact, if $F \colon \mathcal{K} \to \mathcal{L}$ is accessible, then for each object L of $\mathcal{L}$ the comma-category $F \downarrow C_L$, where $C_L \colon \mathcal{L} \to \mathcal{L}$ is the constant functor of value L, is accessible. Any set of generators of $F \downarrow C_L$ is a solution set. □

2.46 Remarks

(1) We will see later that, under some set-theoretical assumptions, the above corollary can be reversed: given accessible categories $\mathcal{K}$ and $\mathcal{L}$, any functor $\mathcal{K} \to \mathcal{L}$ satisfying the solution-set condition is accessible. (See Theorem 6.30.)

(2) The above corollary is a natural generalization of the adjoint functor theorem (1.66).

2.47 Corollary. *The following conditions on a category $\mathcal{K}$ are equivalent:*

(i) *$\mathcal{K}$ is accessible and complete,*

(ii) *$\mathcal{K}$ is accessible and cocomplete,*

(iii) *$\mathcal{K}$ is locally presentable.*

PROOF. Since (ii) $\Rightarrow$ (iii) $\Rightarrow$ (i) is clear (see Corollary 1.28), it remains to prove that (i) $\Rightarrow$ (ii). Let $\mathcal{K}$ be λ-accessible. Then the canonical functor $E \colon \mathcal{K} \to \mathbf{Set}^{\mathcal{A}^{\mathrm{op}}}$, where $\mathcal{A} = \mathbf{Pres}_\lambda \mathcal{K}$, is λ-accessible and preserves limits. By Corollary 2.45, E has a left adjoint and, consequently, $\mathcal{K}$ is cocomplete. □

2.48 Reflection Theorem. *Every accessibly embedded subcategory of a locally presentable category $\mathcal{K}$ closed in $\mathcal{K}$ under limits is reflective.*

PROOF. Let $\mathcal{L}$ be a full subcategory of a locally λ-presentable category $\mathcal{K}$ closed in $\mathcal{K}$ under limits and λ-directed colimits. It follows from Remark 2.31 that $\mathcal{L}$ is closed in $\mathcal{K}$ under λ-pure subobjects. Thus, by Corollary 2.36, $\mathcal{L}$ is accessible. Hence, following Corollary 2.45, $\mathcal{L}$ is a reflective subcategory of $\mathcal{K}$. □

Corollary. *Every full subcategory of a locally λ-presentable category closed under limits and λ-directed colimits is locally λ-presentable.*

This follows from Theorems 2.48 and 1.39. □

2.49 Theorem. *Every accessible category with pushouts is co-wellpowered.*

Remark. The following is, in particular, an elegant proof of the co-wellpoweredness of every locally presentable category (cf. Theorem 1.58). Whether or not *each* accessible category is co-wellpowered depends on set theory as we will see in Corollary 6.8 and Example A.19.

PROOF. Given an accessible category $\mathcal{K}$ with pushouts and an object K of $\mathcal{K}$, we will prove that the full subcategory $\mathcal{E}_K$ of the comma-category $K \downarrow \mathcal{K}$ over all epimorphisms is accessible. Since $\mathcal{E}_K$ is equivalent to a partially ordered class, this implies that $\mathcal{E}_K$ is small (Example 2.3(3)).

Let $\mathcal{A}$ be the following category

By Corollary 2.44 and Theorem 2.39, the comma-category $\mathcal{L} = \left(C_K \downarrow \mathcal{K}^{\mathcal{A}}\right)$, where $C_K \colon \mathcal{A} \to \mathcal{K}$ denotes the constant functor with the value K, is accessible. The objects of the category $\mathcal{L}$ are the commutative squares

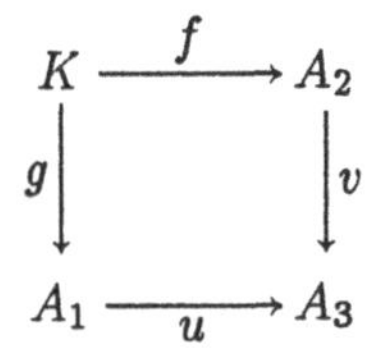

and morphisms are triples (a_1, a_2, a_3) such that the following diagram

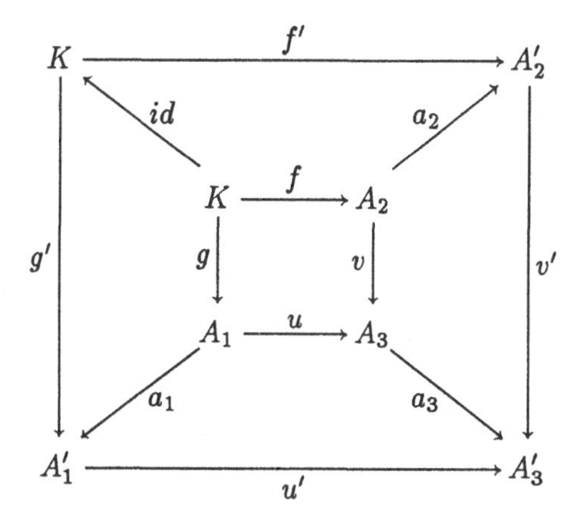

commutes. The comma category $K \downarrow \mathcal{K}$ is isomorphic to the full subcategory $\mathcal{L}_1$ of $\mathcal{L}$ of the following squares

$$\begin{array}{ccc} K & \xrightarrow{\;f\;} & A \\ {\scriptstyle f}\downarrow & & \| \\ A & = & A \end{array}$$

By Corollary 2.44, $\mathcal{L}_1$ is accessible, and it is clearly accessibly embedded into $\mathcal{L}$. Furthermore, the full subcategory $\mathcal{L}_2$ of $\mathcal{L}$ over all pushout squares is ω-accessibly embedded into $\mathcal{L}$ (since pushouts commute with directed colimits), and it is accessible because, obviously, it is isomorphic to the comma-category $(K, K) \downarrow \mathcal{K} \times \mathcal{K}$, and $\mathcal{K} \times \mathcal{K}$ is accessible by Theorem 2.39—take for $\mathcal{A}$ the two-element antichain. By Corollary 2.37, $\mathcal{L}_1 \cap \mathcal{L}_2$ is accessible. Since $f\colon K \to A$ is an epimorphism iff the corresponding square in $\mathcal{L}_1$ is a pushout, $\mathcal{L}_1 \cap \mathcal{L}_2$ is isomorphic to $\mathcal{E}_K$, which concludes the proof. □

2.50 Remark. In Section 2.H below we will study limits of accessible categories. We now just mention a special property which is needed in Section 2.F: if $F\colon \mathcal{K} \to \mathcal{L}$ is an accessible functor, and if $\mathcal{L}_1$ is an accessible, accessibly embedded subcategory of $\mathcal{L}$, then $F^{-1}(\mathcal{L}_1)$ is an accessible, accessibly embedded subcategory of $\mathcal{K}$. Here $F^{-1}(\mathcal{L}_1)$ denotes a full subcategory of $\mathcal{K}$, i.e., the pullback of F and the inclusion functor $\mathcal{L}_1 \hookrightarrow \mathcal{L}$:

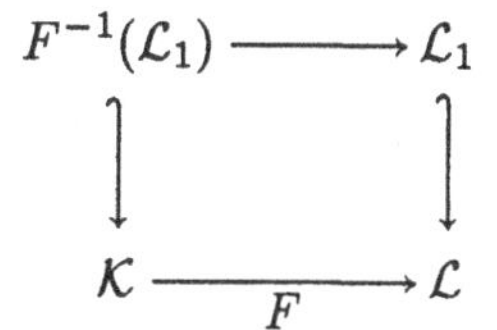

The accessibility of $F^{-1}(\mathcal{L}_1)$ follows from the fact that there exists a regular cardinal λ such that (i) F is λ-accessible and preserves λ-presentable objects, and (ii) $\mathcal{L}_1$ is closed in $\mathcal{L}$ under λ-directed colimits and λ-pure subobjects. By Proposition 2.38 the subcategory $F^{-1}(\mathcal{L}_1)$, which is, obviously, closed under λ-directed colimits in $\mathcal{K}$, is also closed under λ-pure subobjects. Thus, it is accessible by Corollary 2.36. □

Cone-reflective Subcategories

We have seen in Corollary 2.45 that every accessible functor $F: \mathcal{K} \to \mathcal{L}$ satisfies the solution-set condition. In the case when F is the inclusion of $\mathcal{K} \subseteq \mathcal{L}$, this means that $\mathcal{K}$ is a cone-reflective subcategory:

2.51 Definition. Let $\mathcal{A}$ be a subcategory of a category $\mathcal{K}$. By a *cone-reflection* of an object K of $\mathcal{K}$ is meant a cone $\left(K \xrightarrow{a_i} A_i\right)_{i \in I}$ of morphisms of $\mathcal{K}$ with $A_i \in \mathcal{A}^{\mathrm{obj}}$ (for $i \in I$) such that for every morphism $f: K \to A$, $A \in \mathcal{A}^{\mathrm{obj}}$, there exists $i \in I$ and a morphism $f': A_i \to A$ in $\mathcal{A}$ with $f = f' \cdot a_i$.

A subcategory is called *cone-reflective* in $\mathcal{K}$ provided that every object of $\mathcal{K}$ has a cone-reflection in the subcategory.

2.52 Examples

(1) Every reflective subcategory is cone-reflective.

(2) The full subcategory $\mathcal{A}$ of **Pos** formed by posets with a greatest element is cone-reflective: a cone-reflection of a poset is obtained by adding a greatest element to it.

(3) The full subcategory $\mathcal{A}$ of the category $\mathbf{Alg}\,\Sigma$, where Σ consists of one (one-sorted) unary operation, formed by all algebras with a cycle (i.e., algebras (X, α) such that α^n has a fixpoint for some natural number $n \geq 1$) is cone-reflective: a cone-reflection of an algebra K is formed by the cone $(K \hookrightarrow K + C_n)_{n \in \omega}$ where C_n denotes a single n-cycle.

Remark. We have seen in Corollary 2.36 that for each accessible category $\mathcal{K}$, the accessibility of accessibly embedded subcategories is characterized by closedness under pure subobjects. We now prove another result of that sort.

2.53 Theorem. *Let $\mathcal{K}$ be an accessible category, and let $\mathcal{A}$ be an accessibly embedded subcategory of $\mathcal{K}$. Then $\mathcal{A}$ is accessible iff it is cone-reflective in $\mathcal{K}$.*

PROOF. Necessity follows from Corollary 2.45. To prove sufficiency, let $\mathcal{A}$ be an accessibly embedded subcategory of an accessible category $\mathcal{K}$. It follows from Proposition 2.8 and Example 1.41 that $\mathcal{A}$ is also an accessibly embedded subcategory of $\mathbf{Rel}\,\Sigma$ for some finitary signature Σ. Let us prove that there exists a regular cardinal λ such that (1) $\mathcal{A}$ is closed under λ-directed colimits in $\mathbf{Rel}\,\Sigma$ and (2) every λ-presentable object K of $\mathbf{Rel}\,\Sigma$ has a λ-presentable cone-reflection in $\mathcal{A}$, by which we mean a cone-reflection $\left(K \xrightarrow{r_i} A_i\right)_{i\in I}$ such that each A_i, $i \in I$, is λ-presentable in $\mathbf{Rel}\,\Sigma$. Then $\mathcal{A}$ is λ-accessible because for each object A of $\mathcal{A}$ the canonical diagram w.r.t. $\mathbf{Pres}_\lambda\,\mathcal{A}$ is clearly cofinal in the canonical diagram w.r.t. $\mathbf{Pres}_\lambda\,\mathbf{Rel}\,\Sigma$. Since the latter diagram is λ-filtered (because $\mathbf{Rel}\,\Sigma$ is locally finitely presentable), so is the former one, see Exercise 1.o(3).

To prove the existence of such a λ, first let $\delta > \operatorname{card}\Sigma$ be a regular cardinal such that $\mathcal{A}$ is closed in $\mathbf{Rel}\,\Sigma$ under δ-directed colimits. Since there is (up to isomorphism) only a set of λ-presentable objects in $\mathbf{Rel}\,\Sigma$, it is sufficient to find a regular cardinal $\lambda \geq \delta$ such that any morphism $f\colon K \to A$, A in $\mathcal{A}$, factorizes through some A^* in $\mathcal{A}$ which is λ-presentable in $\mathbf{Rel}\,\Sigma$.

We know from Example 1.14(2) that for each regular cardinal $\lambda > \operatorname{card}\Sigma$, an object of $\mathbf{Rel}\,\Sigma$ is λ-presentable iff the power of its underlying set is smaller than λ. We proceed by defining a chain λ_i ($i \leq \delta$) of cardinals by the following transfinite induction:

$\lambda_0 = \delta$;

λ_{i+1} is the smallest regular cardinal such that every λ_i-presentable object of $\mathbf{Rel}\,\Sigma$ has a λ_{i+1}-presentable cone-reflection in $\mathcal{A}$, and $\lambda_{i+1} \geq \lambda_i$;

$\lambda_i = \bigvee_{j<i} \lambda_j$ for all limit ordinals j.

If $\lambda_i = \lambda_{i+1}$, put $\lambda = \lambda_i$. In the case $\lambda_i < \lambda_{i+1}$ for each $i < \delta$, let λ be the cardinal successor of λ_δ. Denote by $\mathcal{L}$ the collection of all objects which are λ'-presentable for some regular cardinal $\lambda' < \lambda_\delta$. Each object in $\mathcal{L}$ has, obviously, a cone-reflection in $\mathcal{A}$ consisting of morphisms to $\mathcal{L}$-objects. It remains to prove that every λ-presentable object $K \notin \mathcal{L}$ has a λ-presentable cone-reflection in $\mathcal{A}$. The underlying many-sorted set $X = |K|$ has power precisely λ_δ: since K is λ-presentable, its power is at most λ_δ, and it cannot be less than λ_δ since $K \notin \mathcal{L}$. In the category of many-sorted sets there is a smooth (see Remark 1.7) chain of subobjects Y_i of X ($i < \lambda_\delta$) with $X = \bigcup_{i<\lambda_\delta} Y_i$ such that each Y_i has power less than λ_i. (In fact, if all points of all sorts of X are labeled as $\{x_j \mid j < \lambda_\delta\}$, then Y_i is formed by the points $\{x_j \mid j < i\}$ distributed to their sorts.) For each i we have a strong subobject K_i of K (with the inclusion $K_i \xrightarrow{k_i} K$) whose underlying set is Y_i. Then $K_i \in \mathcal{L}$ and the inclusion morphisms form a smooth chain $k_{i,j}: K_i \to K_j$ $(i \le j < \lambda_\delta)$ with a colimit $(K_i \xrightarrow{k_i} K)_{i<\lambda_\delta}$ in $\mathbf{Rel}\,\Sigma$.

Consider any object A in $\mathcal{A}$ and any morphism $f: K \to A$. We define a chain $k^*_{i,j}: K^*_i \to K^*_j$ $(i \le j < \lambda_\delta)$ with $K^*_i \in \mathcal{L}$, and compatible collections $r_i: K_i \to K^*_i$ and $f^*_i: K^*_i \to A$ of the chains $k_{i,j}$, $k^*_{i,j}$ $(i \le j < \lambda_\delta)$ with $f^*_i \cdot r_i = f \cdot k_i$ $(i < \lambda_\delta)$, by the following transfinite induction:

First step: $r_0: K_0 \to K^*_0$ and $f^*_0: K^*_0 \to A$ are given by a λ_1-presentable cone-reflection of $f \cdot k_0: K_0 \to A$.

Isolated step: We form the pushout of r_i and $k_{i,i+1}$:

$$\begin{array}{ccc} K_{i+1} & \xrightarrow{\overline{r}_i} & \overline{K}_{i+1} \\ {\scriptstyle k_{i,i+1}}\uparrow & & \uparrow{\scriptstyle \overline{k}_{i,i+1}} \\ K_i & \xrightarrow[r_i]{} & K^*_i \end{array}$$

Since K_i, K^*_i and K_{i+1} lie in $\mathcal{L}$, so does $\overline{K}_{i+1}$ (see Proposition 1.16). Since

$$f \cdot k_{i+1} \cdot k_{i,i+1} = f \cdot k_i = f^*_i \cdot r_i,$$

there exists $\overline{f}_{i+1}: \overline{K}_{i+1} \to A$ with

$$\overline{f}_{i+1} \cdot \overline{r}_i = f \cdot k_{i+1} \qquad \text{and} \qquad \overline{f}_{i+1} \cdot \overline{k}_{i,i+1} = f^*_i .$$

Since $\overline{K}_{i+1} \in \mathcal{L}$, there is a factorization

$$\overline{f}_{i+1}: \overline{K}_{i+1} \xrightarrow{r^*_{i+1}} K^*_{i+1} \xrightarrow{f^*_{i+1}} A$$

with $K^*_{i+1} \in \mathcal{L}$. We define

$$\begin{aligned} k^*_{i,i+1} &= r^*_{i+1} \cdot \overline{k}_{i,i+1}: K^*_i \to K^*_{i+1}, \\ r_{i+1} &= r^*_{i+1} \cdot \overline{r}_{i+1}: K_{i+1} \to K^*_{i+1}. \end{aligned}$$

Limit step: Given a limit ordinal $i < \lambda_\delta$, we form a colimit

$$\left(K^*_j \xrightarrow{\overline{k}_{j,i}} \overline{K}_i\right)_{j<i}$$

of the previously defined chain. Since $i < \lambda_\delta$ and $K^*_j \in \mathcal{L}$ for each $j < i$, it follows that $\overline{K}_i \in \mathcal{L}$ (see Proposition 1.16). There is a unique $\overline{f}_i: \overline{K}_i \to A$ with $f^*_j = \overline{f}_i \cdot \overline{k}_{j,i}$ for all $j < i$ and we have a factorization

$$\overline{f}_i: \overline{K}_i \xrightarrow{r^*_i} K^*_i \xrightarrow{f^*_i} A$$

with $K_i^* \in \mathcal{L}$. We define

$$k_{j,i}^* = r_i^* \cdot \overline{k}_{j,i} : K_j^* \to K_i^*,$$
$$r_i = r_i^* \cdot \operatorname*{colim}_{j<i} r_j : K_i \to K_i^*.$$

The colimit of the chain $k_{i,j}^*$, say $(K_i^* \xrightarrow{k_i^*} K^*)_{i<\lambda_\delta}$, defines an object K^* in $\mathcal{A}$ (since $\mathcal{A}$ is closed under δ-directed colimits). Since $K_i^* \in \mathcal{L}$ for each i, it follows that K^* is λ-presentable in $\mathbf{Rel}\,\Sigma$ (see Proposition 1.16). For $f^* = \operatorname{colim} f_i^* : K^* \to A$ and $r = \operatorname{colim} r_i : K \to K^*$ we have

$$f \cdot k_i = f_i^* \cdot r_i = f^* \cdot k_i^* \cdot r_i = f^* \cdot r \cdot k_i$$

for $i < \lambda_\delta$, thus, $f = f^* \cdot r$.

This concludes the proof. □

2.54 Corollary. *A category is accessible iff it is equivalent to an accessibly embedded, cone-reflective subcategory of* $\mathbf{Set}^{\mathcal{A}}$ *for some small category* $\mathcal{A}$.

In fact, any λ-accessible category $\mathcal{K}$ is equivalent to a full, cone-reflective subcategory of $\mathbf{Set}^{\mathcal{A}}$ closed under λ-directed colimits, where $\mathcal{A} = \mathbf{Pres}_\lambda\,\mathcal{K}$ (see Proposition 2.8 and Corollary 2.45). Therefore the assertion follows from Theorem 2.53. □

2.F Accessible Categories and Sketches

We have introduced the concept of a limit sketch in Definition 1.49, and we have proved that locally presentable categories are precisely the categories "sketchable" by limit sketches (i.e., equivalent to categories of models of limit sketches). We now introduce the concept of a mixed limit-colimit sketch, and we show that accessible categories are precisely those sketchable by the mixed sketches. Recall that we have actually proved a result of that kind already: λ-accessible categories are equivalent to categories of functors from $\mathbf{Set}^{\mathcal{A}}$ to $\mathbf{Set}$ preserving λ-small limits and all colimits (Theorem 2.26). However, there we worked with a large category $\mathbf{Set}^{\mathcal{A}}$, whereas in a sketch everything is required to be small. We will show how the above "large sketch" can be reduced to a small one.

2.55 Definition

(1) By a *sketch* is understood a quadruple $\mathscr{S} = (\mathcal{A}, \mathbf{L}, \mathbf{C}, \sigma)$ consisting of

a small category $\mathcal{A}$,

a set **L** of diagrams in $\mathcal{A}$ (called limit diagrams),

a set **C** of diagrams in $\mathcal{A}$ (called colimit diagrams),

and

a function σ assigning to each diagram in **L** a cone, and to each diagram in **C** a cocone.

(2) The case $\mathbf{C} = \emptyset$ is called a *limit sketch* (see Definition 1.49) and the case $\mathbf{L} = \emptyset$ a *colimit sketch*.

(3) A sketch is called *normal* if σ assigns to each diagram in **L** a limit cone, and to each diagram in **C** a colimit cocone.

Notation. We denote by

$$\mathbf{Mod}\,\mathscr{S}$$

the category of *models* of the sketch $\mathscr{S}$, i.e., functors F from $\mathcal{A}$ to **Set** which for each diagram D in **L** map the cone $\sigma(D)$ to a limit of $F \cdot D$ and for each diagram D in **C** map the cocone $\sigma(D)$ to a colimit of $F \cdot D$. Thus, $\mathbf{Mod}\,\mathscr{S}$ is a full subcategory of $\mathbf{Set}^{\mathcal{A}}$.

2.56 Remarks

(1) If $\mathscr{S}$ is a normal sketch, then models are set-valued functors which preserve limits of **L**-diagrams and colimits of **C**-diagrams. For the theory of accessible categories, normal sketches are actually sufficient.

(2) A more general concept of a model (in other categories than **Set**) is introduced in 2.60 below.

(3) Some authors consider a broader concept of a sketch (which, however, does not make the concept of categories of models any broader): instead of working with a category $\mathcal{A}$, they work with a collection of objects and arrows (called a graph) and a set of (commutativity) diagrams. However, each such presentation defines a free category $\mathcal{A}$ determined by those objects, arrows, and the commutativity of the specified diagrams. The corresponding category $\mathbf{Mod}\,\mathscr{S}$ is equivalent to the category of models of that more general sketch.

2.57 Examples

(1) Let $\mathcal{A}$ be the category with objects 1, a, and a unique non-identity morphism $a \xrightarrow{f} 1$, let **L** consist of the unique diagram which is empty, and let **C** consist of the span $1 \xleftarrow{f} a \xrightarrow{f} 1$. Suppose σ assigns to the

empty diagram the (empty) cone with domain 1, and to the **C**-span the following cocone

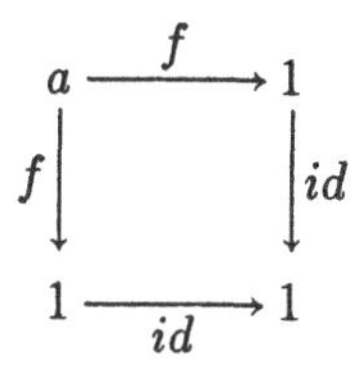

Then $\mathbf{Mod}\,\mathscr{S}$ consists of functors $F\colon \mathcal{A} \to \mathbf{Set}$ such that $F1$ is a singleton set and Fa is a set for which the unique map $Fa \to F1$ is an epimorphism. Thus, $\mathbf{Mod}\,\mathscr{S}$ is equivalent to the category of all non-empty sets and functions.

(2) The sketch for "lists of data". The category $\mathcal{A}$ has objects

$$T \text{ (terminal)}, \quad d \text{ (data)}, \quad l \text{ (lists)}, \quad \text{and } l^+ \text{ (non-empty lists)},$$

and morphisms

$$\text{head: } l^+ \to d, \quad \text{tail: } l^+ \to l, \quad i\colon l^+ \to l, \quad \text{empty: } T \to l$$

(plus the identity morphisms). **L** consists of the empty diagram with $\sigma(\emptyset) = T$ and the discrete diagram $\{d, l\}$ to which σ assigns the cone

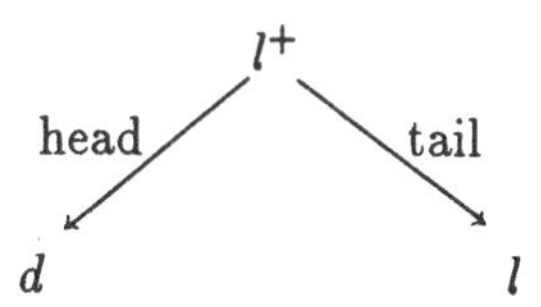

C consists of the single discrete diagram $\{T, l^+\}$ to which σ assigns the cocone

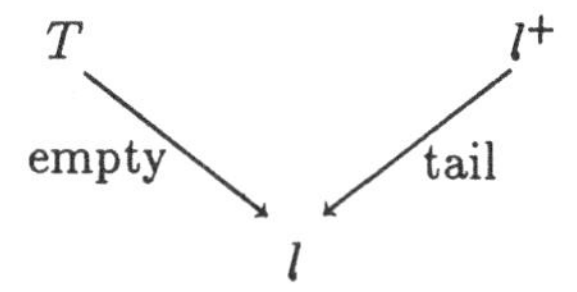

A model of this sketch is a functor $F\colon \mathcal{A} \to \mathbf{Set}$ such that FT is a singleton set (say, $\{\emptyset\}$), $Fl^+ = Fd \times Fl$, and $Fl = FT + Fl^+$. Thus, F is given by the set $D = Fd$ (of data) with Fl representing lists of data and Fl^+ non-empty lists of data.

(3) If we add to the sketch for graphs (see Example 1.50(5)) the diagram

$$\begin{array}{ccc} r & \xrightarrow{p_1 \cdot i} & a \\ {\scriptstyle p_1 \cdot i}\Big\downarrow & & \\ a & & \end{array}$$

to which σ assigns the cocone

$$\begin{array}{ccc} r & \xrightarrow{p_1 \cdot i} & a \\ {\scriptstyle p_1 \cdot i}\Big\downarrow & & \Big\downarrow{\scriptstyle id} \\ a & \xrightarrow[id]{} & a \end{array}$$

we get the sketch for graphs such that any vertex is a source of an arrow.

(4) The sketch for linearly ordered sets and order-preserving maps arises from the sketch for posets (Example 1.50(6)) by adding a morphism $s \colon a^2 \to a^2$ with $p_1 \cdot s = p_2$, $p_2 \cdot s = p_1$ and letting $\mathbf{C}$ consist of the single diagram

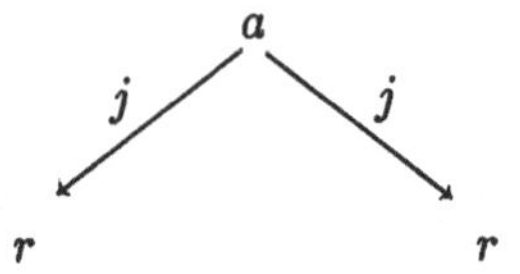

to which σ assigns the cocone

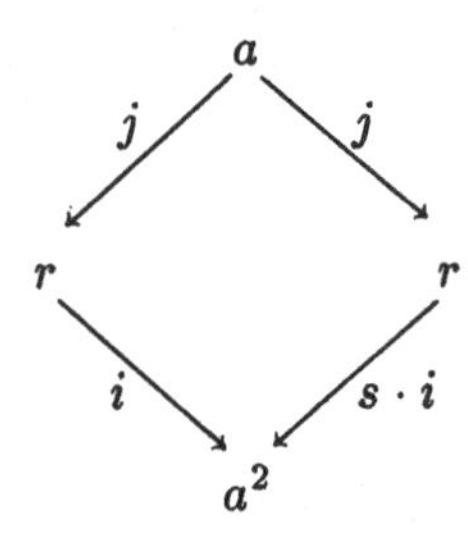

2.58 Theorem. *Every accessible category is normally sketchable, i.e., equivalent to the category of models of a normal sketch.*

Remark. Explicitly, if $\mathcal{K}$ is a λ-accessible category and $\mathcal{B} = \mathbf{Pres}_\lambda \mathcal{K}$ then we will prove that $\mathcal{K}$ is equivalent to $\mathbf{Mod}\,\mathscr{S}$ for the following sketch $\mathscr{S}$ (obtained by "reducing the large sketch" of Theorem 2.26(iv)):

(1) Choose a set $\mathbf{L}$ of representative λ-small diagrams in $Y(\mathcal{B}^{\mathrm{op}})$, where $Y: \mathcal{B}^{\mathrm{op}} \to \mathbf{Set}^{\mathcal{B}}$ is the Yoneda embedding.

(2) For each $D \in \mathbf{L}$ choose a limit cone $\sigma(D)$ with a domain $A(D)$.

(3) Put $\mathscr{S} = (\mathcal{A}, \mathbf{L}, \mathbf{C}, \sigma)$ where

$$\mathcal{A} = Y(\mathcal{B}^{\mathrm{op}}) \cup \{\, A(D) \mid D \in \mathbf{L} \,\}$$

is a full subcategory of $\mathbf{Set}^{\mathcal{B}}$, $\mathbf{L}$ is as in (1), $\mathbf{C}$ is the set of all canonical diagrams of $A(D)$ w.r.t. $Y(\mathcal{B}^{\mathrm{op}})$ for $D \in \mathbf{L}$, and σ assigns to each of the canonical diagrams the canonical colimit.

PROOF. This is an easy corollary of Theorem 2.26 and Lemma 2.24: $\mathcal{K}$ is equivalent to the category of all functors $F^*\colon \mathbf{Set}^{\mathcal{B}} \to \mathbf{Set}$ preserving colimits and λ-small limits of hom-functors. Now, a model $G\colon \mathcal{A} \to \mathbf{Set}$ of the above sketch $\mathscr{S}$ is, due to the $\mathbf{C}$-cocones, determined by its domain-restriction $F\colon \mathcal{B}^{\mathrm{op}} \to \mathbf{Set}$ (more precisely, by its domain-restriction $Y(\mathcal{B}^{\mathrm{op}}) \to \mathbf{Set}$ composed with $\mathcal{B}^{\mathrm{op}} \cong Y(\mathcal{B}^{\mathrm{op}})$) whose Kan-extension $F^*\colon \mathbf{Set}^{\mathcal{B}} \to \mathbf{Set}$ preserves λ-small limits of hom-functors (due to the $\mathbf{L}$). Conversely, if $F^*\colon \mathbf{Set}^{\mathcal{B}} \to \mathbf{Set}$ preserves colimits and λ-small limits, then its domain-restriction $G\colon \mathcal{A} \to \mathbf{Set}$ is a model of $\mathscr{S}$. This establishes an equivalence between $\mathbf{Mod}\,\mathscr{S}$ and $\mathcal{K}$. □

2.59 Remark. We have seen in Remark 1.52 that for a given regular cardinal λ a category is locally λ-presentable iff it is sketchable by a limit sketch containing λ-small limit diagrams only. This does not generalize to accessible categories. The sketch $\mathscr{S}$ of Example 2.57(3) contains finite cones and cocones only; however, we will now show that $\mathbf{Mod}\,\mathscr{S}$ is not finitely accessible.

The object $(\omega, <)$ of natural numbers with their usual irreflexive ordering is, evidently, a model of $\mathscr{S}$. Let us prove that $(\omega, <)$ is not a colimit of a diagram of graphs which are finitely presentable in $\mathbf{Mod}\,\mathscr{S}$. Let (A, α) be such a graph. Suppose the existence of a homomorphism $f\colon (A,\alpha) \to (\omega, <)$. Then $f\colon (A,\alpha) \to (\omega, \leq)$ is also a homomorphism, and since $(\omega, \leq)$ is a directed colimit of all of its finite strong subgraphs (which are models of $\mathscr{S}$, of course), we see that the image of f is finite. Thus, there exists an element x of A with $f(x)$ an upper bound of $f(A)$. This is impossible: we find $x' \in A$ with $(x, x') \in \alpha$, and then $f(x) < f(x')$. Consequently, no homomorphism leads from any finitely presentable model of $\mathscr{S}$ into $(\omega, <)$. □

2.60. So far, we have only considered set-valued models of a sketch. We can also consider, for each category $\mathcal{K}$, $\mathcal{K}$-valued models. That is, given a sketch $\mathscr{S} = (\mathcal{A}, \mathbf{L}, \mathbf{C}, \sigma)$, then

$$\mathbf{Mod}(\mathscr{S}, \mathcal{K})$$

denotes the category of all functors $F\colon \mathcal{A} \to \mathcal{K}$ which for each diagram D in $\mathbf{L}$ map the cone $\sigma(D)$ to a limit of $F \cdot D$ and for each diagram in $\mathbf{C}$ map the cocone $\sigma(D)$ to a colimit of $F \cdot D$. (This is a full subcategory of $\mathcal{K}^{\mathcal{A}}$.)

Theorem. *For each locally presentable category $\mathcal{K}$ and each sketch $\mathscr{S}$ the category $\mathbf{Mod}(\mathscr{S}, \mathcal{K})$ is accessible.*

Remark. Whether, more generally, $\mathbf{Mod}(\mathscr{S}, \mathcal{K})$ is accessible for each accessible category $\mathcal{K}$ depends on set theory—see Example A.19.

PROOF. We can express $\mathbf{Mod}(\mathscr{S}, \mathcal{K})$ as an intersection

$$\mathbf{Mod}(\mathscr{S}, \mathcal{K}) = \mathcal{L} \cap \bigcap_{D \in \mathbf{C}} \mathscr{C}_D$$

where $\mathcal{L} = \mathbf{Mod}(\mathscr{S}_1, \mathcal{K})$ is the category of all models of the limit part $\mathscr{S}_1 = (\mathcal{A}, \mathbf{L}, \sigma)$ of $\mathscr{S}$, and $\mathscr{C}_D$ is the full subcategory of $\mathcal{K}^{\mathcal{A}}$ over all functors F which map $\sigma(D)$ to a colimit of $F \cdot D$. It is sufficient to prove that $\mathcal{L}$ and each $\mathscr{C}_D$ are accessible and accessibly embedded into $\mathcal{K}^{\mathcal{A}}$, see Corollary 2.37.

I. $\mathcal{L}$ is closed in $\mathcal{K}^{\mathcal{A}}$ under λ-directed colimits for each regular cardinal λ such that every $D \in \mathbf{L}$ is λ-small—this follows from the fact that λ-directed colimits commute with λ-small limits in $\mathcal{K}^{\mathcal{A}}$, see Proposition 1.59. By Proposition 1.53, the category $\mathcal{L}$ is locally presentable.

II. To prove that $\mathscr{C}_D$, $D \in \mathbf{C}$, is accessible and accessibly embedded, we use Remark 2.50: we will find an accessible functor G from $\mathcal{K}^{\mathcal{A}}$ to $\mathcal{K}^{\mathbf{2}}$ (the category of $\mathcal{K}$-morphisms) such that for the full embedding $E\colon \mathbf{Iso}\,\mathcal{K} \hookrightarrow \mathcal{K}^{\mathbf{2}}$ (of the subcategory of all isomorphisms) we have the following pullback:

$$\begin{array}{ccc} \mathscr{C}_D & \xrightarrow{G_0} & \mathrm{Iso}\,\mathcal{K} \\ {\scriptstyle E_0}\downarrow & & \downarrow{\scriptstyle E} \\ \mathcal{K}^{\mathcal{A}} & \xrightarrow[G]{} & \mathcal{K}^{\mathbf{2}} \end{array}$$

Since $\mathbf{Iso}(\mathcal{K})$ is accessible (being equivalent to $\mathcal{K}$) and E preserves directed colimits, $\mathscr{C}_D$ is accessible by Remark 2.50. Put

$$\sigma(D) = \big(Dd \xrightarrow{k_d} K\big),$$

and define $G\colon \mathcal{K}^{\mathcal{A}} \to \mathcal{K}^{\mathbf{2}}$ on objects $F\colon \mathcal{A} \to \mathcal{K}$ of $\mathcal{K}^{\mathcal{A}}$ as follows:

$$G(F)\colon \operatorname{colim} FD \to FK$$

is the unique factorization of $F\sigma(D)$ through the colimit of FD. For each morphism $h\colon F \to F'$ in $\mathcal{K}^{\mathcal{A}}$ we denote by $G(h)$ the following morphism of $\mathcal{K}^{\mathbf{2}}$:

$$\begin{array}{ccc} \operatorname{colim} FD & \xrightarrow{G(F)} & FK \\ {\scriptstyle \operatorname{colim} hD}\downarrow & & \downarrow{\scriptstyle h_K} \\ \operatorname{colim} F'D & \xrightarrow[G(F')]{} & F'K \end{array}$$

This yields a well-defined functor $G\colon \mathcal{K}^{\mathcal{A}} \to \mathcal{K}^{\mathbf{2}}$ which is clearly accessible. It is easy to verify that the above square is a pullback for the inclusion functor E_0 and for G_0, the domain-codomain restriction of G. □

2.61 Corollary. *A category is accessible iff it is sketchable.*

This follows from Theorems 2.58 and 2.60. □

2.62 Corollary. *For each sketch $\mathscr{S}$ there exists a normal sketch $\overline{\mathscr{S}}$ with* $\mathbf{Mod}\,\mathscr{S} \approx \mathbf{Mod}\,\overline{\mathscr{S}}$.

2.63 Remark. If $\mathcal{K}$ is locally presentable and $\mathscr{S}$ a limit sketch, then $\mathbf{Mod}(\mathscr{S}, \mathcal{K})$ is complete, i.e., locally presentable (Corollary 2.47). We get the result of Proposition 1.53. Analogously, if $\mathcal{K}$ is locally presentable and $\mathscr{S}$ a colimit sketch then $\mathbf{Mod}(\mathscr{S}, \mathcal{K})$ is cocomplete, i.e., locally presentable.

Consequently, any accessible category $\mathcal{L}$ is an intersection of two locally presentable categories (given by the limit and colimit part of a sketch of $\mathcal{L}$).

2.G Accessible Embeddings Into the Category of Graphs

In this section we will prove that the category **Gra** of graphs has the universal property that every accessible category can be accessibly embedded into it.

We first prove the existence of a *rigid graph*, i.e., a graph whose only endomorphism in **Gra** is the identity, on every set:

2.64 Lemma. *For every set X there exists a rigid graph (X, σ).*

Remark. The rigid graphs constructed in the proof are without cycles.

PROOF. If $X = \{x_1, \dots, x_n\}$, put $\sigma = \{(x_i, x_{i+1}) \mid i = 1, \dots, n-1\}$. If X is infinite, we can assume that $X = \lambda + 2 = \{\alpha \in \mathrm{Ord} \mid \alpha \leq \lambda + 1\}$ for some ordinal λ. For each ordinal $\beta \leq \lambda$ such that $\mathrm{cf}\,\beta = \omega$ (i.e., $\beta = \bigvee \beta_n$ for an increasing sequence $\beta_0 < \beta_1 < \cdots$ of ordinals) choose an increasing sequence β_n $(n = 1, 2, \dots)$ of ordinals with $\beta = \bigvee \beta_n$ and such that $\beta_n = \overline{\beta}_n + n$, where $\overline{\beta}_n$ is a limit ordinal for each n (if $\beta \neq \omega$) or $\overline{\beta}_n = 0$ (if $\beta = \omega$). We will prove that the relation σ consisting of the following pairs (α, β) is rigid:

(i) α, β are limit ordinals, $0 < \alpha < \beta$, $\mathrm{cf}\,\beta > \omega$,

(ii) β is a limit ordinal, $\mathrm{cf}\,\beta = \omega$ and $\alpha = \beta_n$ for some n,

(iii) $\beta = \alpha + 1$.

(iv) $\beta = \lambda + 1$ and α is isolated, $\alpha \neq \lambda + 1$,

(v) $\alpha = 0$, $\beta = 2$.

We proceed by proving that each endomorphism f of (X, σ) has the following properties:

(a) $\alpha < \beta$ implies $f(\alpha) < f(\beta)$. We prove this by induction on β, observing first that each $(\alpha, \beta) \in \sigma$ fulfils $\alpha < \beta$. The case $\beta = 0$ is clear, and the isolated step follows from

$$(\beta, \beta + 1) \in \sigma \implies (f(\beta), f(\beta + 1)) \in \sigma \implies f(\beta) < f(\beta + 1).$$

For the limit step observe that whenever β is a limit ordinal in X, then for each $\alpha < \beta$ there exists $(\gamma, \beta) \in \sigma$ with $\alpha < \gamma$. By the induction hypothesis, $\alpha < \gamma$ $(< \beta)$ implies $f(\alpha) < f(\gamma)$; since $(\gamma, \beta) \in \sigma \implies (f(\gamma), f(\beta)) \in \sigma \implies f(\gamma) < f(\beta)$, we conclude that $f(\alpha) < f(\beta)$.

(b) f is one-to-one and $f(\alpha) \geq \alpha$. This is clear from (a).

(c) $f(\lambda) = \lambda$ and $f(\lambda + 1) = \lambda + 1$. This follows from (a), (b).

(d) $f(\alpha)$ is isolated if α is isolated. Use (c) and $(\alpha, \lambda + 1) \in \sigma$.

(e) $f(\beta)$ is a limit ordinal if $\beta > 0$ is a limit ordinal; moreover, $\mathrm{cf}\,\beta = \omega$ implies $\mathrm{cf}\,f(\beta) = \omega$. In fact, if $\mathrm{cf}\,\beta = \omega$, then the $f(\beta_n)$ are isolated ordinals with $(f(\beta_n), f(\beta)) \in \sigma$; by the definition of σ this implies $f(\beta) = \lambda + 1$ or $\mathrm{cf}\,\beta = \omega$. The former cannot occur by (b), (c). If $\mathrm{cf}\,\beta > \omega$, we have uncountably many γ's with $\mathrm{cf}\,\gamma = \omega$ and $(\gamma, f(\beta)) \in \sigma$ (viz, $\gamma = f(\alpha)$ for any $\alpha < \beta$ with $\mathrm{cf}\,\alpha = \omega$). Thus, $\mathrm{cf}\,f(\beta) > \omega$.

(f) For each $n < \omega$, $f(n) = n$ and $f(\alpha + n) = f(\alpha) + n$. In fact, $f(2)$ is an isolated ordinal, $f(2) \neq \lambda + 1$. Using the fact that σ contains $(0,2)$, $(0,1)$ and $(1,2)$, we conclude that $f(2) = 2$. Consequently, $f(0) = 0$ and $f(1) = 1$. It remains to prove $f(\alpha+1) = f(\alpha)+1$ for each α. We know this if $f(\alpha+1)$ is 2 or $\lambda+1$; if $f(\alpha+1) \neq 2, \lambda+1$, then since $f(\alpha+1)$ is isolated, the unique γ with $(\gamma, f(\alpha+1)) \in \sigma$ is the solution of $f(\alpha+1) = \gamma+1$. The statement follows from $(f(\alpha), f(\alpha+1)) \in \sigma$.

(g) $f(\beta_n) = f(\beta)_n$ for each limit ordinal β cofinal with ω. This follows from (e), (f): $f(\beta_n) = f(\overline{\beta}_n) + n$.

(h) $f(\alpha) = \alpha$ for each α. Assume the contrary and choose the smallest α with $f(\alpha) \neq \alpha$. From (a), (b) we know that $\alpha < f(\alpha) < f^2(\alpha) < \cdots$. Put $\beta = \bigvee f^n(\alpha)$, then $\operatorname{cf} \beta = \omega$. For each β_k there exists $f^n(\alpha)$ with $\beta_k < f^n(\alpha) < \beta$—thus, $f(\beta)_k = f(\beta_k) < f^{n+1}(\alpha) < \beta$, i.e., $f(\beta) = \beta$. From (g) we get $f(\beta_k) = \beta_k$. There exists k with $\alpha < \beta_k$, then $f(\alpha) < \beta_k$, $f^2(\alpha) < \beta_k$, etc. Thus, $\bigvee f^n(\alpha) < \beta_k$, i.e., $\beta < \beta_k$—a contradiction. □

2.65 Theorem. *Every accessible category has an accessible, full embedding into* **Gra**.

Remark. The graphs used in the proof are all connected (i.e., non-empty and indecomposable w.r.t. coproducts). Thus, **Gra** can be substituted by the full subcategory $\mathbf{Gra}_0$ of connected graphs.

PROOF. By Example 1.41 and Proposition 2.8, every λ-accessible category has a λ-accessible, full embedding into $\mathbf{Rel}\,\Sigma$ for some binary (many-sorted) signature Σ. We shall find an accessible, full embedding of $\mathbf{Rel}\,\Sigma$ into **Gra**.

(1) For every signature Σ there exists a one-sorted signature Σ' and a finitely accessible, full embedding $E\colon \mathbf{Rel}\,\Sigma \to \mathbf{Rel}\,\Sigma'$. In fact, let $\Sigma' = \Sigma + S$ where S is the set of sorts of Σ, $\operatorname{ar}(s) = 1$ for $s \in S$, and given $\sigma \in \Sigma$ of arity $(n_s)_{s \in S}$, then σ has arity $\sum_{s\in S} n_s$ in Σ'. For each $\sigma \in \Sigma$ we have a canonical embedding

$$d_\sigma\colon \prod_{s\in S} X_s^{n_s} \longrightarrow \Big(\coprod_{t\in S} X_t\Big)^{\sum n_s} \left[\cong \prod_{s\in S}\Big(\coprod_{t\in S} X_t\Big)^{n_s}\right].$$

Define $E\colon \mathbf{Rel}\,\Sigma \to \mathbf{Rel}\,\Sigma'$ on objects by

$$E(A) = \Big(\coprod_{s\in S} X_s, (d_\sigma(\sigma_A))_{\sigma\in\Sigma}, (X_s)_{s\in S}\Big),$$

where $A = \big((X_s)_{s\in S}, (\sigma_A)_{\sigma\in\Sigma}\big)$, and on morphisms by $E(f_s) = \coprod_{s\in S} f_s$.

(2) For every one-sorted signature Σ there exists a binary signature Σ' (i.e., with all arities equal to 2) and an accessible, full embedding

$$E\colon \mathbf{Rel}\,\Sigma \to \mathbf{Rel}\,\Sigma'.$$

In fact, let n be a cardinal larger or equal to $\operatorname{ar}\sigma$ for all $\sigma \in \Sigma$, and let $\Sigma' = \Sigma + n$ be a binary signature. For each $\sigma \in \Sigma$ choose a one-to-one mapping $h_\sigma = \operatorname{ar}\sigma \to n$. Define $E\colon \mathbf{Rel}\,\Sigma \to \mathbf{Rel}\,\Sigma'$ on objects by

$$E(A) = \big(X^n, (\sigma'_A)_{\sigma\in\Sigma} \cup (r_{A,i})_{i\in n}\big)$$

where $A = \big(X, (\sigma_A)_{\sigma\in\Sigma}\big)$ and

$$\sigma'_A = \{\, (u,v) \mid u,v\colon n \to X, u \cdot h_\sigma \in \sigma_A \,\},$$
$$r_{A,i} = \{\, (u,v) \mid u,v\colon n \to X, v(j) = u(i) \text{ for each } j \in n \,\},$$

and on morphisms by $Ef = f^n$.

It is easy to see that E is an embedding which preserves λ-directed colimits for $\lambda > n$ (by Proposition 1.59). Let us verify that E is full. Let $g\colon E(A) \to E(B)$ be a homomorphism. For the constant n-tuple $[x]$ with value x we have $\big([x],[x]\big) \in r_{A,i}$, thus $\big(g\cdot[x], g\cdot[x]\big) \in r_{B,i}$. Therefore $g\cdot[x]$ is constant. Define $f\colon |A| \to |B|$ by $g\cdot[x] = [f(x)]$. Then $g = f^n$ because for each $u \in X^n$ and each $i \in n$ we have $(u,[u(i)]) \in r_{A,i}$; thus, $\big(g(u),[f(u(i))]\big) \in r_{B,i}$ and consequently $g(u)(i) = f(i)$. To prove that f is a homomorphism in $\mathbf{Rel}\,\Sigma$, consider any $a \in \sigma_A$, and choose $u \in X^n$ with $u \cdot h_\sigma = a$. Then $(u,u) \in \sigma'_A$ implies $(f\cdot u, f\cdot u) \in \sigma'_B$, thus $f\cdot a = f\cdot u \cdot h_\sigma \in \sigma_B$.

(3) For any binary signature Σ there exists a finitely accessible, full embedding $E\colon \mathbf{Rel}\,\Sigma \to \mathbf{Gra}$. Let τ be a rigid binary relation on the set Σ from Lemma 2.64. Define $E\colon \mathbf{Rel}\,\Sigma \to \mathbf{Gra}$ as follows:

$$E(A) = \big(X + (X\times X) + \Sigma + \{\,0,1,2,3\,\}, \varrho_A\big)$$

where $X = |A|$,

$$\varrho_A = \gamma \cup \delta_A \cup \big(\{0\}\times X\big) \cup \big(X \times \{1\}\big) \cup \\ \cup \big(\{1\}\times\Sigma\big) \cup \big(\Sigma\times\{2\}\big) \cup \tau \cup \bigcup_{\sigma\in\Sigma} \big(\sigma_A \times \{\sigma\}\big),$$

$$\gamma = \{\, (0,1),(1,2),(2,0),(0,3),(3,0) \,\}$$

and

$$\delta_A = \big\{\big((x,y),x\big)\big\}_{x,y\in X} \cup \big\{\big(y,(x,y)\big)\big\}_{x,y\in X}.$$

For morphisms, $Ef = f + (f \times f) + id_\Sigma + id_{\{0,1,2,3\}}$. It is easy to check that E is an embedding preserving directed colimits. Let us verify that E is full. Let $g\colon E(A) \to E(B)$ be a homomorphism. Since g preserves n-cycles for each n, and since $E(A)$ has the following n-cycles (τ is without cycles by Remark 2.64):

$n = 1$: none

$n = 2$: $(0, 3)$ and $\big((x, x), x\big)$

$n = 3$: $(0, 1, 2)$,

it is clear that $g(0) = 0$. Thus $g(i) = i$ for $i = 0, 1, 2, 3$. From the path $0 \to x \to 1$ we conclude that $g(|A|) \subseteq |B|$; let $f\colon |A| \to |B|$ be the restriction of g. From the path $y \to (x, y) \to x$ we conclude $g(x, y) = (f(x), f(y))$; since the path $1 \to \sigma \to 2$ implies $g(\Sigma) \subseteq \Sigma$ and since τ is rigid, we have $g(\sigma) = \sigma$ for each $\sigma \in \Sigma$. Consequently, $g = Ef$. Finally, f preserves σ_A because $(x, y) \in \sigma_A$ implies $((x, y), \sigma) \in \varrho_A$, thus $((f(x), f(y)), \sigma) \in \varrho_B$, i.e., $\big(f(x), f(y)\big) \in \sigma_B$. □

2.66 Remark. The category **Sgr** of semigroups also has the universal property that every accessible category has an accessible, full embedding into it. This follows from the fact that **Gra** has a finitely accessible, full embedding into **Sgr**. This result, and a number of others of this kind, can be found in the monograph [Pultr, Trnková 1980].

2.H Limits of Accessible Categories

It is a remarkable property of accessible categories that a limit of a (small) diagram of accessible categories is accessible. This, however, does not hold for "ordinary" limits, but for lax limits, a concept from the theory of 2-categories which we explain in detail below. We do not assume any preliminary knowledge of the theory of 2-categories in our book. The reader can skip this section without breaking the continuity of the text.

We denote by **Cat** the category of small categories and functors, and by **CAT** the quasicategory of all categories and functors. (See 0.13.) The sub-quasicategory of all accessible categories and all accessible functors is denoted by **ACC**.

When talking about limits of accessible categories we mean limits (of small diagrams) in **ACC**. The quasicategory **CAT** is well known to be complete, thus, limits can be always computed via products and equalizers. Products of accessible categories do not present problems, but equalizers (and limits in general) do:

2.67 Proposition. *A product of accessible categories is accessible. More precisely,* **ACC** *is closed under products in* **CAT**.

PROOF. Let $\mathcal{K}_i$, $i \in I$, be accessible categories. There exists a regular cardinal λ such that $\mathcal{K}_i$ is λ-accessible for each $i \in I$, see Remark 2.14. Then the category $\prod_{i \in I} \mathcal{K}_i$ is λ-accessible and the projections are λ-accessible functors. In fact, $\prod_{i \in I} \mathcal{K}_i$ has λ-directed colimits computed coordinatewise. Therefore, an object of $\prod \mathcal{K}_i$ whose components are λ-presentable is λ-presentable. These objects form, up to isomorphism, a set, and every object $\prod_{i \in I} \mathcal{K}_i$ is a λ-directed colimit of those λ-presentable objects. □

2.68 Example of a non-accessible equalizer of two accessible functors. Let $\mathcal{K}$ be a full subcategory of **Set** which is not accessible, and let

$$F\colon \mathbf{Set} \to \mathbf{Set}$$

be a functor naturally isomorphic to Id such that $FX = X$ iff X is in $\mathcal{K}$ (for any set X). Then $F, Id\colon \mathbf{Set} \to \mathbf{Set}$ are finitely accessible functors whose equalizer is not accessible.

2.69 Definition. By the *lax limit* of a diagram $D\colon \mathcal{D} \to \mathbf{CAT}$ is meant the following cone

$$P_d\colon \mathbf{Lax}\, D \to Dd \qquad (d \in \mathcal{D}^{\mathrm{obj}})$$

in **CAT**:

The objects of the category $\mathbf{Lax}\, D$ are the pairs

$$\big((K_d)_{d \in \mathcal{D}^{\mathrm{obj}}}, (k_f)_{f \in \mathcal{D}^{\mathrm{mor}}}\big)$$

where each K_d is an object of the category Dd, and for each morphism $f\colon d \to d'$ in $\mathcal{D}$ we have a morphism $k_f\colon (Df)K_d \to K_{d'}$ in Dd' subject to the following conditions:

(i) if $f = id_d$, then $k_f = id_{K_d}$

(ii) if $f = g \cdot h$, then $k_f = k_g \cdot (Dg)(k_h)$.

Morphisms of $\mathbf{Lax}\, D$ from $\big((K_d), (k_f)\big)$ to $\big((K_d^*), (k_f^*)\big)$ are families $(r_d)_{d \in \mathcal{D}^{\mathrm{obj}}}$ of morphisms $r_d\colon K_d \to K_d^*$ in Dd such that the square

$$\begin{array}{ccc} (Df)K_d & \xrightarrow{\;k_f\;} & K_{d'} \\ {\scriptstyle (Df)r_d}\downarrow & & \downarrow{\scriptstyle r_{d'}} \\ (Df)K_d^* & \xrightarrow[\;k_f^*\;]{} & K_{d'}^* \end{array}$$

commutes for each $f\colon d \to d'$ in $\mathcal{D}$.

The functor $P_d\colon \mathbf{Lax}\,D \to Dd$ is the projection assigning the d-component to each object and each morphism.

2.70 Examples

(1) Lax product and "ordinary" product coincide.

(2) A *lax equalizer* of functors $F_1, F_2\colon \mathcal{K} \to \mathcal{L}$ is the category

$$\mathbf{Lax}\,\mathbf{Eq}(F_1, F_2)$$

whose objects are pairs $\big((K, L), (k_1, k_2)\big)$ of morphisms $k_i\colon F_iK \to L$ in $\mathcal{L}$ and whose morphisms are all morphisms (r, r') of $\mathcal{K} \times \mathcal{L}$ such that the squares

$$\begin{array}{ccc} F_iK & \xrightarrow{k_i} & L \\ {\scriptstyle F_ir}\downarrow & & \downarrow{\scriptstyle r'} \\ F_iK' & \xrightarrow[k_i']{} & L' \end{array}$$

commute ($i = 1, 2$), together with the obvious projection functors

$$P_1\colon \mathbf{Lax}\,\mathbf{Eq}(F_1, F_2) \to \mathcal{K} \quad \text{and} \quad P_2\colon \mathbf{Lax}\,\mathbf{Eq}(F_1, F_2) \to \mathcal{L}.$$

2.71. In order to prove the limit theorem, which states that accessible categories are closed in **CAT** under lax limits, we introduce categories of inserters and equifiers.

Given functors $F_1, F_2\colon \mathcal{K} \to \mathcal{L}$, the *inserter category*

$$\mathbf{Ins}(F_1, F_2)$$

is the subcategory of the comma-category $F_1 \downarrow F_2$ (see Notation 2.42) consisting of all objects $F_1K \xrightarrow{f} F_2K$ and all morphisms

$$\begin{array}{ccc} F_1K & \xrightarrow{f} & F_2K \\ {\scriptstyle F_1k}\downarrow & & \downarrow{\scriptstyle F_2k} \\ F_1K' & \xrightarrow[f']{} & F_2K' \end{array}$$

We denote by $P\colon \mathbf{Ins}(F_1, F_2) \to \mathcal{K}$ the natural forgetful functor.

Remark. There is an "inserted" natural transformation $\psi\colon F_1P \to F_2P$ given by

$$\psi_{(f\colon F_1K \to F_2K)} = f.$$

2.72 Theorem. *The category* $\mathbf{Ins}(F_1, F_2)$ *is accessible for arbitrary accessible functors* $F_1, F_2\colon \mathcal{K} \to \mathcal{L}$.

PROOF. There is a regular cardinal λ such that F_1 and F_2 are λ-accessible functors. By Remark 2.19 and by Theorems 2.34 and 2.43 there exists a regular cardinal $\mu > \lambda$ such that

(a) the categories $\mathcal{K}$, $\mathcal{L}$, $\mathbf{Pure}_\lambda\,\mathcal{K}$ and $F_1 \downarrow F_2$ are μ-accessible

and

(b) the functors $F_1, F_2\colon \mathcal{K} \to \mathcal{L}$, $P_1, P_2\colon F_1 \downarrow F_2 \to \mathcal{K}$ and the embedding $\mathbf{Pure}_\lambda\,\mathcal{K} \to \mathcal{K}$ are μ-accessible and preserve μ-presentable objects.

Therefore, an object $F_1K_1 \xrightarrow{f} F_2K_2$ is μ-presentable in $F_1 \downarrow F_2$ iff K_1 and K_2 are μ-presentable in $\mathcal{K}$.

We are going to prove that $\mathbf{Ins}(F_1, F_2)$ is a μ-accessible category. Since it is, obviously, closed under μ-directed colimits in $F_1 \downarrow F_2$, it is sufficient to prove that each object $F_1K \xrightarrow{h} F_2K$ of $\mathbf{Ins}(F_1, F_2)$ is a μ-filtered colimit of the (μ-presentable) objects $F_1\overline{K} \to F_2\overline{K}$ such that $\overline{K}$ is μ-presentable in $\mathcal{K}$.

I. First, consider a morphism in $F_1 \downarrow F_2$:

$$\begin{array}{ccc} F_1K_1 & \xrightarrow{f} & F_2K_2 \\ {\scriptstyle F_1g_1}\downarrow & & \downarrow{\scriptstyle F_2g_2} \\ F_1K & \xrightarrow[h]{} & F_2K \end{array}$$

such that $F_1K_1 \xrightarrow{f} F_2K_2$ is μ-presentable in $F_1 \downarrow F_2$. We will prove that it has a factorization

$$\begin{array}{ccc} F_1K_1 & \xrightarrow{f} & F_2K_2 \\ {\scriptstyle F_1\overline{g}_1}\downarrow & & \downarrow{\scriptstyle F_2\overline{g}_2} \\ F_1\overline{K} & \xrightarrow{\overline{f}} & F_2\overline{K} \\ {\scriptstyle F_1t}\downarrow & & \downarrow{\scriptstyle F_2t} \\ F_1K & \xrightarrow[h]{} & F_2K \end{array}$$

with $\overline{K}$ a μ-presentable object and $t\colon \overline{K} \to K$ a λ-pure morphism in $\mathcal{K}$.

Since $\mathbf{Pure}_\lambda\,\mathcal{K}$ is μ-accessible and the embedding $\mathbf{Pure}_\lambda\,\mathcal{K} \to \mathcal{K}$ preserves μ-presentable objects, there exists a μ-directed diagram D of μ-presentable objects and λ-pure morphisms in $\mathcal{K}$ with a colimit $(D_i \xrightarrow{k_i} K)_{i\in I}$. The object K_1 is μ-presentable, thus, there is a morphism $u_0\colon K_1 \to D_{i_0}$ ($i_0 \in I$) with $k_{i_0} \cdot u_0 = g_1$. Since the objects K_2 and $F_1 D_{i_0}$ are μ-presentable, there exist $i_0 \leq i_1$ and morphisms $f_0\colon F_1 D_{i_0} \to F_2 D_{i_1}$ and $v_0\colon K_2 \to D_{i_1}$ such that

$$k_{i_1} \cdot v_0 = g_2 \qquad \text{and} \qquad h \cdot F_1 k_{i_0} = F_2 k_{i_1} \cdot f_0.$$

Since $F_2 k_{i_1} \cdot f_0 \cdot F_1 u_0 = F_2 k_{i_1} \cdot F_2 v_0 \cdot f$, we can assume without loss of generality that

$$f_0 \cdot F_1 u_0 = F_2 v_0 \cdot f.$$

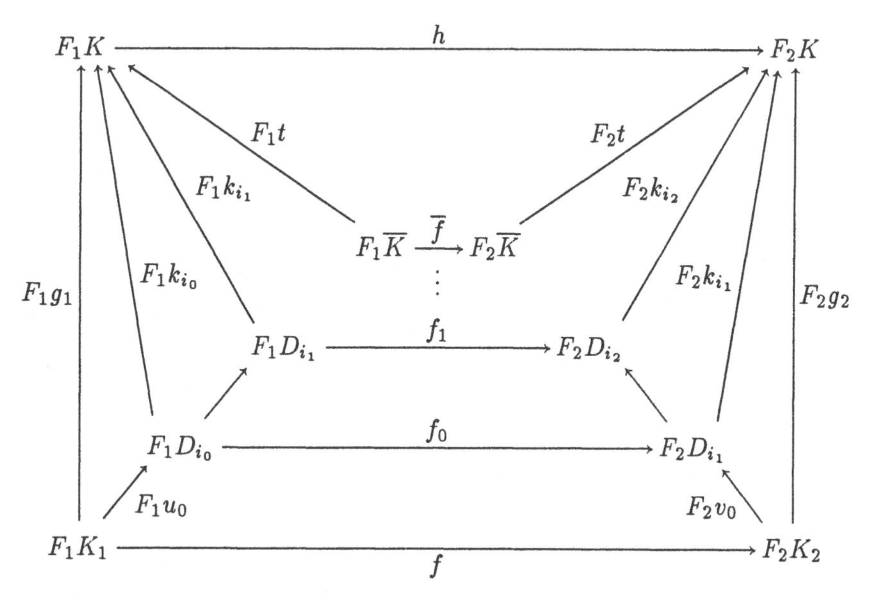

By transfinite induction we will construct morphisms

$$f_\alpha\colon F_1 D_{i_\alpha} \to F_2 D_{i_{\alpha+1}}$$

for $\alpha < \lambda$ (where $i_\alpha \in I$ satisfies $i_\alpha \leq i_\beta$ for $\alpha < \beta < \lambda$) such that $\alpha < \beta$ implies that

$$\big(D(i_\alpha \to i_\beta), D(i_{\alpha+1} \to i_{\beta+1})\big)\colon f_\alpha \to f_\beta$$

and

$$(k_{i_\alpha}, k_{i_{\alpha+1}})\colon f_\alpha \to h$$

are morphisms in $F_1 \downarrow F_2$ (see the diagram above).

Isolated step: let $f_\alpha\colon F_1 D_{i_\alpha} \to F_2 D_{i_{\alpha+1}}$ be given; $FD_{i_{\alpha+1}}$ is μ-presentable, thus, there exist $i_{\alpha+1} \le i_{\alpha+2}$ and $f_{\alpha+1}\colon F_1 D_{i_{\alpha+1}} \to F_2 D_{i_{\alpha+2}}$ such that

$$h \cdot F_1 k_{i_{\alpha+1}} = F_2 k_{i_{\alpha+2}} \cdot f_{\alpha+1}.$$

As above, we can assume without loss of generality that

$$f_{\alpha+1} \cdot F_1 D(i_\alpha \to i_{\alpha+1}) = F_2 D(i_{\alpha+1} \to i_{\alpha+2}) \cdot f_\alpha.$$

Limit step: let $f_\alpha\colon F_1 D_{i_\alpha} \to F_2 D_{i_{\alpha+1}}$ be given for $\alpha < \beta$, where $0 < \beta$ is a limit cardinal. There exist $i_\beta \le i_{\beta+1}$ such that $i_\alpha \le i_\beta$ for any $\alpha < \beta$, and $f_\beta\colon F_1 D_{i_\beta} \to F_2 D_{i_{\beta+1}}$ such that $h \cdot F_1 k_{i_\beta} = F_2 k_{i_{\beta+1}} \cdot f_\beta$. Since

$$\begin{aligned} F_2 k_{i_{\beta+1}} \cdot f_\beta \cdot D(i_\alpha \to i_\beta) &= h \cdot F_1 k_{i_\beta} \cdot D(i_\alpha \to i_\beta) \\ &= h \cdot F_1 k_{i_\alpha} = F_2 k_{i_{\alpha+1}} \cdot f_\alpha \\ &= F_2 k_{i_{\beta+1}} \cdot D(i_{\alpha+1} \to i_{\beta+1}) \cdot f_\alpha \end{aligned}$$

for any $\alpha < \beta$, we can assume without loss of generality that for any $\alpha < \beta$

$$f_\beta \cdot D(i_\alpha \to i_\beta) = D(i_{\alpha+1} \to i_{\beta+1}) \cdot f_\alpha.$$

Since $\mathcal{K}$ is λ-accessible, we can take a colimit $\overline{K}$ of D_{i_α}, $\alpha < \lambda$. The morphisms f_α, $\alpha < \lambda$ induce the morphism $\overline{f}\colon F_1\overline{K} \to F_2\overline{K}$ such that $\overline{f} \cdot F_1 \overline{k}_\alpha = F_2 \overline{k}_{\alpha+1} \cdot f_\alpha$ where $\overline{k}_\alpha\colon D_{i_\alpha} \to \overline{K}$ is the colimit cocone. Then $\overline{f}$ is an object in $\mathbf{Ins}(F_1, F_2)$ which is μ-presentable in $\mathbf{Ins}(F_1, F_2)$ (because $\overline{K}$ is μ-presentable in $\mathcal{K}$ by Proposition 1.16). Since the $k_{i_\alpha}\colon D_{i_\alpha} \to K$ form a compatible cocone, there is a unique $t\colon \overline{K} \to K$ satisfying

$$t \cdot \overline{k}_\alpha = k_{i_\alpha}$$

for any $\alpha < \lambda$. Since, for any $\alpha < \lambda$

$$h \cdot F_1 t \cdot F_1 \overline{k}_\alpha = h \cdot F_1 k_{i_\alpha} = F_2 k_{i_{\alpha+1}} \cdot f_\alpha = F_2 t \cdot F_2 \overline{k}_{\alpha+1} \cdot f_\alpha = F_2 t \cdot \overline{f} \cdot F_1 \overline{k}_\alpha,$$

we have

$$h \cdot F_1 t = F_2 t \cdot \overline{f}.$$

Therefore we obtain the factorization

$$(g_1, g_2)\colon f \xrightarrow{(\overline{g}_1, \overline{g}_2)} \overline{f} \xrightarrow{t} h$$

where $\overline{g}_1 = \overline{k}_0 \cdot u_0$ and $\overline{g}_2 = \overline{k}_1 \cdot v_0$. Moreover, $t\colon \overline{K} \to K$ is λ-pure in $\mathcal{K}$ (by Remark 2.34).

II. Let $F_1K \xrightarrow{h} F_2K$ be in $\mathbf{Ins}(F_1, F_2)$. Let $\mathcal{D}_0$ be the comma-category in $\mathbf{Ins}(F_1, F_2)$ of h w.r.t. objects $F_1\overline{K} \xrightarrow{\overline{f}} F_2\overline{K}$ having $\overline{K}$ μ-presentable, $\mathcal{D}$ the comma-category in $F_1 \downarrow F_2$ of h w.r.t. μ-presentable objects in $F_1 \downarrow F_2$, and $H\colon \mathcal{D}_0 \to \mathcal{D}$ the inclusion. We will prove that $\mathcal{D}_0$ is μ-filtered and H cofinal, which will complete the proof.

The condition 0.11(a) follows by I. Moreover, if we take from I the factorization

$$g\colon f \xrightarrow{(g_1,g_2)} \overline{f} \xrightarrow{t} h$$

of a morphism g belonging to $\mathbf{Ins}(F_1, F_2)$ then $g_1 = g_2$, i.e., (g_1, g_2) belongs to $\mathbf{Ins}(F_1, F_2)$ too. Indeed, since $t \cdot g_1 = g = t \cdot g_2$ and t is λ-pure, it follows by Proposition 2.29. Now, this property together with I implies that H is cofinal and $\mathcal{D}_0$ λ-filtered (see Exercise 1.o(3)). □

2.73 Remark. It follows from the proof that the embedding

$$\mathbf{Ins}(F_1, F_2) \to F_1 \downarrow F_2$$

is accessible. Hence the natural forgetful functor $P\colon \mathbf{Ins}(F_1, F_2) \to \mathcal{K}$ is accessible (see Remark 2.43).

2.74 Notation. Let $T\colon \mathcal{K} \to \mathcal{K}$ be a functor. Denote by $\mathbf{Alg}\,T$ the category of *T-algebras*, i.e., arrows $TK \xrightarrow{k} K$ in $\mathcal{K}$, and T-homomorphisms, i.e., morphisms $f\colon K \to K'$ such that the square

$$\begin{array}{ccc} TK & \xrightarrow{k} & K \\ {\scriptstyle Tf}\downarrow & & \downarrow{\scriptstyle f} \\ TK' & \xrightarrow[k']{} & K' \end{array}$$

commutes.

2.75 Corollary. *For each accessible functor $T\colon \mathcal{K} \to \mathcal{K}$ the category $\mathbf{Alg}\,T$ is accessible.*

In fact, $\mathbf{Alg}\,T = \mathbf{Ins}(T, Id_{\mathcal{K}})$.

Remark. Moreover, if $\mathcal{K}$ is locally λ-presentable and T is λ-accessible then $\mathbf{Alg}\,T$ is locally λ-presentable. In fact, the forgetful functor $P\colon\ \mathbf{Alg}\,T \to \mathcal{K}$ has a left adjoint by Theorem 1.66, and free T-algebras over λ-presentable $\mathcal{K}$-objects are λ-presentable in $\mathbf{Alg}\,T$ and form a strong generator. It suffices to apply Theorem 1.20.

2.76 Lemma. *Let $F_1, F_2\colon \mathcal{K} \to \mathcal{L}$ be accessible functors. Then for each pair $\varphi, \psi\colon F_1 \to F_2$ of natural transformations the equifier of φ, ψ, i.e., the full subcategory* $\mathbf{Eq}(\varphi, \psi)$ *of $\mathcal{K}$ over all objects K with $\varphi_K = \psi_K$, is accessible.*

PROOF. By Remark 2.19 there exists a regular cardinal λ such that $\mathcal{K}$ and $\mathcal{L}$ are λ-accessible categories, and the F_i are λ-accessible functors preserving λ-presentable objects. It follows immediately that $\mathbf{Eq}(\varphi, \psi)$ is closed under λ-directed colimits. Moreover, since the F_i preserve λ-pure subobjects by Proposition 2.38, it is clear that $\mathbf{Eq}(\varphi, \psi)$ is closed under λ-pure subobjects. Therefore, it is λ-accessible by Corollary 2.36. □

Remark. The above lemma holds, more generally, for any set of pairs

$$F_1^t, F_2^t\colon \mathcal{K} \to \mathcal{L}_t \qquad (t \in T)$$

of accessible functors, and any set of pairs

$$\varphi^t, \psi^t : F_1^t \to F_2^t$$

of natural transformations: the joint equifier, i.e., the full subcategory of $\mathcal{K}$ consisting of objects K with $\varphi_K^t = \psi_K^t$ for all $t \in T$, is accessible. (The proof is the same.)

2.77 Limit Theorem. *A lax limit of accessible categories is accessible. More precisely,* **ACC** *is closed under lax limits in* **CAT**.

PROOF. Let $D\colon \mathcal{D} \to \mathbf{ACC}$ be a diagram, and let

$$F_1, F_2\colon \prod_{d \in \mathcal{D}^{\mathrm{obj}}} Dd \to \prod_{\substack{f\colon d \to d' \in \mathcal{D}^{\mathrm{mor}} \\ f \neq id_d}} Dd'$$

be the following functors: given $f\colon d \to d'$ in $\mathcal{D}^{\mathrm{mor}}$, $f \neq id_d$, the f-component of F_1 is the projection $P_{d'}$ of the first product to Dd', and the f-component of F_2 is $Df \cdot P_d$. The inserter category $\mathbf{Ins}(F_1, F_2)$ is accessible by Theorem 2.72. It is, obviously, isomorphic to the category $\mathbf{Lax}^* D$ defined precisely as $\mathbf{Lax}\, D$ (see Definition 2.69) except that the condition (ii) is deleted. (In fact, the isomorphism from $\mathbf{Ins}(F_1, F_2)$ into $\mathbf{Lax}^* D$ "enriches" each object by defining $k_f = id_{K_d}$ for $f = id_d$.) From Remark 2.73 and Proposition 2.67 we know that the projection functors $P_d^*\colon \mathbf{Lax}^* D \to Dd$ are accessible. Every morphism $f\colon d \to d'$ in $\mathcal{D}$ defines a natural transformation

$$f^* : (Df) \cdot P_d \to P_{d'}$$

which is the projection of each object to k_f.

For each composition

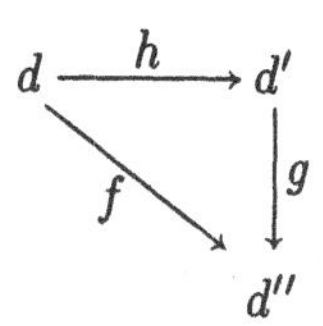

in $\mathcal{D}$ consider the equifier of the natural transformations

$$f^*,\ g^* \cdot D(g)h^* : (Df) \cdot P_d \to P_{d''}.$$

The equifier is accessible (by Lemma 2.76) and consists of all objects in $\mathbf{Lax}^*\, D$ which fulfil the condition (ii) for this instance of composition. By Remark 2.76 the joint equifier for all instances of compositions is also accessible. This joint equifier is isomorphic to $\mathbf{Lax}\, D$. □

2.78. As another application of the accessibility of equifiers we will prove that for each *accessible monad*, i.e., a monad $\mathcal{T} = (T, \mu, \eta)$ over a category $\mathcal{K}$ such that the functor T (and hence, the category $\mathcal{K}$) is accessible, the Eilenberg–Moore category $\mathcal{K}^{\mathcal{T}}$ of $\mathcal{T}$-algebras is also accessible.

Recall that a *monad* over a category $\mathcal{K}$ is a triple $\mathcal{T} = (T, \mu, \eta)$ consisting of a functor $T: \mathcal{K} \to \mathcal{K}$ and natural transformations $\mu: T^2 \to T$ and $\eta: Id \to T$ such that the following diagrams

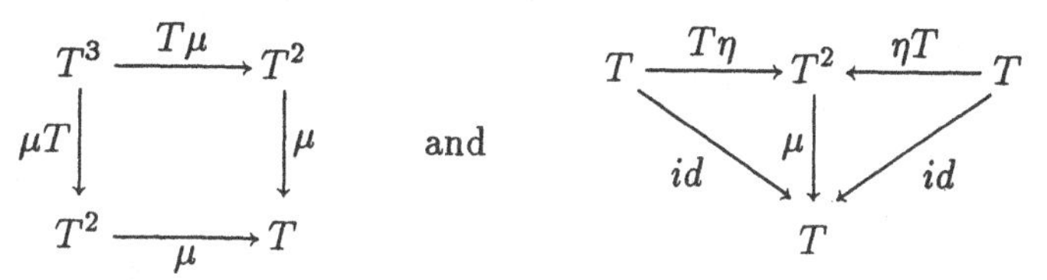

commute. We denote by $\mathcal{K}^{\mathcal{T}}$ the *Eilenberg–Moore category*: the objects are $\mathcal{T}$-*algebras*, i.e., arrows $TK \xrightarrow{k} K$ satisfying (i) $k \cdot \eta_K = id_K$ and (ii) $k \cdot Tk = k \cdot \mu_K$. Morphisms $f: (K, k) \to (L, l)$, called homomorphisms, are $\mathcal{K}$-morphisms $f: K \to L$ such that $f \cdot k = l \cdot Tf$.

Theorem. *The Eilenberg–Moore category of an accessible monad is accessible.*

PROOF. We know that $\mathbf{Alg}\, T$ is an accessible category (Corollary 2.75). Let P: $\mathbf{Alg}\, T \to \mathcal{K}$ be the (accessible) forgetful functor, and let $\varphi: TU \to U$ be the natural transformation with $\varphi_{(K,k)} = k$. The Eilenberg–Moore category $\mathcal{K}^{\mathcal{T}}$ is the full subcategory of $\mathbf{Alg}\, T$ over T-algebras $TK \xrightarrow{k} K$ satisfying the above given conditions (i) and (ii). These conditions can be expressed by equifiers of the following natural transformations:

(i) $id, \varphi \cdot \eta U : U \to U$

and

(ii) $\varphi \cdot T\varphi, \varphi \cdot \mu U : T^2U \to U$.

The theorem thus follows from Remark 2.76. □

Remark. If $\mathcal{T} = (T, \mu, \eta)$ is a λ-accessible monad (i.e., T preserves λ-directed colimits) on a locally λ-presentable category $\mathcal{K}$ then $\mathcal{K}^{\mathcal{T}}$ is locally λ-presentable. In fact, $\mathbf{Alg}\, T$ is locally λ-presentable by Remark 2.75 and $\mathcal{K}^{\mathcal{T}}$ is closed in $\mathbf{Alg}\, T$ under limits and λ-directed colimits. Therefore, we can apply the reflection theorem, 2.48.

Exercises

2.a Accessible Categories

(1) Verify that the category of sets and partial injective maps is finitely accessible.

(2) Verify that the full subcategory of **Lin** consisting of dense linear orderings without endpoints is ω_1-accessible, but not finitely accessible.

(3) Is the full subcategory of **Pos** over all posets with a greatest element accessible?

(4) Let $\mathcal{K}$ be a category whose objects are pairs (A_0, A) of sets $A_0 \subseteq A$ and whose morphisms $f\colon (A_0, A) \to (B_0, B)$ are maps $f\colon A \to B$ with

$$f(a) \in B_0 \iff a \in A_0.$$

Show that $\mathcal{K}$ is finitely accessible but the full subcategory of $\mathcal{K}$ over all objects (A_0, A) with $\operatorname{card} A_0 = \operatorname{card}(A - A_0)$ is not accessible.

2.b Splitting of Idempotents

(1) Prove that a category which has either equalizers or coequalizers has split idempotents.

(2) For a category $\mathcal{K}$, let $\widetilde{\mathcal{K}}$ be the category whose objects are (A, f), where $A \in \mathcal{K}^{\mathrm{obj}}$ and $f\colon A \to A$ is an idempotent in $\mathcal{K}$, and whose morphisms $(A, f) \to (A', f')$ are $\mathcal{K}$-morphisms $h\colon A \to A'$ satisfying $f' \cdot h \cdot f = h$. Show that $E\colon \mathcal{K} \to \widetilde{\mathcal{K}}$ given by

$$E(K) = (K, id_K) \qquad \text{and} \qquad E(h) = h$$

is a universal extension with split idempotents in the sense of Remark 2.5(2).

(3) Prove that for a small category $\mathcal{K}$, the universal extension with split idempotents is equivalent to the closure of $Y(\mathcal{K})$ under split subobjects in $\mathbf{Set}^{\mathcal{K}^{\mathrm{op}}}$ (where $Y\colon \mathcal{K} \to \mathbf{Set}^{\mathcal{K}^{\mathrm{op}}}$ is the Yoneda embedding).

(4) Prove that the dual of a category with split idempotents has split idempotents.

(5) Show that an accessibly embedded subcategory of an accessible category has split idempotents. (Hint: use Observation 2.4.)

2.c Accessibility of the Category of Morphisms

(1) Prove that for each λ-accessible category $\mathcal{K}$ the category $\mathcal{K}^{\mathbf{2}}$ of $\mathcal{K}$-morphisms (see Example 1.55) is λ-accessible.

(2) Prove that an arrow is λ-presentable in $\mathcal{K}^2$ iff both its domain and codomain are λ-presentable in $\mathcal{K}$.

2.d Directed Colimits in Accessible Categories. Prove that λ-accessible categories have the property of Exercise 1.o(1).

2.e Raising the Index of Accessibility

(1) Prove that if $\lambda \leq \mu$, then $\lambda \lhd (2^{<\mu})^+$.

(2) Show that if $\lambda < \mu$ and $\mathrm{cf}\,\mu \leq \lambda$, then $\lambda^+ \lhd \mu^+$ does not hold.

2.f Accessible Functors. Prove that for each accessible category $\mathcal{K}$ and for each functor $H: \mathcal{A} \to \mathcal{B}$ between small categories the induced functor $\mathcal{K}^H: \mathcal{K}^{\mathcal{B}} \to \mathcal{K}^{\mathcal{A}}$ is finitely accessible.

2.g Cone-reflective Subcategories. Prove that if $\mathcal{A}$ is a cone-reflective subcategory of a category $\mathcal{B}$, and $\mathcal{B}$ is a cone-reflective subcategory of a category $\mathcal{C}$, then $\mathcal{A}$ is cone-reflective in $\mathcal{B}$.

2.h Comma-categories. Prove that if λ-accessible functors $F_1: \mathcal{K}_1 \to \mathcal{L}$ and $F_2: \mathcal{K}_2 \to \mathcal{L}$ between locally λ-presentable categories preserve limits, then the comma-category $F_1 \downarrow F_2$ is locally λ-presentable.

2.i Equivalence and Isomorphism

(1) Prove that a category which is equivalent to a full subcategory of an accessible category is also isomorphic to a full subcategory of an accessible category. (Hint: By Remark 2.65, it is sufficient to prove that the category $\mathbf{Gra}_0$ of connected graphs has the property that each category equivalent to a full subcategory of $\mathbf{Gra}_0$ is also isomorphic to a full subcategory of $\mathbf{Gra}_0$. This follows from the fact that every object of $\mathbf{Gra}_0$ is non-empty and thus has a proper class of isomorphic copies in $\mathbf{Gra}_0$.)

(2) Prove that (1) holds if locally presentable categories replace accessible categories.

2.j Inserters

(1) Prove that if $F_1, F_2: \mathcal{K} \to \mathcal{L}$ are λ-accessible functors between locally λ-presentable categories and F_2 preserves limits, then the inserter category $\mathbf{Ins}(F_1, F_2)$ is locally λ-presentable.

(2) Prove that if $F_1, F_2: \mathcal{K} \to \mathcal{L}$ are accessible functors between locally presentable categories and F_1 preserves colimits, then $\mathbf{Ins}(F_1, F_2)$ is locally presentable.

In particular, for any accessible functor $T: \mathcal{K} \to \mathcal{K}$ on a locally presentable category $\mathcal{K}$, the category of *T-coalgebras* $\mathbf{Coalg}\,T = \mathbf{Ins}(Id_{\mathcal{K}}, T)$ is locally presentable.

2.k Fixpoints. For a functor $T\colon \mathcal{K} \to \mathcal{K}$, the category $\mathbf{Fix}\,T$ of fixpoints of T is defined as a full subcategory of $\mathbf{Alg}\,T$ over all T-algebras $TK \xrightarrow{k} K$ such that k is an isomorphism.

Show that $\mathbf{Fix}\,T$ is accessible for any accessible functor T.

2.l Limits. Prove that a lax limit of locally λ-presentable categories and limit-preserving λ-accessible functors is locally λ-presentable.

2.m Inverters. Let $F_1, F_2\colon \mathcal{K} \to \mathcal{L}$ be accessible functors. For each natural transformation $\varphi\colon F_1 \to F_2$ prove that the *inverter* of φ, i.e., the full subcategory $\mathbf{Inv}(\varphi)$ of $\mathcal{K}$ over all objects K such that φ_K is an isomorphism, is accessible.

2.n Pseudolimits. By the *pseudolimit* of a diagram $D\colon \mathcal{D} \to \mathbf{CAT}$ is meant the full subcategory $\mathbf{Plim}\,D$ of $\mathbf{Lax}\,D$ over all objects

$$\Big((K_d)_{d\in\mathcal{D}^{\mathrm{obj}}}, (k_f)_{f\in\mathcal{D}^{\mathrm{mor}}}\Big)$$

such that each k_f, $f \in \mathcal{D}^{\mathrm{mor}}$, is an isomorphism. Prove that a pseudolimit of accessible categories is accessible (i.e., $\mathbf{ACC}$ is closed under pseudolimits in $\mathbf{CAT}$). (Hint: combine Theorem 2.77 and Exercise 2.m.)

2.o Orthogonality Hulls. Let $\mathcal{A}$ be a full subcategory of a locally presentable category $\mathcal{K}$. For each regular cardinal λ, denote by $\mathcal{A}_\lambda$ the closure of $\mathcal{A}$ under λ-directed colimits in $\mathcal{K}$ (i.e., $\mathcal{A}_\lambda$ is the smallest full subcategory of $\mathcal{K}$ containing $\mathcal{A}$ and closed under λ-directed colimits).

Prove that

$$\mathcal{A}_\infty = \bigcap_{\lambda\in\mathrm{Ord}} \mathcal{A}_\lambda$$

is the *orthogonality hull* of $\mathcal{A}$, i.e., the smallest orthogonality class in $\mathcal{K}$ containing $\mathcal{A}$. (Hint: use Exercise 1.j(2) for proving that $\mathcal{A}_\infty$ is contained in the orthogonality hull of $\mathcal{A}$, and Theorem 2.48 (and Exercise 1.j(1)) for the converse.)

2.p Sketches. Prove that $\mathbf{Mod}(\mathscr{S}, \mathcal{K})$ is accessible whenever $\mathcal{K}$ is an accessible category and $\mathscr{S}$ a limit sketch. (Hint: let $\mathscr{S} = (\mathcal{A}, \mathbf{L}, \sigma)$ be a limit sketch and $\mathcal{K}$ be λ-accessible. For any $D \in \mathbf{L}$ denote by $\mathcal{L}_D$ the full subcategory of $\mathcal{K}^{\mathcal{A}}$ over all functors F for which $\lim FD$ exists. Then $\mathcal{L}_D \approx \overline{\mathcal{L}}_D$ in the following pseudo-pullback $\overline{\mathcal{L}}_D$

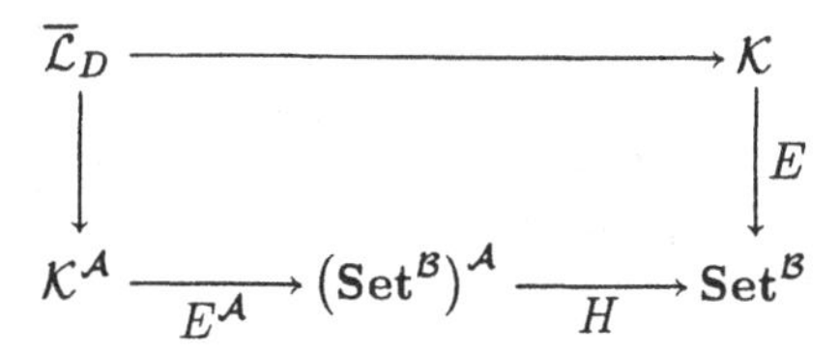

where $\mathcal{B} = (\mathbf{Pres}_\lambda \mathcal{K})^{\mathrm{op}}$, $E\colon \mathcal{K} \to \mathbf{Set}^{\mathcal{B}}$ is the canonical embedding, and H assigns to each $G\colon \mathcal{A} \to \mathbf{Set}^{\mathcal{B}}$ the value $\lim GD$. Use Exercise 2.n and the argument of the proof of Theorem 2.60.

Historical Remarks

The first concept closely related to that of accessible category was used by [Artin, Grothendieck, Verdier 1972]. They also defined λ-accessible functors in much the same way as used here. Related concepts were investigated by [Banaschewski, Herrlich 1976], [Smyth 1978] and [Johnstone, Joyal 1982]. An important step was the introduction of locally multipresentable categories, see Chapter 4, by [Diers 1980a]. Categories axiomatizable in infinitary logic were characterized in terms of accessibility by [Rosický 1981b].

The present definition of an accessible category is due to [Lair 1981], where the fundamental result that accessible categories are precisely those sketchable in the sense of [Ehresmann 1968] was proved. The name which C. Lair, used was "catégories modelables". Accessible categories were rediscovered by M. Makkai and R. Paré, who gave them their present name and who later published a substantial treatise on the theory of accessible categories [Makkai, Paré 1989], whose ten-page introduction is a valuable source of further historical comments. Accessible categories were, independently, introduced in the thesis [Rosický 1983] where the (unpublished) proofs of results announced in [Rosický 1981b] are presented.

Our treatment in sections 2.A–2.C closely follows [Makkai, Paré 1989]. The application of pure subobjects in 2.D is new. The concept stems from model theory (see Chapter 5), and its categorical formulation was presented by [Fakir 1975] in terms similar to our Definition 2.27. The fact that accessible categories have enough pure subobjects (Theorem 2.33) is new, but it could be derived from results on first-order logic such as Proposition 3.2.8 of [Makkai, Paré 1989]. The main result, that closedness under λ-directed colimits and λ-pure subobjects implies accessibility (Corollary 2.36), is new. Related concepts of purity were used by [Ulmer 1975] and [Bird 1984] in an investigation of constructions performed with locally presentable categories. The fact that a full subcategory of a locally λ-presentable category, closed under limits and λ-directed colimits, is reflective (Theorem 2.48) was proved for $\lambda = \aleph_0$ by [Makkai, Pitts 1987] in a manner which cannot be generalized to arbitrary λ; the general theorem was proved in an entirely different manner (using pure subobjects) in [Adámek, Rosický 1989].

A nice example of the influence accessible categories have on the theory of locally presentable categories is the elegant proof that locally presentable categories are co-wellpowered (Theorem 2.49), due to [Makkai, Paré 1989].

The characterization of accessible categories by means of cone-reflectivity (Theorem 2.53 and Corollary 2.54) was presented for categories with products in [Adámek, Rosický 1993], and then generalized to all categories (using the same technique) by [Hu, Makkai 1994].

The results of section 2.G were proved by the "Prague School" in the 1960's, and our treatment closely follows that of [Pultr, Trnková 1980].

The material of section 2.H is from [Makkai, Paré 1989].

Chapter 3

Algebraic Categories

The topic of the present chapter is varieties and quasivarieties, i.e., equational and implicational classes of algebras (with a set of many-sorted operations). Whereas locally presentable categories are characterized as categories of models of limit sketches, varieties can be characterized as categories of models of product sketches. These are limit sketches with discrete diagrams; in fact, to sketch a given variety of algebras, the Lawvere–Linton algebraic theory of that variety, considered as a product-sketch, can be used (Theorems 3.16 and 3.30). Moreover, varieties are precisely the accessibly monadic categories over many-sorted sets (Theorem 3.31). Quasivarieties can be abstractly characterized as precisely the locally presentable categories with a dense set of regular projectives; and varieties are then characterized as precisely the quasivarieties with effective equivalence relations (Theorem 3.33).

The name "presentable" stems from algebra: an algebra is a λ-presentable object of a finitary variety iff it can be presented by less than λ generators and less than λ equations in the usual algebraic sense [Theorem 3.12].

The chapter is concluded by a characterization of locally presentable categories which, although known as folklore, has never been published before: locally presentable categories are precisely the essentially algebraic categories, i.e., varieties of partial algebras in which the domain of definition of each partial operation is described by equations involving total operations only (Theorem 3.36).

Since all the results in the present chapter are quite analogous in the finitary case and in the general case, we present all the details for the (notationally simpler) finitary algebras, and then mention the general results more briefly.

3.A Finitary Varieties

Algebras and Free Algebras

We will show that varieties of finitary, many-sorted algebras are locally finitely presentable categories. More precisely: finitary varieties are precisely the categories sketchable by *FP sketches*, i.e., sketches in which the choice of limits is that of finite products.

Let S be a set of sorts. By an *S-sorted signature of finitary algebras* is understood a set Σ (of *operation symbols*) together with an arity function assigning to each operation symbol σ a finite word $(s_1, s_2, \ldots, s_n)$ in the alphabet S (determining the sorts of variables of σ) and an element $s \in S$ (determining the sort of the result of σ). Notation:

$$\sigma \colon s_1 \times s_2 \times \cdots \times s_n \to s.$$

In the case $n = 0$ we just write $\sigma \colon \to s$. An *algebra* A of the signature Σ is an S-sorted set $|A| = (A_s)_{s \in S}$ together with an operation

$$\sigma_A \colon A_{s_1} \times A_{s_2} \times \cdots \times A_{s_n} \to A_s$$

for each $\sigma \in \Sigma$ of arity $s_1 \times s_2 \times \cdots \times s_n \to s$. In particular, nullary operations $\sigma \colon \to s$ correspond to constants $\sigma_A \in A_s$, and unary operations $\sigma \colon s_1 \to s_2$ to functions $\sigma_A \colon A_{s_1} \to A_{s_2}$.

A *homomorphism* from an algebra A to an algebra B (of the same signature Σ) is an S-sorted function $f \colon |A| \to |B|$ preserving the operations in the usual sense: given $\sigma \colon s_1 \times \cdots \times s_n \to s$ in Σ, then

$$f_s(\sigma_A(x_1, \ldots, x_n)) = \sigma_B(f_{s_1}(x_1), \ldots, f_{s_n}(x_n))$$

for all $x_i \in A_{s_i}$, $i = 1, \ldots, n$. This yields the category $\mathbf{Alg}\,\Sigma$ of Σ-algebras and homomorphisms.

The set of *terms* over an (S-sorted) set X of variables is defined in a standard manner, except that for each term τ we must define a (value) *sort* of τ. Thus, the set

$$T_\Sigma(X)$$

of all terms over an S-sorted set X (of variables) is the smallest S-sorted set such that

(a) each variable of sort s is a term of sort s,

and

(b) for each operation $\sigma \colon s_1 \times s_2 \times \cdots \times s_n \to s$ in Σ and each n-tuple of terms τ_i of sort s_i $(i = 1, \ldots, n)$, we conclude that $\sigma(\tau_1, \ldots, \tau_n)$ is a term of sort s.

For reasons of clarity, we assume that the set Σ is disjoint with the set X_s of variables of sort s (for each $s \in S$). Observe that terms $\sigma(\tau_1, \ldots, \tau_n)$ are just formal expressions, thus, two terms $\sigma(\tau_1, \ldots, \tau_n)$ and $\sigma'(\tau'_1, \ldots, \tau'_m)$ are equal iff $\sigma = \sigma'$, $n = m$, and $\tau_1 = \tau'_1$, ..., $\tau_n = \tau'_n$.

We can consider $T_\Sigma(X)$ as an algebra: its underlying set is the S-sorted set of all terms, and its operations are given by (b) above.

3.1 Examples

(1) Let Σ be the one-sorted signature of one binary symbol σ. Terms over $X = \{x, y, z\}$ are:

$$x, \quad \sigma(x,y), \quad \sigma\big(\sigma(x,y),z\big), \quad \sigma\big(x,\sigma(y,z)\big), \quad \text{etc.}$$

(2) Let S = {ring, module} and let Σ be the signature of the following eight operations:

$$\begin{aligned} +&: \text{ring} \times \text{ring} \to \text{ring} & +_m&: \text{module} \times \text{module} \to \text{module} \\ 0&: \to \text{ring} & 0_m&: \to \text{module} \\ -&: \text{ring} \to \text{ring} & -_m&: \text{module} \to \text{module} \\ \times&: \text{ring} \times \text{ring} \to \text{ring} & \cdot&: \text{ring} \times \text{module} \to \text{module} \end{aligned}$$

Let $X = (X_{\text{ring}}, X_{\text{module}})$ be a 2-sorted set of variables, and suppose that we denote the variables in X_{ring} by r, r_1, etc., and those in X_{module} by x, x_1, etc. The following expressions are examples of terms:

$$r, \quad r_1 + r_2, \quad -(0 \times r), \quad x, \quad x_1 + x_2, \quad r.x, \quad -_m(0.0_m), \ldots.$$

(3) Sequential deterministic automata are algebras of three sorts:

$$S = \{\text{state, input, output}\}$$

and of three operations:

$$\begin{aligned} i&: \ \to \ \text{state} && \text{(initial state)} \\ n&: \ \text{state} \times \text{input} \ \to \ \text{state} && \text{(next state)} \\ o&: \ \text{state} \ \to \ \text{output} && \text{(output)} \end{aligned}$$

Denote by $\sigma_1, \sigma_2, \ldots$ variables of sort input. Here are some terms representing the reaction to words (of input symbols) inside the automaton:

$$i, \quad n(i,\sigma_1), \quad n\big(n(i,\sigma_1),\sigma_2\big), \quad n\big(n\big(n(i,\sigma_1),\sigma_2\big),\sigma_3\big), \ldots$$

and outside:

$$o(i), \quad o\big(n(i,\sigma_1)\big), \quad o\big(n\big(n(i,\sigma_1),\sigma_2\big)\big), \ldots.$$

3.2 Proposition. *The term-algebra $T_\Sigma(X)$ is a free Σ-algebra generated by X. That is, for each Σ-algebra A and each S-sorted function $f\colon X \to |A|$ there exists a unique extension of f to a homomorphism $f^\# : T_\Sigma(X) \to A$.*

PROOF. For each variable x of sort s we put $f^\#_s(x) = f_s(x)$. For each term $\sigma(\tau_1, \ldots, \tau_n)$ of sort s we put

$$f^\#_s(\sigma(\tau_1, \ldots, \tau_n)) = \sigma_A(f^\#_{s_1}(\tau_1), \ldots, f^\#_{s_n}(\tau_n)).$$

The unique $f^\#$ defined by this rule is a homomorphism. □

3.3 Corollary. *The natural forgetful functor from* $\mathbf{Alg}\,\Sigma$ *to* $\mathbf{Set}^S$ *is a right adjoint.*

3.4 Remark: Properties of categories of algebras. The categories of finitary many-sorted algebras have the following properties:

(1) $\mathbf{Alg}\,\Sigma$ is complete, and the natural forgetful functor $U\colon \mathbf{Alg}\,\Sigma \to \mathbf{Set}^S$ creates limits. This means that for each diagram $D\colon \mathcal{D} \to \mathbf{Alg}\,\Sigma$, given a limit $(L \xrightarrow{l_d} UDd)_{d\in\mathcal{D}^{\mathrm{obj}}}$ of $U \cdot D$ in $\mathbf{Set}^S$, there exists a unique algebra A with $|A| = L$ and such that $l_d\colon A \to Dd$ is a homomorphism for $d \in \mathcal{D}^{\mathrm{obj}}$; moreover, the latter cone is a limit of D. (In fact, the operations of A are defined coordinate-wise, i.e., by $(l_d)_s[\sigma_A(x_1, \ldots, x_n)] = \sigma_{Dd}((l_d)_{s_1}(x_1), \ldots, (l_d)_{s_n}(x_n))$ for each d and each $\sigma\colon s_1 \times \cdots \times s_n \to s$.)

(2) $\mathbf{Alg}\,\Sigma$ has (regular epi, mono)-factorizations of morphisms:

(i) Monomorphisms are precisely the homomorphisms which are injective in each sort. (In fact, U is a faithful right adjoint, thus, it preserves and reflects monomorphisms.)

(ii) Every homomorphism bijective in each sort is an isomorphism. (In fact, since f and $f^{-1} \cdot f = id$ preserve the operations, so does f^{-1}.)

(iii) Given an algebra A, by a *congruence* $\sim$ is meant a collection $(\sim_s)_{s\in S}$ of equivalence relations $\sim_s$ on A_s such that each Σ-operation $\sigma\colon s_1 \times \cdots \times s_n \to s$ fulfils

$$\sigma_A(x_1, \ldots, x_n) \sim_s \sigma_A(y_1, \ldots, y_n) \qquad \text{whenever} \quad x_i \sim_{s_i} y_i \text{ for } i = 1, \ldots, n.$$

For each homomorphism $f\colon A \to B$ the equivalences defined by $x \sim_s y$ iff $f_s(x) = f_s(y)$ form a congruence on A, called the *kernel congruence* on A.

(iv) Let $\sim$ be a congruence on an algebra A. The *quotient algebra* $A/\!\sim$ is the algebra whose s-sort underlying set is $(A/\!\sim)_s = A_s/\!\sim_s$, i.e., the set of all equivalence classes $[x]$ of elements $x \in A_s$ (under the equivalence $\sim_s$) and whose operations are given by

$$\sigma_{A/\sim}([x_1], \ldots, [x_n]) = [\sigma_A(x_1, \ldots, x_n)].$$

The *canonical homomorphism* $c\colon A \to A/\!\sim$ assigning to each element its congruence class is surjective in each sort.

(v) For each homomorphism $f\colon A \to B$ we have a factorization $f = i \cdot c$ where $c\colon A \to A/\!\sim$ is the canonical homomorphism of the kernel equivalence of f, and $i\colon A/\!\sim \to B$ is the monomorphism defined by $i_s([x]) = f_s(x)$.

(vi) Regular epimorphisms in **Alg** Σ are precisely the homomorphisms which are surjective in each sort, and each epimorphism represents the same quotient as the corresponding canonical homomorphism. In fact, let $f\colon A \to B$ be a homomorphism surjective in each sort. The kernel congruence $\sim$ of f represents a subalgebra C of the product $A \times A$. The two restricted projections $\pi_1, \pi_2\colon C \to A$ have a coequalizer $c\colon A \to A/\!\sim$ and, since f is surjective, its canonical factorization $f = i \cdot c$ has $i\colon A/\!\sim \to B$ bijective in each sort. Now use (ii).

(vii) Every epimorphism is regular in **Alg** Σ. See Exercise 3.b.

(3) **Alg** Σ is wellpowered [see (2.i)] and co-wellpowered [see (2.vii)].

(4) **Alg** Σ is cocomplete, and the forgetful functor creates directed colimits. In fact:

(i) Coproducts are "free products". Given Σ-algebras A_t $(t \in T)$, the *coproduct* is the quotient algebra of the free algebra $T_\Sigma(\coprod_{t\in T} |A_t|)$ generated by the coproduct in $\mathbf{Set}^S$ under the smallest congruence which merges the term $\sigma(x_1, \ldots, x_n)$ with the variable

$$x = \sigma_{A_t}(x_1, \ldots, x_n)$$

for each $\sigma\colon s_1 \times \cdots \times s_n \to s$ in Σ, each $t \in T$, and each n-tuple $x_i \in (A_t)_{s_i}$, $i = 1, \ldots, n$.

(ii) A coequalizer of two homomorphisms $f, g\colon A \to B$ is given by the canonical homomorphism $c\colon B \to B/\!\sim$, where $\sim$ is the smallest congruence on B with $f_s(a) \sim_s g_s(a)$ for every element a of A_s.

(iii) Let $D\colon (I, \leq) \to \mathbf{Alg}\,\Sigma$ be a directed diagram, and let $(UD_i \xrightarrow{c_i} C)_{i \in I}$ be a colimit of $U \cdot D$ in $\mathbf{Set}^S$. The unique algebra A with $|A| = C$ such that the $c_i\colon D_i \to A$ are homomorphisms is defined as follows. Given $\sigma\colon s_1 \times \cdots \times s_n \to s$ in Σ and given elements $x_t \in C_{s_t}$ $(t = 1, \ldots, n)$, there exists $i \in I$ with x_t lying in the image of $(c_i)_{s_t}$ for $t = 1, \ldots, n$, say $x_t = (c_i)_{s_t}(y_t)$. Then in order that c_i be a homomorphism, we must define

$$\sigma_A(x_1, \ldots, x_n) = (c_i)_s\big(\sigma_{D_i}(y_1, \ldots, y_n)\big).$$

The above equation describes $\sigma_A\colon C_{s_1} \times \cdots \times C_{s_n} \to C_s$ such that the resulting algebra A and the homomorphisms $c_i\colon D_i \to A$ form a colimit of D in $\mathbf{Alg}\,\Sigma$.

(5) The free algebras $T_\Sigma(X)$, where X is a finite set of variables, are finitely presentable in $\mathbf{Alg}\,\Sigma$, and they form a dense (essentially small) collection. The former follows from the creation of directed colimits by U, and the latter from the fact that for each Σ-algebra A, in order to verify that a map $f\colon |A| \to |B|$ is a homomorphism in $\mathbf{Alg}\,\Sigma$, i.e., that $f_s\big(\sigma_A(a_1, \ldots, a_n)\big) = \sigma_B\big(f_{s_1}(a_1), \ldots, f_{s_n}(a_n)\big)$ (for $\sigma\colon s_1 \times \cdots \times s_n \to s$ in Σ and for $a_i \in A_{s_i}$), it is sufficient to show that $f \cdot h\colon T_\Sigma\big(\{x_1^{s_1}, \ldots, x_n^{s_n}\}\big) \to B$ is a homomorphism, where $h\colon T_\Sigma\big(\{x_1^{s_1}, \ldots, x_n^{s_n}\}\big) \to A$ is the unique homomorphism with $h_{s_1}(x_1^{s_1}) = a_1, \ldots, h_{s_n}(x_n^{s_n}) = a_n$.

Moreover, each of the free algebras is a *regular projective*, i.e., an object K such that for every regular epimorphism $e\colon A' \to A$ and each morphism $f\colon K \to A$ there exists a morphism $f'\colon K \to A'$ with $f = e \cdot f'$. In fact, for every homomorphism $e\colon A' \to A$ surjective in each sort and each homomorphism $f\colon T_\Sigma(X) \to A$ we can clearly choose a many-sorted function $\overline{e}\colon |A| \to |A'|$ with $e \cdot \overline{e} = id_{|A|}$, and we can extend the function $g = \overline{e} \cdot f \cdot \eta_X\colon X \to |A'|$ to a homomorphism $g^\#\colon T_\Sigma(X) \to A'$. The equality $e \cdot g^\# = f$ then follows from the fact that

$$\big(e \cdot g^\#\big) \cdot \eta_X = e \cdot \overline{e} \cdot f \cdot \eta_X = f \cdot \eta_X.$$

(6) $\mathbf{Alg}\,\Sigma$ is a locally finitely presentable category. This follows from (4) and (5), see Theorem 1.20.

(7) The forgetful functor $U\colon \mathbf{Alg}\,\Sigma \to \mathbf{Set}^S$ *creates absolute coequalizers.* That is, given homomorphisms $f, g\colon A \to B$ in $\mathbf{Alg}\,\Sigma$ and given a coequalizer

$$UA \underset{Ug}{\overset{Uf}{\rightrightarrows}} UB \xrightarrow{c} X$$

in $\mathbf{Set}^S$ which is absolute (i.e., preserved by every functor $F\colon \mathbf{Set}^S \to \mathcal{L}$), then there exists a unique Σ-algebra C such that $UC = X$ and $c\colon B \to C$ is a homomorphism; moreover, c is a coequalizer of f and g in $\mathbf{Alg}\,\Sigma$.

In fact, for each operation $\sigma\colon s_1 \times \cdots \times s_n \to s$ in Σ we have

$$\begin{aligned} c_s \cdot \sigma_B \cdot (f_{s_1} \times \cdots \times f_{s_n}) &= c_s \cdot f_s \cdot \sigma_A \\ &= c_s \cdot g_s \cdot \sigma_A \\ &= c_s \cdot \sigma_B \cdot (g_{s_1} \times \cdots \times g_{s_n}). \end{aligned}$$

The functor $\mathbf{Set}^S \to \mathbf{Set}$ given by $(X_s)_{s \in S} \mapsto X_{s_1} \times \cdots \times X_{s_n}$ preserves the coequalizer of f and g (since it is absolute); thus, there exists a unique map

$$\sigma_C\colon C_{s_1} \times \cdots \times C_{s_n} \to C$$

with

$$c_s \cdot \sigma_B = \sigma_C \cdot (c_{s_1} \times \cdots \times c_{s_n}).$$

This equips the S-sorted set X with operations σ_C ($\sigma \in \Sigma$) such that $c\colon B \to C$ is a homomorphism. It is easy to verify that c is a coequalizer of f and g in $\mathbf{Alg}\,\Sigma$ (see Exercise 3.b(2)).

(8) $\mathbf{Alg}\,\Sigma$ has *effective equivalence relations*, i.e., every equivalence relation is a kernel pair of some morphism.

Recall that a *kernel pair* of a morphism $f\colon K \to K'$ in a category is a pair $e_1, e_2\colon E \to K$ such that the square

$$\begin{array}{ccc} E & \xrightarrow{e_1} & K \\ {\scriptstyle e_2}\downarrow & & \downarrow{\scriptstyle f} \\ K & \xrightarrow[f]{} & K' \end{array}$$

is a pullback.

Recall that a *relation* on an object K is a subobject of $K \times K$ (usually represented by a pair $e_1, e_2\colon E \to K$ of morphisms such that the morphism $(e_1, e_2)\colon E \to K \times K$ is a monomorphism). We call $e_1, e_2\colon E \to K$ an *equivalence relation* provided that it is

(i) *reflexive*, i.e., the diagonal of $K \times K$ is contained in the subobject represented by (e_1, e_2),

(ii) *symmetric*, i.e., the monomorphisms (e_1, e_2) and (e_2, e_1) represent the same subobject, and

(iii) *transitive*, i.e., when we form the pullback of e_2 and e_1:

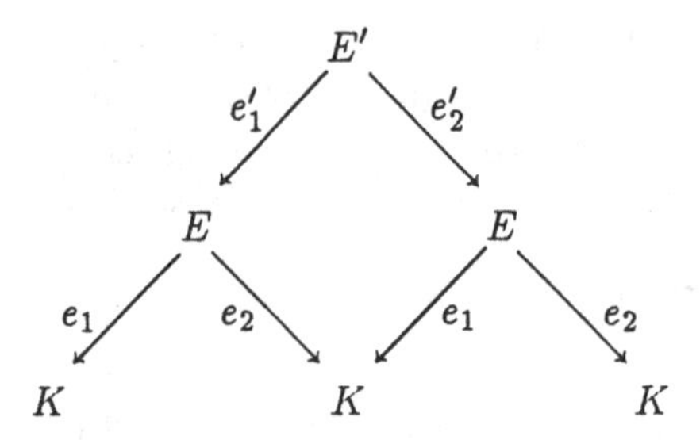

then the subobject represented by $(e_1 \cdot e_1', e_2 \cdot e_2')$ is contained in that represented by (e_1, e_2).

To verify that $\mathbf{Alg}\,\Sigma$ has effective equivalence relations, it is sufficient to show that every equivalence relation in $\mathbf{Alg}\,\Sigma$ is a kernel pair of its coequalizer. This follows from the fact that $U\colon\ \mathbf{Alg}\,\Sigma \to \mathbf{Set}^S$ creates coequalizers of equivalence relations (because they are absolute, see Exercise 3.h), and that $\mathbf{Set}^S$ has effective equivalence relations.

Equational Presentation

We now assume that a "standard" many-sorted set V of variables has been chosen such that for each sort s the set V_s is denumerable.

An *equation* is a pair (τ_1, τ_2) of terms in $T_\Sigma(V)$ of the same sort; notation:

$$\tau_1 = \tau_2.$$

A Σ-algebra A *satisfies* this equation if for each interpretation of the standard variables, i.e., each S-sorted function $f\colon V \to |A|$, we have $f_s^\#(\tau_1) = f_s^\#(\tau_2)$. An *equational class* or *variety* of Σ-algebras is a class $\mathcal{C}$ for which there exists a set

$$E \subseteq T_\Sigma(V) \times T_\Sigma(V)$$

of equations such that a Σ-algebra lies in $\mathcal{C}$ iff it satisfies all equations in E. The pair (Σ, E) is called an *equational presentation* of the class $\mathcal{C}$, and $\mathcal{C}$-algebras are also called (Σ, E)-algebras. Each equational class is considered as a full subcategory $\mathbf{Alg}(\Sigma, E)$ of $\mathbf{Alg}\,\Sigma$.

3.5 Examples

(1) Abelian groups form an equational class of one-sorted algebras of signature $\{+, -, 0\}$ with arities $s \times s \to s$, $s \to s$, and $\to s$, respectively,

presented by the following equations:

$$(x+y)+z = x+(y+z)$$
$$x+y = y+x$$
$$x+0 = x$$
$$x+(-x) = 0.$$

(2) Modules over arbitrary rings form an equational class of two-sorted algebras of the signature of Example 3.1(2) presented by the following set of equations

$$E = E_r \cup E_m \cup E_0$$

where E_r are the four equations of (1), E_m are the corresponding four equations for $+_m$, 0_m, and $-_m$, and E_0 are the following equations

$$r \times (r' \times r'') = (r \times r') \times r''$$
$$r \times (r' + r'') = (r \times r') + (r \times r'')$$
$$(r' + r'') \times r = (r' \times r) + (r'' \times r)$$
$$r \times (x +_m x') = (r \cdot x) +_m (r \cdot x')$$
$$(r + r') \cdot x = (r \cdot x) +_m (r' \cdot x)$$
$$(r \times r') \cdot x = r \cdot (r' \cdot x).$$

(3) For each small category $\mathcal{A}$, the functor-category $\mathbf{Set}^{\mathcal{A}}$ is isomorphic to an equational class of unary algebras (besides being a small-orthogonality class of binary relations, see Example 1.41). The objects of $\mathcal{A}$ play the role of sorts and the morphisms the role of operation symbols: Let

$$S = \mathcal{A}^{\text{obj}} \qquad \text{and} \qquad \Sigma = \mathcal{A}^{\text{mor}};$$

each morphism

$$f: s \to s'$$

has arity, as indicated, $s \to s'$. Let

$$E = E_{\text{com}} \cup E_{id}$$

be the following set of equations: E_{com} consists of all the equations

$$h(x) = f(g(x))$$

where $h = f \cdot g: s \to s'$ in $\mathcal{A}^{\text{mor}}$ and x is a variable of sort s, and E_{id} consists of all the equations

$$id_s(x) = x$$

for $s \in \mathcal{A}^{\text{obj}}$, where x is a variable of sort s. Then each (Σ, E)-algebra A defines a functor $\widehat{A}\colon \mathcal{A} \to \mathbf{Set}$ by $\widehat{A}(s) = A_s$ $(s \in \mathcal{A}^{\text{obj}})$ and $\widehat{A}(f) = f_A$ (for f in $\Sigma = \mathcal{A}^{\text{mor}}$). In this way we obtain an isomorphism

$$(\hat{-})\colon \mathbf{Alg}(\Sigma, E) \to \mathbf{Set}^{\mathcal{A}}$$

defined on morphisms by $\hat{r} = (r_s)_{s \in \mathcal{A}^{\text{obj}}}$.

3.6 Remark: Properties of Varieties. Every variety $\mathbf{Alg}(\Sigma, E)$ of Σ-algebras has the following properties.

(1) $\mathbf{Alg}(\Sigma, E)$ is closed under subalgebras in $\mathbf{Alg}\,\Sigma$: if an algebra A satisfies all E-equations, then so do all subalgebras $B \subseteq A$ because terms are computed in B as in A.

(2) $\mathbf{Alg}(\Sigma, E)$ is closed under products in $\mathbf{Alg}\,\Sigma$ (because terms are computed in a product coordinate-wise).

(3) $\mathbf{Alg}(\Sigma, E)$ is a (regular epi)-reflective subcategory of $\mathbf{Alg}\,\Sigma$. This follows from (1) and (2) since $\mathbf{Alg}\,\Sigma$ is a complete, co-wellpowered category with (regular epi, mono)-factorizations (see Remark 3.4).

(4) For each many-sorted set X of variables there exists the smallest congruence $\sim$ on the term-algebra $T_\Sigma(X)$ for which the quotient algebra

$$T_{\Sigma,E}(X) = T_\Sigma(X)/\sim$$

lies in $\mathbf{Alg}(\Sigma, E)$. Moreover, $T_{\Sigma,E}(X)$ is a free (Σ, E)-algebra with the universal map $\eta\colon X \to T_{\Sigma,E}(X)$ assigning to each variable x its congruence class $[x]$. (That is, for each (Σ, E)-algebra A and each many-sorted function $f\colon X \to |A|$ there exists a unique Σ-homomorphism $f^\# : T_{\Sigma,E}(X) \to A$ with $f = f^\# \cdot \eta$.) All this follows from (3): consider the reflection of $T_\Sigma(X)$ in $\mathbf{Alg}(\Sigma, E)$.

(5) $\mathbf{Alg}(\Sigma, E)$ is closed under quotients (that is, under homomorphic images) in $\mathbf{Alg}\,\Sigma$. In fact, given (Σ, E)-algebra A and a Σ-homomorphism $h\colon A \to B$ which is surjective in each sort, then there is a many-sorted function $k\colon |B| \to |A|$ (not a homomorphism, in general) with $h \cdot k = id_B$. It follows that each equation $\tau = \tau'$ satisfied by A is also satisfied by B: given a function $f\colon V \to |B|$, we have

$$h \cdot (k \cdot f)^\# = (h \cdot k \cdot f)^\# = f^\# : T_\Sigma(V) \to B;$$

thus, if $(k \cdot f)^\#(\tau) = (k \cdot f)^\#(\tau')$, then $f^\#(\tau) = f^\#(\tau')$.

(6) $\mathbf{Alg}(\Sigma, E)$ is closed under directed colimits in $\mathbf{Alg}\,\Sigma$. In fact, let $(A_i \xrightarrow{h_i} A)_{i \in I}$ be a directed colimit of algebras A_i satisfying an equation $\tau = \tau'$ in the variables $x_1, \ldots, x_n$. For each many-sorted function $f\colon \{x_1, \ldots, x_n\} \to |A|$ there exists a factorization $f = h_i \cdot g$ for some $i \in I$, and then $f^\# = h_i \cdot g^\#$. Thus, $g_s^\#(\tau) = g_s^\#(\tau')$ implies $f_s^\#(\tau) = f_s^\#(\tau')$.

(7) $\mathbf{Alg}(\Sigma, E)$ has effective equivalence relations. This follows from Remark 3.4(8): from the closedness under (finite) products and subobjects it follows that $\mathbf{Alg}(\Sigma, E)$ is closed under equivalence relations in $\mathbf{Alg}\,\Sigma$, and from the closedness under quotient algebras it follows that $\mathbf{Alg}(\Sigma, E)$ is closed under coequalizers of equivalence relations in $\mathbf{Alg}\,\Sigma$ (which are created by the forgetful functor).

(8) $\mathbf{Alg}(\Sigma, E)$ has a dense subcategory consisting of finitely presentable regular projectives. The argument is analogous to that of Remark 3.4(5), it is just necessary to substitute $T_\Sigma(X)$ by $T_{\Sigma,E}(X)$.

3.7 Corollary. *Every variety of finitary algebras is a locally finitely presentable category.*

In fact, this follows from Theorem 1.11, since every variety is cocomplete (see Remarks 3.4(4) and 3.6(3)) and has a dense set of finitely presentable objects (by Remark 3.6(8)). □

3.8 Example of a locally finitely presentable category which is not equivalent to a many-sorted variety: the category **Pos** of posets does not have effective equivalence relations. In fact, let $K = (X, \leq)$ be any non-discrete poset, and let E be the set $X \times X$ with discrete order. Then the projections $e_1, e_2\colon E \to K$ represent an equivalence which is not a kernel pair of any morphism. The same argument holds for the category **Gra**.

3.9 The Birkhoff Variety Theorem. *A class of Σ-algebras is a variety iff it is closed under products, subalgebras, and homomorphic images (= quotients) in $\mathbf{Alg}\,\Sigma$.*

PROOF. The necessity has been verified in Remark 3.6.

Let $\mathcal{A}$ be a class of Σ-algebras closed under products, subalgebras, and homomorphic images. Then $\mathcal{A}$ is a full, epireflective subcategory of $\mathbf{Alg}\,\Sigma$ (see 0.8 and Remark 3.4(6)). Let us prove that $\mathcal{A}$ is closed under directed colimits in $\mathbf{Alg}\,\Sigma$. Given a directed diagram $D\colon (I, \leq) \to \mathbf{Alg}\,\Sigma$ with $D_i \in \mathcal{A}$ for each i, we form a colimit $(D_i \xrightarrow{c_i} C)_{i \in I}$ in $\mathbf{Alg}\,\Sigma$. By the description of directed colimits (see Remark 3.4(4) and Exercise 1.a), C is the algebra whose elements of sort s are equivalence classes $[x]$ of elements x of sort s in D_i, $i \in I$, where x is equivalent to $x' \in \left(D_{i'}\right)_s$ iff $(d_{i,j})_s(x) = (d_{i',j})_s(x)$ for some upper bound j of i and i'. It is sufficient to find a subalgebra B of the algebra $\prod_{i \in I} D_i$ such that C is a quotient of B. Let B be the subalgebra of all tuples $(x_i)_{i \in I}$ in $\left|\prod D_i\right|_s$ for which there exists $i_0 \in I$ with $x_i = \left(d_{i_0,i}\right)_s(x_{i_0})$ for all $i \geq i_0$. (Since the $d_{i,j}$ are homomorphisms, such tuples form a subalgebra of $\prod_{i \in I} D_i$.) Then the following is an epimorphism in $\mathbf{Alg}\,\Sigma$:

$$f : B \to C, \quad f_s(x_i)_{i \in I} = [x_{i_0}] \quad \text{whenever } x_i = \left(d_{i_0,i}\right)_s(x_{i_0}) \text{ for all } i \geq i_0.$$

Let us consider all finite subsets Z of the standard set V of variables (i.e., many-sorted subsets of a finite power), and let $c_Z : T_\Sigma(Z) \to T_\Sigma(Z)/{\sim}$ be the canonical map forming a reflection of the free Σ-algebra in $\mathcal{A}$. We claim that $\mathcal{A}$ is presented by the set E of all equations $\tau = \tau'$ where $Z \subseteq V$ is finite and (τ, τ') lies in the kernel of c_Z. Observe that a Σ-algebra satisfies all equations in E iff it is orthogonal to each c_Z. Thus, we need to prove that $\mathcal{A} = \{c_Z \mid Z \subseteq V, Z \text{ finite}\}^{\perp}$. By Theorem 1.38, $\mathcal{A}$ is the orthogonality class of all reflection maps of finitely presentable Σ-algebras. Thus, it is sufficient to prove that orthogonality to each c_Z implies orthogonality to the reflection map of any finitely presentable algebra A. Let $Z \subseteq |A|$ be a finite set of generators, and let $k : T_\Sigma(Z) \to A$ be the corresponding epimorphism. Let us form a pushout of k and c_Z in $\mathbf{Alg}\,\Sigma$:

$$\begin{array}{ccc} T_\Sigma(Z) & \xrightarrow{c_Z} & T_\Sigma(Z)/{\sim} \\ {\scriptstyle k}\big\downarrow & & \big\downarrow{\scriptstyle k'} \\ A & \xrightarrow[r]{} & A' \end{array}$$

Since k' is an epimorphism, and $\mathcal{A}$ is closed under quotients, we have A' in $\mathcal{A}$—thus, r is a reflection of A in $\mathcal{A}$. It is obvious that a Σ-algebra orthogonal to c_Z is orthogonal to r. □

3.10 Presentation and Generation of Algebras. We are now going to prove that an algebra in $\mathbf{Alg}\,\Sigma$ is

(1) λ-generated (see Definition 1.67) iff it has less than λ generators in the usual algebraic sense,

and

(2) λ-presentable iff it can be presented by less than λ generators and less than λ equations in the usual algebraic sense.

(These results explain the terminology used throughout our book.)

Recall that for each algebra A in Σ-**Alg**, a (many-sorted) subset X of $|A|$ is said to *generate* the algebra A provided that every subalgebra of A containing X is equal to A. This is, obviously, equivalent to A's being isomorphic to a quotient algebra of the free algebra $T_\Sigma(X)$ w.r.t. the extension of the inclusion $X \hookrightarrow |A|$ to a homomorphism $T_\Sigma(X) \to A$. We say that A has *less than λ generators* provided that it can be generated by a subset X of power smaller than λ.

By a *presentation* of a Σ-algebra A is meant a set X of (many-sorted) variables, not necessarily standard, and a set $\tau_i = \tau_i'$ $(i \in I)$ of equations in $T_\Sigma(X)$ such that A is isomorphic to the quotient of $T_\Sigma(X)$ modulo the smallest congruence $\sim$ with $\tau_i \sim \tau_i'$ for each $i \in I$. We say that A is *presentable by less than λ generators and less than λ equations* provided that we can choose a presentation with X and I both of power smaller than λ.

More in general, for each variety $\mathcal{V}$ of Σ-algebras a *presentation of an algebra A in the variety $\mathcal{V}$* is a set X of variables and a set of equations $\tau_i = \tau_i'$ in $T_\Sigma(X)$ such that A is isomorphic to the algebra $T_\Sigma(X)/\!\sim$ for the smallest congruence $\sim$ such that $T_\Sigma(X)/\!\sim$ lies in $\mathcal{V}$ and $\tau_i \sim \tau_i'$ for each i.

3.11 Proposition. *For each regular cardinal λ, an algebra is a λ-generated object in* $\mathbf{Alg}\,\Sigma$ *iff it has less than λ generators.*

PROOF. I. If A is λ-generated, consider the diagram D whose objects are all subalgebras of A on less than λ generators, and whose morphisms are the inclusion morphisms $B \hookrightarrow B'$ for $B \subseteq B'$ in D. Then A is a canonical colimit of D, and $\hom(A, -)$ preserves that colimit. Thus, id_A factorizes through $B \hookrightarrow A$ for some algebra B in D; it follows that $B = A$.

II. If A has less than λ generators, it is obvious that $\hom(A, -)$ preserves λ-directed unions (= colimits) of subalgebras. □

3.12 Theorem. *For each regular cardinal λ, an algebra is a λ-presentable object of* $\mathbf{Alg}\,\Sigma$ *iff it is presentable by less than λ generators and less than λ equations.*

PROOF. I. Let A be an algebra presentable in $\mathbf{Alg}\,\Sigma$ by less than λ generators and less than λ equations. We can assume that $A = T_\Sigma(X)/\!\sim$ where $\#X < \lambda$ and $\sim$ is the smallest congruence containing a set E of equations with $\operatorname{card} E < \lambda$. Denote by $k_0 \colon X \to A$ the domain restriction of the canonical homomorphism $k \colon T_\Sigma(X) \to A$. Let D be a λ-directed diagram in $\mathbf{Alg}\,\Sigma$ with a colimit $(D_i \xrightarrow{c_i} C)_{i \in I}$. Given a homomorphism $f \colon A \to C$, there exists $i_0 \in I$ with $f(|A|) \subseteq c_{i_0}(|D_{i_0}|)$: this follows from the fact that, since I is directed, $C_0 = \bigcup_{i \in I} c_i(|D_i|)$ is a subalgebra of C, hence C_0 is equal to C. From the λ-directedness of I we get i_0 with $f(k(X)) \subseteq c_{i_0}(|D_{i_0}|)$. Thus, there is a mapping $h\colon X \to |D_{i_0}|$ such that $c_{i_0} \cdot h = f \cdot k_0$, i.e., $c_{i_0} \cdot h^{\#} = f \cdot k$ for the homomorphism $h^{\#}\colon T_\Sigma(X) \to D_{i_0}$ extending h.

For each $(u, v) \in E$ we have $c_{i_0} \cdot h^{\#}(u) = c_{i_0} \cdot h^{\#}(v)$, and since I is directed, it follows that there exists $i \geq i_0$ such that the homomorphism $d_{i_0,i}\colon D_{i_0} \to D_i$ of D merges $h^{\#}(u)$ and $h^{\#}(v)$. Since $\operatorname{card} E < \lambda$, there exists $i_1 \geq i_0$ such that d_{i_0,i_1} merges $(h^{\#})^2(E)$, thus, it merges $(h^{\#})^2(\sim)$. Then f factorizes through c_{i_1}: there exists a mapping $f'\colon |A| \to |D_{i_1}|$ with $f' \cdot k = d_{i_0,i_1} \cdot h^{\#}$, and this is a homomorphism $f'\colon A \to D_{i_1}$ (see Exercise 3.b(2)) with $f = c_{i_1} \cdot f'$. To show that the factorization is essentially unique, let $f''\colon A \to D_{i_1}$ be a homomorphism with $f = c_{i_1} \cdot f''$. For each $x \in X_s$, since $(c_{i_1})_s(f'_s(x)) = (c_{i_1})_s(f''_s(x))$ it follows that there exists $i \geq i_1$ with $d_{i_1,i}(f'_s(x)) = d_{i_1,i}(f''_s(x))$. Since $\#X < \lambda$, there exists $i_2 \geq i_1$ with $d_{i_1,i_2} \cdot f' \cdot k_0 = d_{i_1,i_2} \cdot f'' \cdot k_0$. This implies $d_{i_1,i_2} \cdot f' \cdot k = d_{i_1,i_2} \cdot f'' \cdot k$, and since k is an epimorphism, we conclude that $d_{i_1,i_2} \cdot f' = d_{i_1,i_2} \cdot f''$.

II. Let A be a λ-presentable object of $\mathbf{Alg}\,\Sigma$. By Proposition 3.11 the algebra A has less than λ generators. Let $X \subseteq |A|$ be a set of generators of power less than λ, and let $k\colon T_\Sigma(X) \to A$ be the canonical homomorphism. Let $\ker k$ denote the many-sorted kernel set of k:

$$(\ker k)_s = \{\, (t, t') \mid t, t' \in T_\Sigma(X)_s \text{ and } k_s(t) = k_s(t') \,\}.$$

Consider the set $\{E_i \mid i \in I\}$ of all many-sorted subsets $E_i \subseteq \ker k$ of power less than λ. We have a λ-directed diagram D' whose objects are the quotient algebras $D'_i = T_\Sigma(X)/{\sim_i}$, where $\sim_i$ is the congruence generated by E_i, and for $E_i \subseteq E_j$ the D'-connecting morphism is the homomorphism $k_{i,j}\colon D'_i \to D'_j$ defined via the quotient maps $k_i\colon T_\Sigma(X) \to D'_i$ by

$$k_{i,j} \cdot k_i = k_j.$$

It is obvious that a colimit of D' is formed by the canonical homomorphisms $k'_i\colon D'_i \to A$ $(i \in I)$ defined by $k'_i \cdot k_i = k$.

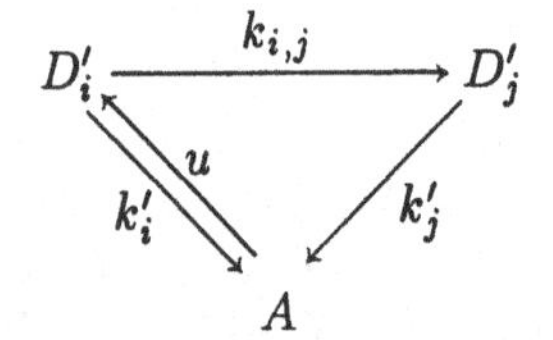

Since A is λ-presentable and the diagram D' is λ-directed, id_A factorizes through some k'_i, say

$$k'_i \cdot u = id_A.$$

Next we use the fact, established in I above, that each D'_i is λ-presentable: since D' is a λ-directed diagram, and since k'_i merges $id_{D'_i}$ with $u \cdot k'_i$ (we have $k'_i \cdot u \cdot k'_i = k'_i$), there exists $j \geq i$ with

$$k_{i,j} \cdot u \cdot k'_i = k_{i,j}.$$

We claim that k'_j is an isomorphism with the inverse $k_{i,j} \cdot u$. In fact,

$$k'_j \cdot (k_{i,j} \cdot u) = k'_i \cdot u = id_A,$$

and to prove $(k_{i,j} \cdot u) \cdot k'_j = id_{D'_j}$, we use the fact that $k_{i,j}$ is an epimorphism:

$$\left[(k_{i,j} \cdot u) \cdot k'_j\right] \cdot k_{i,j} = k_{i,j} \cdot u \cdot k'_i = k_{i,j}.$$

Thus, $A \cong D'_j$ has a presentation by less than λ generators and less than λ equations. □

3.13 Corollary. *In every variety $\mathcal{V}$ of finitary algebras, λ-presentable objects are precisely the algebras presentable by less than λ generators and less than λ equations in $\mathcal{V}$ (for each regular cardinal λ).*

For $\lambda >$ card Σ these are precisely the algebras whose underlying set has power less than λ.

PROOF. Every variety is locally finitely presentable by Corollary 3.7. $\mathcal{V}$ is closed under directed colimits in $\mathbf{Alg}\,\Sigma$, and every Σ-algebra A has a reflection $A \to A/\!\sim$ in $\mathcal{V}$ for some congruence $\sim$, see Remark 3.6. If an algebra A of $\mathcal{V}$ is presentable by less than λ generators and less than λ equations in $\mathcal{V}$, then it is a reflection of a Σ-algebra A_0 presentable by less than λ generators and less than λ equations in $\mathbf{Alg}\,\Sigma$. Since A_0 is λ-presentable in $\mathbf{Alg}\,\Sigma$, it follows that A is λ-presentable in $\mathcal{V}$. Conversely, let A be a λ-presentable object of $\mathcal{V}$. The proof that A can be presented by less than λ generators and less than λ equations is analogous to the above proof in the case $\mathcal{V} = \mathbf{Alg}\,\Sigma$. □

Remark. For each λ-presentable algebra A of $\mathcal{V}$ there exists a coequalizer in $\mathcal{V}$

$$T_{\Sigma,E}(I) \rightrightarrows T_{\Sigma,E}(X) \to A$$

with X and I of power less than λ. In fact, let us assume that $A = T_\Sigma(X)/\!\sim$ for a set X of generators of power less than λ and for a congruence $\sim$ generated by less than λ equations $\tau_i = \tau'_i$, $i \in I$. We can consider I as an S-sorted set: the sort of $i \in I$ is the sort of τ_i. Denote by $e\colon T_\Sigma(X) \to T_{\Sigma,E}(X)$ the canonical homomorphism. Then the maps $f, f'\colon I \to |T_{\Sigma,E}(X)|$ given by $f_s(i) = e_s(\tau_i)$ and $f'_s(i) = e_s(\tau'_i)$ have extensions to a pair of homomorphisms $f^\#, (f')^\#\colon T_{\Sigma,E}(I) \to T_{\Sigma,E}(X)$ whose coequalizer is the canonical homomorphism $T_{\Sigma,E}(X) \to A$.

Algebraic Theories

With each variety $\mathcal{V}$ of many-sorted algebras we associate a category $\mathbf{Th}\,\mathcal{V}$, called the *theory of $\mathcal{V}$-algebras*, such that $\mathcal{V}$ is equivalent to the category of all set functors on $\mathbf{Th}\,\mathcal{V}$ preserving finite products. This presentation of $\mathcal{V}$ via its theory has the fundamental property of being independent of signatures and equations.

Recall that we assume that a standard set of variables is given; in each sort s the variables are denoted by $x_1^s, x_2^s, x_3^s, \dots$. We have shown that the forgetful functor of $\mathcal{V}$ has a left adjoint $F\colon \mathbf{Set}^S \to \mathcal{V}$ (see Remark 3.6(4)).

3.14 Definition. Let $\mathcal{V}$ be a variety of finitary S-sorted algebras. The *theory of $\mathcal{V}$-algebras*, denoted by $\mathbf{Th}\,\mathcal{V}$, is the dual category of the category of all $\mathcal{V}$-free algebras $F\{x_1^{s_1}, \dots, x_n^{s_n}\}$ for all n-tuples of sorts $s_1, \dots, s_n \in S$ ($n = 0, 1, 2, \dots$), as a full subcategory of $(\mathbf{Alg}\,\Sigma)^{\mathrm{op}}$.

3.15 Remarks

(1) In the one-sorted case the objects of $\mathbf{Th}\,\mathcal{V}$ are just $F\{x_1, \dots, x_n\}$. Thus, they can be identified with the natural numbers n. Given two objects n and m, morphisms from n to m in $\mathbf{Th}\,\mathcal{V}$ are the homomorphisms $f\colon F\{x_1, \dots, x_m\} \to F\{x_1, \dots, x_n\}$. Since f is fully determined by its values $f(x_i) \in F\{x_1, \dots, x_n\}$, we can describe $\hom(n, m)$ as the set of all m-tuples of terms in the variables $x_1, \dots, x_n$.

(2) In the many-sorted case, the objects $F\{x_1^{s_1}, \dots, x_n^{s_n}\}$ of $\mathbf{Th}\,\mathcal{V}$ can be identified with the expressions $s_1 \times \cdots \times s_n$. Morphisms from $s_1 \times \cdots \times s_n$ to $t_1 \times \cdots \times t_m$ are m-tuples $(\tau_1, \dots, \tau_m)$ where τ_i is an element of sort t_i of $F\{x_1^{s_1}, \dots, x_n^{s_n}\}$. For a detailed description see Exercise 3.i.

(3) The category $\mathbf{Th}\,\mathcal{V}$ has finite products: the free algebra $F\emptyset$ is a terminal object, and every object $F\{x_1^{s_1}, \dots, x_n^{s_n}\}$ is a product of the $F\{x_i^{s_i}\}$ ($i = 1, \dots, n$) with the ith projection given by the homomorphism $F\{x_i^{s_i}\} \to F\{x_1^{s_1}, \dots, x_n^{s_n}\}$ mapping $x_i^{s_i}$ to itself.

3.16 FP sketches. Recall from Definition 1.49 that a limit sketch $\mathscr{S} = (\mathcal{A}, \mathbf{L}, \sigma)$ consists of a small category $\mathcal{A}$, a choice $\mathbf{L}$ of diagrams in $\mathcal{A}$, and a function σ assigning a cone $\sigma(D)$ to each diagram D in $\mathbf{L}$. If all diagrams in $\mathbf{L}$ are finite and discrete, then $\mathscr{S}$ is called a *finite-product sketch*, briefly an FP sketch. Thus, a model of an FP sketch is a set-functor turning specified cones to product cones.

In particular, each small category $\mathcal{A}$ with finite products can be considered as an FP sketch with $\mathbf{L}$ consisting of all finite, discrete diagrams to which σ assigns product-cones. Then the category $\mathbf{Mod}_{\mathrm{FP}}\,\mathcal{A}$ of models of that sketch is the full subcategory of $\mathbf{Set}^{\mathcal{A}}$ over functors preserving finite products.

Theorem. *Each finitary variety $\mathcal{V}$ is equivalent to the category of models of its theory, considered as an FP sketch, i.e.,*

$$\mathcal{V} \approx \mathbf{Mod}_{\mathrm{FP}}\, \mathbf{Th}\, \mathcal{V}.$$

PROOF. Since $\mathcal{V}$ is a locally finitely presentable category (Corollary 3.7), it is equivalent to $\mathbf{Mod}\,\overline{\mathscr{S}}$ for the limit sketch $\overline{\mathscr{S}}$ given by all finite limits in the category $(\mathbf{Pres}_\omega\, \mathcal{V})^{\mathrm{op}}$—see Theorem and Remark 1.46. We are going to show that $\mathbf{Mod}\,\overline{\mathscr{S}}$ is equivalent to $\mathbf{Mod}\,\mathscr{S}$ for the FP sketch $\mathscr{S}$ on the category $\mathbf{Th}\,\mathcal{V}$. For each model $\overline{H}\colon (\mathbf{Pres}_\omega\, \mathcal{V})^{\mathrm{op}} \to \mathbf{Set}$ of $\overline{\mathscr{S}}$ the domain restriction to $\mathbf{Th}\,\mathcal{V}$ is, obviously, a model of $\mathscr{S}$. Thus, we have a natural functor

$$G\colon \mathbf{Mod}\,\overline{\mathscr{S}} \to \mathbf{Mod}\,\mathscr{S}, \qquad \overline{H} \mapsto \overline{H}/_{\mathbf{Th}\,\mathcal{V}}$$

(assigning to each morphism the domain-codomain restriction). This functor is full and faithful—to verify this, recall from Remark 3.13 that every object of $\mathbf{Pres}_\omega\, \mathcal{V}$ can be expressed by a coequalizer (in $\mathbf{Alg}\,\Sigma$) of two homomorphisms between objects of $\mathbf{Th}\,\mathcal{V}$. To prove that G is an equivalence (0.12), it remains to verify that each model $H\colon \mathbf{Th}\,\mathcal{V} \to \mathbf{Set}$ of $\mathscr{S}$ is naturally isomorphic to a domain-restriction of a model of $\overline{\mathscr{S}}$. We are going to find an algebra A such that H is naturally isomorphic to the domain restriction of $\hom(-, A)$ (and the latter functor, restricted to $(\mathbf{Pres}_\omega\, \mathcal{V})^{\mathrm{op}}$, is a model of $\overline{\mathscr{S}}$, of course).

Let $\mathcal{V} = \mathbf{Alg}(\Sigma, E)$ be an equational presentation. For each set X of variables the free $\mathcal{V}$-algebra FX is a quotient of the term-algebra $T_\Sigma X$. Thus, given a term τ in $(T_\Sigma X)_s$, we have the unique Σ-homomorphism $[\tau]\colon F\{x_1^s\} \to FX$ assigning to x_1^s the congruence class of τ in $(FX)_s$. Consequently, for each term τ in $\big(T_\Sigma\{x_1^{s_1}, \ldots, x_n^{s_n}\}\big)_s$ we get a morphism

$$[\tau]\colon F\{x_1^{s_1}, \ldots, x_n^{s_n}\} \to F\{x_1^s\} \qquad \text{in } \mathbf{Th}\,\mathcal{V}.$$

Observe that the cone

$$\Big(F\{x_1^{s_1}, \ldots, x_n^{s_n}\} \xrightarrow{[x_i^{s_i}]} F\{x_1^{s_i}\}\Big)_{i=1}^{n}$$

is a product in $\mathbf{Th}\,\mathcal{V}$ (Remark 3.15(3)).

Since $H\colon \mathbf{Th}\,\mathcal{V} \to \mathbf{Set}$ preserves finite products, we can assume that

$$HF\{x_1^{s_1}, \ldots, x_n^{s_n}\} = \prod_{i=1}^{n} HF\{x_1^{s_i}\}$$

is the cartesian product with projections $H[x_i^{s_i}]$ in **Set**. Define a Σ-algebra A as follows. The underlying set of sort s is

$$A_s = HF\{x_1^s\} \qquad (s \in S).$$

For each operation symbol $\sigma\colon s_1 \times \cdots \times s_n \to s$ put

$$\sigma_A = H\left[\sigma(x_1^{s_1}, \ldots, x_n^{s_n})\right] : A_{s_1} \times \cdots \times A_{s_n} \to A_s.$$

In order to show that A lies in $\mathcal{V}$, we will prove that for each term τ in $T_\Sigma\{x_1^{s_1}, \ldots, x_n^{s_n}\}$ of sort s the value of the map $H[\tau]$ at a given n-tuple $(a_1, \ldots, a_n)$ in $A_{s_1} \times \cdots \times A_{s_n}$ (or, equivalently, at each many-sorted map $a\colon \{x_1^{s_1}, \ldots, x_n^{s_n}\} \to |A|$) is the value

$$H[\tau](a) = a_s^\#(\tau)$$

which the homomorphism $a^\# : T_\Sigma\{x_1^{s_1}, \ldots, x_n^{s_n}\} \to A$ assigns to τ. It then follows that, whenever all $\mathcal{V}$-algebras satisfy an equation $\tau = \tau'$ in variables $x_1^{s_1}, \ldots, x_n^{s_n}$, then A satisfies it too, because $[\tau] = [\tau']$. Since Σ is finitary, we conclude that A satisfies all equations in E, thus, $A \in \mathcal{V}$. We prove the rule $H[\tau](a) = a_s^\#(\tau)$ by induction on the complexity of τ:

(a) If $\tau = x_i^{s_i}$, then $H[\tau]$ is the ith projection (by our hypothesis on H), and $a_{s_i}^\#(x_i^{s_i}) = a_i$ too.

(b) If $\tau = \sigma(\tau_1, \ldots, \tau_m)$ and the rule holds for each of the terms $\tau_k \in (T\{x_1^{s_1}, \ldots, x_n^{s_n}\})_{t_k}$ $(k = 1, \ldots, m)$, we prove the rule for τ. In fact, the morphisms $[\tau_k]\colon F\{x_1^{s_1}, \ldots, x_k^{s_k}\} \to F\{x_1^{t_k}\}$ yield a morphism

$$([\tau_1], \ldots, [\tau_m]) : F\{x_i^{s_i}\}_{i=1}^n \longrightarrow \prod_{k=1}^m F\{x_1^{t_k}\} = F\{x_1^{t_1}, \ldots, x_m^{t_k}\}.$$

Composing it with $[\sigma(x_i^{s_i})]$ in $\mathbf{Th}\,\mathcal{V}$, we get $[\tau]$. Thus, by our hypothesis on H we have $H[\tau] = H\left[\sigma(x_i^{s_i}\right]\left(H[\tau_1], \ldots, H[\tau_k]\right)$, therefore

$$\begin{aligned}
H[\tau](a) &= H\left[\sigma(x_i^{s_i})\right]\left(H[\tau_1](a), \ldots, H[\tau_m](a)\right) && \text{(definition of } \sigma_A) \\
&= \sigma_A\left(H[\tau_1](a), \ldots, H[\tau_m](a)\right) && \text{(induction)} \\
&= \sigma_A\left(a_{t_1}^\#(\tau_1), \ldots, a_{t_m}^\#(\tau_m)\right) && (a^\# \text{ is a homomorphism}) \\
&= a_s^\#\left(\sigma_A(\tau_1, \ldots, \tau_m)\right) \\
&= a_s^\#(\tau).
\end{aligned}$$

Let us verify that H is naturally isomorphic to the domain restriction of $\hom(-, A)$. Since both functors preserve finite products, and every object of $\mathrm{Th}(\mathcal{V})$ is a finite product of the objects $F\{x_1^s\}$, $s \in S$, it is sufficient to observe that

$$\hom(F\{x_1^s\}, A) \cong A_s = HF\{x_1^s\}. \qquad \square$$

3.17 Remark: FP sketches define finitary varieties. We have seen that each variety can be described by an FP sketch. Conversely, we will now show that for each FP sketch $\mathscr{S} = (\mathcal{A}, \mathbf{L}, \sigma)$ the category $\mathbf{Mod}\,\mathscr{S}$ of models is a variety.

Let us extend the signature of Example 3.5(3) to Σ^* by adding, for each σ-cone $c = (s \xrightarrow{\pi_i} s_i)_{i=1,\ldots,n}$, an operation $\hat{c}$ of arity

$$\hat{c}\colon s_1 \times \cdots \times s_n \to s,$$

and by expanding the equations E to a set E^* by the equations

$$\begin{aligned} \hat{c}(\pi_1(x), \ldots, \pi_n(x)) &= x \\ \pi_i(\hat{c}(x_1, \ldots, x_n)) &= x_i \qquad \text{for each } i = 1, \ldots, n \end{aligned}$$

for each σ-cone $c = (s \xrightarrow{\pi_i} s_i)$ (including the case $n = 0$, where the first line reads $\hat{c} = x$, and the second line is empty).

The variety $\mathbf{Alg}(\Sigma^*, E^*)$ is equivalent to the category of models of $(\mathcal{A}, \mathbf{L}, \sigma)$. In fact, each algebra A defines a functor $F\colon \mathcal{A} \to \mathbf{Set}$ as in 3.5(3) ($Fs = A_s$, $Ff = f_A$) which preserves $\mathbf{L}$-products because the above equations guarantee that the function $A_s \to A_{s_1} \times \cdots \times A_{s_n}$ with components $F\pi_i$ is a bijection (whose inverse is $\hat{c}_A$). That is, $(Fs \xrightarrow{F\pi_i} Fs_i)$ is a product in $\mathbf{Set}$. Conversely, for any functor $F\colon \mathcal{A} \to \mathbf{Set}$ preserving the specified products we define an algebra A in $\mathbf{Alg}(\Sigma^*, E^*)$ as follows: $A_s = Fs$, $h_A = Fh$ for $h \in \mathcal{A}^{\mathrm{mor}}$, and given $c = (s \xrightarrow{\pi_i} s_i)$ in $\mathbf{L}$, then $c_A\colon Fs \to Fs_1 \times \cdots \times Fs_n$ is the canonical bijection.

3.18 Finitary Monads. Another way of describing varieties of algebras independently of signatures and equations is via monads (see 2.78).

A monad $\mathbf{T}$ is called *finitary* provided that T preserves directed colimits. A category isomorphic to $\mathcal{K}^{\mathbf{T}}$ for a finitary monad $\mathbf{T}$ is said to be *finitary monadic* over $\mathcal{K}$.

Theorem. *Varieties of S-sorted finitary algebras are precisely the finitary monadic categories over* $\mathbf{Set}^S$.

PROOF. I. For each variety $\mathbf{Alg}(\Sigma, E)$ of finitary S-sorted algebras the forgetful functor

$$U\colon \mathbf{Alg}(\Sigma, E) \to \mathbf{Set}^S$$

has the following properties:

(1) U has a left adjoint F (see Remark 3.6(4)),

(2) U creates absolute coequalizers (see Remarks 3.4(7) and 3.6(5)),

(3) U preserves directed colimits (see Remarks 3.4(4) and 3.6(6)).

By Beck's theorem (see e.g. [MacLane 1971]), conditions (1) and (2) guarantee that $\mathbf{Alg}(\Sigma, E)$ is isomorphic to $(\mathbf{Set}^S)^{\mathbf{T}}$ for the monad $\mathbf{T}$ generated by the adjoint situation $F \dashv U$. Condition (3) guarantees that $T = UF$ preserves directed colimits (since F preserves colimits), thus, $\mathbf{T}$ is finitary.

II. Each finitary monadic category $(\mathbf{Set}^S)^{\mathbf{T}}$ is isomorphic to a variety of S-sorted algebras. In fact, let Σ be the following signature: for each word $s_1 \times \cdots \times s_n$ in S and each element $s \in S$ the Σ-operations

$$\sigma\colon s_1 \times \cdots \times s_n \to s$$

are precisely the elements of sort s in the free algebra over the variables $x_1^{s_1}, \ldots, x_n^{s_n}$, i.e., $\sigma \in \big(T\{x_1^{s_1}, \ldots, x_n^{s_n}\}\big)_s$. For every $\mathbf{T}$-algebra $(TA \xrightarrow{\alpha} A)$ we define a Σ-algebra A^* on the underlying set of A as follows: given $\sigma\colon s_1 \times \cdots \times s_n \to s$ in Σ, the value of $\sigma_{A^*}\colon A_{s_1} \times \cdots \times A_{s_n} \to A_s$ in an n-tuple $(a_1, \ldots, a_n) \in A_{s_1} \times \cdots \times A_{s_n}$ is defined by forming the corresponding S-sorted map

$$a\colon \{\, x_1^{s_1}, \ldots, x_n^{s_n} \,\} \to |A|$$

and putting

$$\sigma_A(a_1, \ldots, a_n) = (Ta)_s(\sigma).$$

This yields a functor

$$G\colon (\mathbf{Set}^S)^{\mathbf{T}} \to \mathbf{Alg}\,\Sigma$$

with $Gf = f$ on morphisms. This functor is an embedding, since $\mathbf{T}$ is finitary: each $\mathbf{T}$-algebra is a canonical directed colimit of the free $\mathbf{T}$-algebras $T\{x_1^{s_1}, \ldots, x_n^{s_n}\}$, and free $\mathbf{T}$-algebras are free Σ-algebras. Thus, the category of $\mathbf{T}$-algebras is isomorphic to the image of G. It is sufficient to verify that $\operatorname{im} G$ is a variety. Since G obviously preserves products and subalgebras, $\operatorname{im} G$ is closed under products and subalgebras. It remains to prove that $\operatorname{im} G$ is closed under quotients.

Let A be a Σ-algebra in $\operatorname{im} G$, and let $h\colon A \to B$ be a surjective homomorphism. The kernel pair

$$p_1, p_2\colon A_0 \to A$$

of h (formed by the projections of the subalgebra A_0 of $A \times A$ where $(x_1, x_2) \in (A_0)_s$ iff $h_s(x_1) = h_s(x_2)$) lies in $\operatorname{im} G$, since G preserves products and subalgebras. Let $U\colon \mathbf{Alg}\,\Sigma \to \mathbf{Set}^S$ be the natural forgetful functor. Then

$$UA_0 \underset{Up_2}{\overset{Up_1}{\rightrightarrows}} UA \xrightarrow{Uh} UB$$

is an absolute coequalizer in $\mathbf{Set}^S$ (see Exercise 3.h). The forgetful functor $U \cdot G$ of the monadic category $(\mathbf{Set}^S)^{\mathbf{T}}$ creates absolute coequalizers. Consequently, B lies in $\operatorname{im} G$. □

3.B Finitary Quasivarieties

Finitary quasivarieties are classes of many-sorted algebras that can be described by implications (rather than equations). An implication is a formula of the form

$$\bigwedge_i (\tau_i = \tau_i') \implies (\tau = \tau')$$

where $\tau_i = \tau_i'$ and $\tau = \tau'$ are equations. There is some flexibility in the concept of implication, related to restriction on the size of the set of premises; in the present section we assume that there are only finitely many premises $\tau_i = \tau_i'$, but we will return to this question later.

Thus, given a many-sorted signature Σ, by an *implication* we understand a formula

$$(*) \qquad (\tau_1 = \tau_1') \wedge \cdots \wedge (\tau_n = \tau_n') \implies (\tau = \tau')$$

where $\tau_i = \tau_i'$ and $\tau = \tau'$ are equations. A Σ-algebra A is said to *satisfy* the implication provided that for each interpretation of the standard variables $f\colon V \to |A|$ such that $f^{\#}(\tau_i) = f^{\#}(\tau_i')$ holds for $i = 1, \ldots, n$ it follows that $f^{\#}(\tau) = f^{\#}(\tau')$.

3.19 Remark. A *finitary quasivariety* of algebras is a class $\mathcal{V}$ of Σ-algebras for which there exists a set of implications $(*)$ such that a Σ-algebra lies in $\mathcal{V}$ iff it satisfies each of the given implications. Every quasivariety $\mathcal{V}$ has the following properties.

(1) $\mathcal{V}$ is closed under subalgebras and products in Σ (i.e., it is an SP-class, or, equivalently, a full, epireflective subcategory, of $\mathbf{Alg}\,\Sigma$).

(2) $\mathcal{V}$ is closed under directed colimits in $\mathbf{Alg}\,\Sigma$.

(3) $\mathcal{V}$ has free algebras.

(4) $\mathcal{V}$ has a dense subcategory of finitely presentable regular projectives.

The proof is analogous to Remark 3.6.

3.20 Examples

(1) Every variety is a quasivariety.

(2) Torsion-free Abelian groups form a quasivariety presented by the equations for Abelian groups (Example 3.5(1)) together with the following countable set of implications:

$$\begin{aligned} x + x = 0 \quad &\Longrightarrow \quad x = 0 \\ x + x + x = 0 \quad &\Longrightarrow \quad x = 0 \\ &\vdots \end{aligned}$$

(3) Permutation automata are precisely the finite sequential automata (see Example 3.1(3)) satisfying the implication

$$n(q, \sigma) = n(q', \sigma) \implies q = q'$$

(i.e., each input σ leads to an injective next-state function $n(-, \sigma)$).

(4) The category of graphs can be presented as a quasivariety of 2-sorted unary algebras. The two sorts are "edges" and "vertices", and the operations assign to each edge its source and target, respectively. That is, we put

$$S = \{\, \text{vertex, edge} \,\} \qquad \text{and} \qquad \Sigma = \{\, s, t \,\},$$

where s, t both have the arity edge $\rightarrow$ vertex. The quasivariety $\mathcal{A}$ presented by the implication

$$\big(s(x) = t(x)\big) \wedge \big(s(y) = t(y)\big) \implies x = y$$

for standard variables x, y of sort edge is equivalent to **Gra**. In fact, let $E\colon \mathbf{Gra} \rightarrow \mathcal{A}$ be the functor assigning to each graph $K = (X, \alpha)$ the algebra A, where

$$\begin{aligned} |A| &= (X, \alpha) \\ s_A(u, v) &= u \\ t_A(u, v) &= v \end{aligned}$$

for each edge $(u, v) \in \alpha$. To every graph homomorphism $f\colon K \rightarrow K'$ we assign the Σ-homomorphism $Ef = (f, f_0)$, where f_0 is the domain-codomain restriction of the map $f \times f$. It is obvious that E is a full embedding. To show that this is an equivalence of categories (see 0.12), consider an algebra A in $\mathcal{A}$, and denote by K the graph $K = (|A|_{\text{vertex}}, \alpha)$, where the relation α consists of all pairs $(s_A(e), t_A(e))$ with $e \in |A|_{\text{edge}}$. Then A is, obviously, isomorphic to EK.

(5) More generally, for each relational signature Σ the category $\mathbf{Rel}\,\Sigma$ is equivalent to a quasivariety of unary algebras.

3.21 Remarks

(1) Unlike **Gra** and $\mathbf{Rel}\,\Sigma$, the category of posets is *not* equivalent to a quasivariety: **Pos** does not have a dense set of regular projectives. In fact, the only regular projectives are the discretely ordered sets. To verify this, consider the following regular epimorphism in **Pos**:

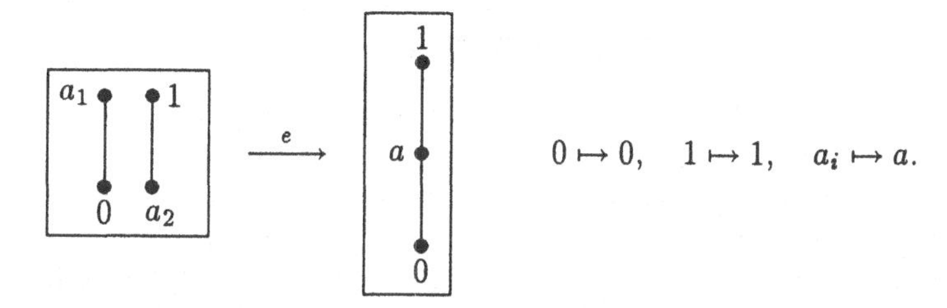

For each poset $(X, \leq)$ and each element $x_0 \in X$ we have a morphism $f\colon (X, \leq) \to (\{0, a, 1\}, \leq)$ defined by

$$f(x) = \begin{cases} 0 & \text{if } x \leq x_0 \\ 1 & \text{otherwise.} \end{cases}$$

If $(X, \leq)$ is a regular projective, then f factorizes through e, which implies that for each $x \in X$ either $x \leq x_0$, or $x_0 \not\leq x$. Since this holds for every $x_0 \in X$, it is clear that the poset $(X, \leq)$ is discrete.

(2) The category **Gra** is equivalent to a quasivariety (Example 3.20(4)), but not to a variety (Example 3.8).

3.22 Theorem. *A class of many-sorted Σ-algebras is a finitary quasivariety iff it is closed in* $\mathbf{Alg}\,\Sigma$ *under products, subobjects, and directed colimits.*

PROOF. Let $\mathcal{K}$ be a class of Σ-algebras closed under products, subobjects, and directed colimits. Then $\mathcal{K}$ is a full, epireflective subcategory of $\mathbf{Alg}\,\Sigma$ (since $\mathbf{Alg}\,\Sigma$ is a complete, (co-)wellpowered category, see Remark 3.4).

Let $\mathcal{A}$ be a set of representatives for all finitely presentable Σ-algebras, and let $c_A : A \to A^*$ be a reflection of A in $\mathcal{K}$ ($A \in \mathcal{A}$). Then, by Remark 1.39, $\mathcal{K}$ is the orthogonality class of these reflections:

$$\mathcal{K} = \{ c_A \}^{\perp}_{A \in \mathcal{A}}.$$

For each algebra $A \in \mathcal{A}$ we have, by Theorem 3.12, a finite set X of (standard, many-sorted) variables and a presentation of A by the equations $\tau_1 = \tau_1', \ldots, \tau_n = \tau_n'$ in X. Thus, there is a surjective homomorphism $h_A : T_\Sigma(X) \to A$ whose kernel congruence is generated by

$$(\tau_1, \tau_1'), \ldots, (\tau_n, \tau_n').$$

Then consider the implications

$$(*) \qquad (\tau_1 = \tau_1' \wedge \cdots \wedge \tau_n = \tau_n') \implies (\tau = \tau')$$

for all $A \in \mathcal{A}$ and for all pairs τ, τ' of terms in $T_\Sigma(X)_s$ with $(c_A \cdot h_A)_s(\tau) = (c_A \cdot h_A)_s(\tau')$. An algebra K satisfies all of these implications iff K is orthogonal to c_A. In fact, a homomorphism $h : A \to K$ is nothing other than an interpretation $h_0 : X \to K$ of the variables above such that the homomorphism $h_0^{\#}$ merges τ_i with τ_i' for $i = 1, 2, \ldots, n$. And h factorizes through c_A iff $h_0^{\#}$ merges τ with τ' whenever $(c_A \cdot h_A)_s(\tau) = (c_A \cdot h_A)_s(\tau')$. Thus, the implications $(*)$, where A ranges through $\mathcal{A}$, determine $\mathcal{K}$.

Conversely, every quasivariety has the mentioned properties—see Remark 3.19. □

3.23 Remarks

(1) The above theorem indicates a close relationship between the concepts of locally finitely presentable category and finitary quasivariety: the first one describes all reflective subcategories of algebras closed under directed colimits (see Theorem 1.46 and Example 3.5(3)), the latter all epireflective subcategories closed under directed colimits. The category of posets demonstrates that these two concepts do not coincide, see Example 3.20(4).

(2) If we admit implications with *infinitely* many premises, we obtain a characterization of epireflective subcategories of algebras. More precisely, a class of Σ-algebras is an SP-class, i.e., is closed under products and subalgebras in $\mathbf{Alg}\,\Sigma$, iff it can be presented by a class of formulas $\bigwedge_{i \in I}(\tau_i = \tau_i') \implies (\tau = \tau')$ where all $\tau_i = \tau_i'$ and $\tau = \tau'$ are equations. The proof is analogous to (in fact, easier than) that of Theorem 3.22: For each Σ-algebra A let $c_A : A \to A^*$ be a reflection in the given class,

and let $h_A : T_\Sigma(|A|) \to A$ be the canonical homomorphism. Consider the implications

$$\bigwedge_{(\varrho,\varrho')\in\ker h_A} (\varrho = \varrho') \implies (\tau = \tau')$$

for all $\tau, \tau' \in T_\Sigma(|A|)_s$ with $(c_A h_A)_s(\tau) = (c_A h_A)_s(\tau')$.

Whether each SP-class can be described by a set of implications is undecidable: see Corollary 6.14.

(3) Epireflective full subcategories need not be closed under directed colimits. For example, in the category **Grp** of groups, all subgroups of products of all countable groups form an epireflective subcategory not closed under directed colimits. The question of whether epireflective, full subcategories are always locally presentable is undecidable, see Corollary 6.24.

(4) Whereas the Birkhoff variety theorem and Theorem 3.22 characterize finitary varieties and quasivarieties as special subcategories of $\mathbf{Alg}\,\Sigma$, we will now present an abstract characterization. Recall the concept of a regular projective from Remark 3.4(5).

3.24 Characterization Theorem of Finitary Quasivarieties. *A category is equivalent to a finitary quasivariety iff it is cocomplete and has a dense subcategory formed by finitely presentable regular projectives.*

PROOF. Every category equivalent to a quasivariety has all the mentioned properties, by Remark 3.19.

Conversely, let $\mathcal{K}$ be a cocomplete category with a dense subcategory $\mathcal{B}$ of finitely presentable regular projectives. It is easy to see that a finite colimit of regular projectives is a regular projective. Thus, the essentially small closure of $\mathcal{B}$ under finite colimits consists of regular projectives; let $\{P_s \mid s \in S\}$ be a set of representatives of that closure. Define a "forgetful" functor

$$U : \mathcal{K} \to \mathbf{Set}^S$$

by

$$UK = \big(\hom(P_s, K)\big)_{s\in S}$$

and

$$Uf = \big(\hom(P_s, f)\big)_{s\in S}.$$

Since each P_s is finitely presentable, the functor U preserves directed colimits (see Proposition 1.26). It is well known that, since $\mathcal{K}$ has coproducts,

U is a right adjoint: a corresponding left adjoint $F\colon \mathbf{Set}^S \to \mathcal{K}$ is given by $FX = \coprod_{s\in S} \coprod_{x\in X_s} P_s$. We denote by $(\varepsilon, \eta)\colon F \dashv U$ the resulting adjoint situation.

Let $\mathcal{A} = \mathbf{Pres}_\omega\, \mathbf{Set}^S$ be a set of representatives of finite S-sorted sets. We define an S-sorted signature Σ to consist of the following operation symbols

$$\sigma_{X,t}\colon s(x_1) \times \cdots \times s(x_n) \to s(t)$$

where $X = \{x_1, \ldots, x_n\}$ ranges through $\mathcal{A}$, $s(x_i)$ denotes the sort of x_i, and t is an element of UFX of sort $s(t)$. Let $U^*\colon \mathbf{Alg}\,\Sigma \to \mathbf{Set}^S$ be the natural forgetful functor. We have an obvious functor $H\colon \mathcal{K} \to \mathbf{Alg}\,\Sigma$ for which the triangle

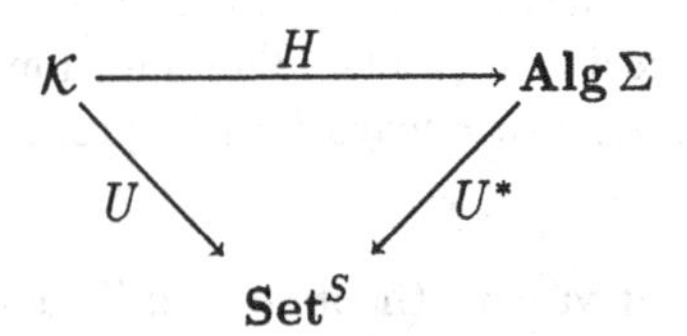

commutes: for each object K of $\mathcal{K}$ we equip the many-sorted set UK with the following operations $\sigma_{X,t}$. Every X-tuple in UK (i.e., every S-sorted map $m\colon X \to UK$) defines an adjoint morphism $\widehat{m}\colon FX \to K$, and we put

$$(\sigma_{X,t})_{HK}(m) = (U\widehat{m})_{s(t)}(t).$$

It is clear that for each morphism $f\colon K \to K'$ in $\mathcal{K}$ we have a homomorphism $Uf\colon HK \to HK'$ in $\mathbf{Alg}\,\Sigma$. We are going to prove that H is a (faithful and) full functor such that the category $H(\mathcal{K})$, which is equivalent to $\mathcal{K}$, is a quasivariety of Σ-algebras.

(1) H preserves directed colimits. In fact, since Σ is a finitary signature, U^* creates directed colimits (see Remark 3.4), and $U^*H = U$ preserves directed colimits.

(2) H is full. Let us first observe that HF is a left adjoint of U^*, i.e., for each object X in $\mathbf{Set}^S$, $HF(X)$ is a free Σ-algebra with the universal map

$$\eta\colon X \to UF(X) = U^*HF(X).$$

Due to (1), it is sufficient to show this for all X in $\mathcal{A}$. And this is clear: for each algebra A and each map $h\colon X \to U^*A$ the unique extension to a homomorphism $h^*\colon HF(X) \to A$ is defined by

$$h^*_s(t) = (\sigma_{X,t})_A(h) \qquad \text{for all } t \in (FX)_s.$$

Now, for each s we have $P_s \cong FX_s$ where X_s is the S-sorted set with all sorts empty except the sort s which is a singleton set. Consequently, the restriction of H yields a bijective correspondence between $\hom_{\mathcal{K}}(P_s, K)$ (isomorphic to $\hom(X_s, UK)$ in $\mathbf{Set}^S$) and $\hom_{\mathbf{Alg}\,\Sigma}(HP_s, HK)$. Since H preserves directed colimits, and since each object of $\mathcal{K}$ is a directed colimit of the objects of $\{ P_s \mid s \in S \}$ (because this is a dense set closed under finite colimits), it follows that H is full.

(3) $H(\mathcal{K})$ is a quasivariety. In fact, by (1), $H(\mathcal{K})$ is closed under directed colimits, thus, it is sufficient to show that $H(\mathcal{K})$ is an epireflective subcategory, see Theorem 3.21. Let (T, μ, η) with $T = HFU^*\colon \mathbf{Alg}\,\Sigma \to \mathbf{Alg}\,\Sigma$ be the monad induced by $HF \dashv U^*$. Then each Σ-algebra A is a coequalizer

$$T^2A \underset{T\varepsilon_A}{\overset{\mu_A}{\rightrightarrows}} TA \xrightarrow{\varepsilon_A} A$$

and, since H is full, there exist morphisms $f, g\colon FU^*TA \to FU^*A$ in $\mathcal{K}$ with $\mu_A = Hf$ and $T\varepsilon_A = Hg$. Let us form a coequalizer in $\mathcal{K}$:

$$FU^*TA \underset{g}{\overset{f}{\rightrightarrows}} FU^*A \xrightarrow{c} K.$$

There exists a unique homomorphism $r_A\colon A \to HK$ with $Hc = r_A \cdot \varepsilon_A$. This is a reflection of A in $H(\mathcal{K})$. In fact, let $r'\colon A \to HK'$ be a homomorphism, then, since H is full and faithful, there exists a unique morphism $k\colon FU^*A \to K'$ in $\mathcal{K}$ with $Hk = r' \cdot \varepsilon_A$. We have $k \cdot f = k \cdot g$, because $Hk \cdot Hf = r' \cdot \varepsilon_A \cdot \mu_A = r' \cdot \varepsilon_A \cdot T\varepsilon_A = Hk \cdot Hg$. Thus, k factorizes uniquely through c, which implies that $Hk = r' \cdot \varepsilon_A$ factorizes uniquely through $Hc = r_A \cdot \varepsilon_A$. Since ε_A is an epimorphism, r' factorizes uniquely through r_A.

To show that r_A is an epimorphism, it is sufficient to verify that U^*r_A is surjective (in each sort), and this follows from the fact that

$$Uc = U^*Hc = U^*r_A \cdot U^*\varepsilon_A\colon UFU^*A \to UK$$

is surjective. In fact, since P_s is a regular projective and c is a regular epimorphism, for each $t \in (UK)_s = \hom(P_s, K)$ there exists

$$t' \in \hom(P_s, FU^*A) = (UFU^*A)_s$$

with $t = c \cdot t'$, i.e., with $t' = (Uc)_s(t)$. ☐

3.25 Corollary (Characterization of Finitary Varieties). *A category is equivalent to a finitary variety iff it is cocomplete, has effective equivalence relations, and has a dense set of finitely presentable regular projectives.*

PROOF. For every quasivariety $\mathcal{V}$ of Σ-algebras with effective equivalence relations we will show that $\mathcal{V}$ is closed under quotients in $\mathbf{Alg}\,\Sigma$, thus, it is a variety (by the Birkhoff variety theorem). To verify this, let $f\colon K \to L$ be a quotient with $K \in \mathcal{V}$, and let $e_1, e_2\colon E \to K$ be a kernel pair of f in $\mathbf{Alg}\,\Sigma$ (i.e., a pullback of f and f). Since $\mathcal{V}$ is closed under subalgebras and (finite) limits in $\mathbf{Alg}\,\Sigma$, it follows that $E \in \mathcal{V}$, and e_1, e_2 is an equivalence relation in $\mathcal{V}$. Consequently, e_1, e_2 is a kernel pair of its coequalizer $f_0\colon K \to L_0$ in $\mathcal{V}$. In other words, f and f_0 induce the same kernel equivalence on K. Since both f and f_0 are surjective homomorphisms, it follows that L and L_0 are isomorphic algebras. Thus, from $L_0 \in \mathcal{V}$ we conclude $L \in \mathcal{V}$. □

3.C Infinitary Varieties and Quasivarieties

Varieties

Infinitary varieties form locally presentable categories whenever they are given by a set of operations and equations (example: σ-complete lattices); we do not consider here equational presentations based on a proper class of operations (example: complete lattices). We will show that infinitary varieties precisely correspond to *product sketches,* i.e., sketches in which the chosen diagrams are discrete.

By an *S-sorted signature of algebras* is understood a set Σ (of operation symbols) together with an arity function assigning to each operation symbol $\sigma \in \Sigma$ a collection $(s_i)_{i<n}$ of sorts for some cardinal number n, and a sort s. Notation:

$$\sigma\colon \prod_{i<n} s_i \to s.$$

The signature is said to be *λ-ary* provided that λ is a regular cardinal larger than n for each operation of arity $\prod_{i<n} s_i \to s$ in Σ. An *algebra* A of signature Σ is an S-sorted set $|A| = (A_s)_{s\in S}$ together with an operation

$$\sigma_A\colon \prod_{i<n} A_{s_i} \to A_s$$

for each $\sigma \in \Sigma$ of arity $\prod s_i \to s$. A *homomorphism* $f\colon A \to B$ is an S-sorted function such that for each $\sigma\colon \prod s_i \to s$ in Σ we have $f_s(\sigma_A(x_i)) = \sigma_B(f_{s_i}(x_i))$. This yields a category $\mathbf{Alg}\,\Sigma$.

The set $T_\Sigma(X)$ of all *terms* in variables from X is defined as in the finitary case: each $x \in X_s$ is a term of sort s, and for $\sigma\colon \prod s_i \to s$ and for terms τ_i of sorts s_i, $\sigma(\tau_i)$ is a term of sort s.

Examples of terms of a one-sorted ω-ary operation σ:

$$x, \quad \sigma(x_0, x_1, x_2, \ldots), \quad \sigma\big(\sigma(x_0, x_1, \ldots), x_0, x_1, \ldots\big), \quad \text{etc.}$$

Again, $T_\Sigma(X)$ is the free Σ-algebra over X.

Assumption. We assume below that Σ is a λ-ary signature, and that a standard set V of variables has been chosen with each sort V_s $(s \in S)$ having cardinality λ.

The concept of an equational presentation is quite analogous to the finitary case: we consider equations $\sigma = \tau$, i.e., pairs (σ, τ) of terms in the standard variables, and then define satisfaction as above. Classes of algebras described by equations are called *λ-ary varieties.*

3.26 Example. An equational presentation of upper σ-semilattices (i.e., posets with countable joins) and countable-join-preserving functions. Let Σ be the one-sorted signature of one ω-ary operation (denoted by $x_0 \vee x_1 \vee x_2 \vee \ldots$) and one nullary operation 0, and let E be the following set of equations:

$$x \vee x \vee x \vee \ldots = x$$
$$0 \vee x \vee x \vee \ldots = x$$
$$x_{i_0} \vee x_{i_1} \vee x_{i_2} \vee \ldots = x_{j_0} \vee x_{j_1} \vee x_{j_2} \ldots \quad \text{whenever } \{i_k\}_{k\in\omega} = \{j_k\}_{k\in\omega}.$$

Each (Σ, E)-algebra A is a poset under the ordering

$$a \leq b \quad \text{iff} \quad a \vee b \vee b \vee \cdots = b$$

with countable joins $a_0 \vee a_1 \vee a_2 \vee \cdots$ and the empty join 0.

3.27 Properties of varieties are quite analogous to the finitary case (Remark 3.6) except that a variety need not be closed under directed colimits. Instead, each λ-ary variety is closed under λ-directed colimits in $\mathbf{Alg}\,\Sigma$.

The Birkhoff variety theorem also holds in the infinitary case.

3.28 Theorem. *Every λ-ary variety is a locally λ-presentable category.*

The proof is analogous to Corollary 3.7. □

3.29 Theory of $\mathcal{V}$-algebras. The construction of a theory $\mathbf{Th}\,\mathcal{V}$ for each λ-ary many-sorted variety $\mathcal{V}$ is quite analogous to the finitary case. We define a category $\mathbf{Th}\,\mathcal{V}$ with products of families of less than λ objects. Then we prove that $\mathcal{V}$ is equivalent to the category of models of the product-sketch $\mathbf{Th}\,\mathcal{V}$. To make this precise, for each small category $\mathcal{A}$ with products of families of less than λ objects we denote by

$$\mathbf{Mod}_{\mathrm{P}_\lambda}\,\mathcal{A}$$

the category of all functors in $\mathbf{Set}^{\mathcal{A}}$ which preserve products of less than λ objects.

The theory $\mathbf{Th}\,\mathcal{V}$ is defined to be the dual of the category of $\mathcal{V}$-free algebras generated by sets $\{ x_i^{s_i} \}_{i<n}$ of standard variables for all cardinals $n < \lambda$ (as a full subcategory of $(\mathbf{Alg}\,\Sigma)^{\mathrm{op}}$).

The category $\mathbf{Th}\,\mathcal{V}$ is easily seen to have products of less than λ objects. The following is proved analogously to Theorem 3.16:

3.30 Theorem. *Every λ-ary variety $\mathcal{V}$ of many-sorted algebras is equivalent to the category of models of its theory, considered as a P_λ-sketch, i.e.,*

$$\mathcal{V} \approx \mathbf{Mod}_{\mathrm{P}_\lambda}\, \mathbf{Th}\,\mathcal{V}.$$

Remark. Analogously to Remark 3.17, for each product sketch $\mathscr{S} = (\mathcal{A}, \mathbf{L}, \sigma)$ the category of models is equivalent to a variety of many-sorted algebras. We extend the signature of 3.5(3) to Σ^* by adding, for each cone $c = (s \xrightarrow{\pi_i} s_i)_{i \in I}$ in σ, an operation $\hat{c}\colon \prod s_i \to s$ and by expanding the equations E to a set E^* with the equations

$$\hat{c}\big(\pi_i(x)\big) = x$$
$$\pi_j\big(\hat{c}(x_i)\big) = x_j$$

for each σ-cone $c = (s \xrightarrow{\pi_i} s_i)$. The proof that $\mathbf{Alg}(\Sigma^*, E^*)$ is equivalent to $\mathbf{Mod}\,\mathscr{S}$ is analogous to Remark 3.17.

3.31. Analogously to the finitary case, varieties correspond to monads over $\mathbf{Set}^S$. Recall that a monad $T = (T, n, \eta)$ over $\mathbf{Set}^S$ is called *accessible* if there exists a regular cardinal λ such that T preserves λ-directed colimits.

Theorem. *Varieties of S-sorted algebras are precisely the accessibly monadic categories over $\mathbf{Set}^S$.*

The proof is analogous to that of Theorem 3.18.

Remark. In more detail, varieties of λ-ary S-sorted algebras are precisely the λ-accessibly monadic categories over $\mathbf{Set}^S$.

Quasivarieties

Classes of Σ-algebras described by a set of *implications*

$$(*) \qquad \bigwedge_{i \in I} (\tau_i = \tau_i') \Longrightarrow (\tau = \tau')$$

are called *quasivarieties*. Here we put no restriction on the number of premises $\tau_i = \tau_i'$ of an implication. *Satisfaction* is defined as in the finitary case: an algebra A satisfies $(*)$ provided that for each interpretation f of the standard variables we have that $f^{\#}(\tau_i) = f^{\#}(\tau_i')$ for all $i \in I$ implies $f^{\#}(\tau) = f^{\#}(\tau')$.

3.32 Remark. A class of Σ-algebras is called a *λ-ary quasivariety* provided that Σ is a λ-ary signature and that the class can be described by a set of implications $(*)$ of less than λ premises (i.e., $\operatorname{card} I < \lambda$). Every λ-ary quasivariety

(1) is closed under products and subalgebras in $\mathbf{Alg}\,\Sigma$ (i.e., it is an epireflective full subcategory of $\mathbf{Alg}\,\Sigma$);

(2) is closed under λ-directed colimits in $\mathbf{Alg}\,\Sigma$;

(3) has free algebras;

(4) has a dense subcategory formed by λ-presentable regular projectives.

The proof is analogous to Remark 3.6.

Conversely, a full subcategory of $\mathbf{Alg}\,\Sigma$ satisfying (1) and (2) is a λ-ary quasivariety. This is proved analogously to Theorem 3.22. The condition (4) yields an abstract characterization of quasivarieties:

3.33 Theorem. *A category is equivalent to*

(i) *a quasivariety iff it is locally presentable and has a dense subcategory formed by regular projectives;*

(ii) *a variety iff it is locally presentable, has effective equivalence relations and has a dense subcategory formed by regular projectives.*

The proof is analogous to that of Theorem 3.24 and Corollary 3.25. The statement can be formulated relative to any regular cardinal λ: a category is equivalent to a λ-ary (quasi)variety of many sorted algebras iff it is locally λ-presentable and satisfies the above additional conditions.

3.D Essentially Algebraic Categories

Equational theories of many-sorted algebras, as studied above, do not suffice to describe all locally presentable categories (e.g. **Pos**, see Remark 3.21). We now turn to equational theories of *partial* algebras. The type of theory of partial algebras used to describe locally presentable categories is very "pleasant": some of the operation symbols are required to be total, and the remaining (partial) symbols have their domain of definition fully described by equations using solely the total symbols. Example: the theory of (small) categories is a two-sorted equational theory, $S = \{\text{obj}, \text{mor}\}$, with two total operations $d, c\colon \text{mor} \to \text{obj}$ (the domain and codomain, resp.) and a partial binary operation $\cdot\colon \text{mor} \times \text{mor} \to \text{mor}$ with $x \cdot y$ defined iff $d(x) = c(y)$.

A *partial algebra* A of signature Σ is an S-sorted set $|A| = (A_s)_{s \in S}$ together with partial functions $\sigma_A : \prod_{i \in I} A_{s_i} \to A_s$ for all operation symbols $\sigma\colon \prod_{i \in I} s_i \to s$ in Σ. We denote by $\mathbf{Palg}\,\Sigma$ the category of all partial algebras of signature Σ and all homomorphisms, where a *homomorphism* from A to B is an S-sorted function $f\colon |A| \to |B|$ such that, for each $\sigma\colon \prod_{i \in I} s_i \to s$ in Σ, whenever $\sigma_A(a_i)$ is defined, then $\sigma_B(f_{s_i}(a_i))$ is defined and is equal to $f_s(\sigma_A(a_i))$.

The concepts of terms and equations are introduced exactly as in the theory of total algebras above. We say that $\tau = \tau'$ is an equation in standard variables x_j ($\in V_{s_j}$) for $j \in J$ if both τ and τ' are terms in

$$T_\Sigma(\{\, x_j \,\}_{j \in J}).$$

A partial algebra A *satisfies that equation in the elements* a_j ($\in A_{s_j}$) provided that both τ and τ' are defined in A under the substitution $x_j \mapsto a_j$ and that they give the same result.

3.34 Definition

(1) An *essentially algebraic theory* is a quadruple

$$\Gamma = (\Sigma, E, \Sigma_t, \text{Def})$$

consisting of a many-sorted signature Σ of algebras, a set E of Σ-equations, a set $\Sigma_t \subseteq \Sigma$ of "total" operation symbols, and a function Def assigning to each operation symbol $\sigma\colon \prod_{i \in I} s_i \to s$ in $\Sigma - \Sigma_t$ a set $\text{Def}(\sigma)$ of Σ_t-equations in the standard variables $x_i \in V_{s_i}$ ($i \in I$).

(2) We say that the theory Γ is λ*-ary*, for a regular cardinal λ, provided that Σ is λ-ary, each of the equations of E and $\text{Def}(\sigma)$ uses less than λ standard variables, and each $\text{Def}(\sigma)$ contains less than λ equations.

(3) By a *model* of an essentially algebraic theory Γ we mean a partial Σ-algebra A such that

(a) A satisfies all equations of E,

(b) for each $\sigma \in \Sigma_t$, the operation σ_A is everywhere defined,

(c) for each $\sigma \in \Sigma - \Sigma_t$ with $\sigma\colon \prod_{j \in J} s_j \to s$ and any $a_j \in A_{s_j}$ ($j \in J$) we have that $\sigma_A(a_j)$ is defined iff A satisfies all equations of $\mathrm{Def}(\sigma)$ in the elements a_j.

The category of all models and homomorphisms is denoted by $\mathbf{Mod}\,\Gamma$. A category is called *essentially algebraic* if it is equivalent to $\mathbf{Mod}\,\Gamma$ for some essentially algebraic theory Γ.

3.35 Examples

(1) Every equational theory is essentially algebraic: we simply put $\Sigma = \Sigma_t$.

(2) The implicational theory of graphs (Example 3.20(4)) can be turned into an essentially algebraic theory by introducing a new operation $\sigma \in \Sigma - \Sigma_t$ whose only role is to translate the implication

$$\big(s(x) = t(x)\big) \wedge \big(s(y) = t(y)\big) \implies x = y$$

into a definability statement. Thus, we put

$$\begin{aligned} S &= \{\,\text{vertex, edge}\,\}, \\ \Sigma &= \{\,s, t, \sigma\,\} \quad \text{with} \quad \Sigma_t = \{\,s, t\,\}, \\ s, t&\colon \text{edge} \to \text{vertex}, \\ \sigma&\colon \text{edge} \times \text{edge} \to \text{edge}, \\ \mathrm{Def}(\sigma) &= \{\, s(x) = t(s), s(y) = t(y)\,\}, \\ E &= \{\,\sigma(x, y) = x, \sigma(x, y) = y\,\}. \end{aligned}$$

A partial Σ-algebra is a model of this essentially algebraic theory Γ iff its total operations s, t satisfy the above implication. Thus, $\mathbf{Mod}\,\Gamma$ is equivalent to $\mathbf{Gra}$.

(3) More generally, every theory given by implications can be easily translated to an essentially algebraic theory: for each implication $\bigwedge_{i \in I}(\tau_i = \tau_i') \implies \tau = \tau'$ we add a new operation symbol $\sigma \in \Sigma - \Sigma_t$ depending on all the variables which appear in all τ_i, τ_i', and then we put

$$\mathrm{Def}(\sigma) = \{\, \tau_i = \tau_i' \mid i \in I\,\},$$

while extending E by the equations $\sigma = \tau$ and $\sigma = \tau'$.

(4) The following is an essentially algebraic theory of posets. We add three operations to the theory of graphs in (2): a total operation

$$\delta\colon \text{vertex} \to \text{edge}$$

expressing the reflexivity (in each poset $\delta(x) = (x, x)$), a partial operation τ (which will translate the implication of antisymmetry

$$\big(t(x) = s(y)\big) \wedge \big(t(y) = s(x)\big) \implies x = y$$

into a definability condition):

$$\tau\colon \text{edge} \times \text{edge} \to \text{edge}$$
$$\operatorname{Def}(\tau) = \{\, t(x) = s(y), t(y) = s(x) \,\}$$

and an operation ϱ (expressing the transitivity, i.e., in each poset with edges $x = (u, v)$ and $y = (v, w)$ we have $\varrho(x, y) = z$ for $z = (u, w)$):

$$\varrho\colon \text{edge} \times \text{edge} \to \text{edge}$$
$$\operatorname{Def}(\varrho) = \{\, t(x) = s(y) \,\}.$$

We extend E by the following equations:

$$\begin{array}{lll} t\delta(x) = x, & s\delta(x) = x & \text{(reflexivity)} \\ \tau(x, y) = x, & \tau(x, y) = y & \text{(antisymmetry)} \\ s\varrho(x, y) = s(x), & t\varrho(x, y) = t(y) & \text{(transitivity)}. \end{array}$$

The resulting essentially algebraic theory Γ has the property that **Pos** is equivalent to $\mathbf{Mod}\,\Gamma$.

Remark. Since **Pos** is not equivalent to a quasivariety of algebras (see Remark 3.20), the last example demonstrates that essentially algebraic theories have a wider expressive power than quasivarieties.

3.36 Theorem. *A category is locally presentable iff it is essentially algebraic.*

Remark. In more detail, a category is locally λ-presentable iff it is equivalent to the category of models of a λ-ary essentially algebraic theory. (Thus, locally finitely presentable categories can always be described by an essentially algebraic theory with all arities finite, and with the sets $\operatorname{Def}(\sigma)$ all finite.)

PROOF. I. Let Γ be an λ-ary essentially algebraic theory, then we will prove that $\mathbf{Mod}\,\Gamma$ is locally λ-presentable. It is sufficient to prove that the category $\mathbf{Palg}\,\Sigma$ is locally λ-presentable, and that $\mathbf{Mod}\,\Gamma$ is closed under limits and λ-directed colimits in $\mathbf{Palg}\,\Sigma$. It then follows that $\mathbf{Mod}\,\Gamma$ is a reflective subcategory of $\mathbf{Palg}\,\Sigma$ (by the reflection theorem, 2.48) and the closedness under λ-directed colimits guarantees that $\mathbf{Mod}\,\Gamma$ is locally λ-presentable, see Theorem 1.39.

To prove that $\mathbf{Palg}\,\Sigma$ is locally λ-presentable, we consider it as a subcategory of the category $\mathbf{Rel}\,\Sigma^*$ of relational structures of the signature $\Sigma^* = \Sigma$ where for each operation symbol $\sigma\colon \prod_{i\in I} s_i \to s$ in Σ we define the arity of the (relation) symbol σ in Σ^* to be $\left(\prod_{i\in I} s_i\right) \times s$. Then we have a full embedding $E\colon \mathbf{Palg}\,\Sigma \to \mathbf{Rel}\,\Sigma^*$ assigning to each partial Σ-algebra A the relational structure EA on the same underlying set, and with each partial operation $\sigma_A\colon \prod_{i\in I} A_{s_i} \to A_s$ interpreted by its graph, i.e., by the corresponding subset of $(\prod_{i\in I} A_{s_i}) \times A_s$. The image of this full embedding is a full subcategory of $\mathbf{Rel}\,\Sigma^*$ characterized by the property that for each symbol $\sigma \in \Sigma^*$ of arity $(\prod_{i\in I} s_i)\times s$ the following holds: if σ_A contains $\left((a_i)_{i\in I}, a\right)$ and $\left((a_i)_{i\in I}, b\right)$, then $a = b$. It is easy to verify that this full subcategory is closed in $\mathbf{Rel}\,\Sigma^*$ under limits and λ-directed colimits. Since $\mathbf{Rel}\,\Sigma^*$ is locally λ-presentable (see Example 1.18), it follows that $\mathbf{Palg}\,\Sigma$ is locally λ-presentable by Theorem 1.39.

The subcategory $\mathbf{Mod}\,\Gamma$ is closed under limits in $\mathbf{Palg}\,\Sigma$ because an equation (in E or in $\mathrm{Def}(\sigma)$) holds in the limit of a diagram of partial algebras A_i iff the corresponding equations hold in each A_i. Analogously, a partial operation is defined in an I-tuple of the limit iff the corresponding partial operations are defined in the corresponding I-tuples in each A_i.

Closedness of $\mathbf{Mod}\,\Gamma$ under λ-directed colimits follows from the fact that all the conditions (a), (b), (c) defining a model of Γ (see Definition 3.34) involve in each instance less than λ variables.

II. Let $\mathcal{K}$ be a locally λ-presentable category. Then $\mathcal{K}$ is equivalent to a full, reflective subcategory of $\mathbf{Set}^{\mathcal{A}}$ ($\mathcal{A}$ a small category), closed under λ-directed colimits, see Theorem 1.46. Without loss of generality, let us assume that $\mathcal{K}$ *is* actually a full, reflective subcategory of $\mathbf{Set}^{\mathcal{A}}$, closed under λ-directed colimits. Recall that $\mathbf{Set}^{\mathcal{A}}$ is isomorphic to a variety of unary algebras (Example 3.5(3)).

Thus, we can consider $\mathcal{K}$ as a full, reflective subcategory of the category $\mathbf{Alg}\,\Sigma$ of unary algebras of signature Σ, closed under λ-directed colimits. Moreover, by Theorem and Remark 1.39, we have a set $\mathcal{M} = \{ A_i \xrightarrow{m_i} A_i' \}_{i\in I}$ of morphisms of $\mathbf{Alg}\,\Sigma$ such that

$$\mathcal{K} = \mathcal{M}^{\perp},$$

each A_i is λ-presentable in $\mathbf{Alg}\,\Sigma$, and $m_i\colon A_i \to A_i'$ is a reflection of A_i in $\mathcal{K}$. Let us choose a presentation of A_i by $\alpha_i < \lambda$ variables, say,

$$\text{variables } y_{i,k} \text{ of sort } s_{i,k} \text{ for } i \in I,\ k < \alpha_i$$

and by $\beta_i < \lambda$ equations (see Theorem 3.12). Denote by $p_i \colon \{y_{i,k}\} \to |A|$ the interpretation of variables used by the presentation.

We now define an essentially algebraic theory

$$\Gamma = (\Sigma^*, E, \Sigma, \mathrm{Def})$$

for which we will prove that $\mathcal{K} \approx \mathbf{Mod}\,\Gamma$. The set Σ of total operation symbols is the above unary signature Σ. This we extend to Σ^* by choosing, for each $i \in I$ and $a \in |A_i'|_s$, an operation

$$h_{i,a} \colon \prod_{k<\alpha_i} s_{i,k} \to s.$$

The set E consists of all of the following equations:

(i) $\sigma h_{i,a} = h_{i,\sigma_{A_i'}(a)}$ for each $\sigma \colon s \to s'$ in Σ, $i \in I$, and $a \in |A_i'|_s$,

and

(ii) $y_{i,k} = h_{i,a_{i,k}}$ for each $i \in I$ and $k < \alpha_i$

with $a_{i,k} = (m_i \cdot p_i)_{s_{i,k}}(y_{i,k})$. Finally, $\mathrm{Def}(h_{i,a})$ is the above set of β_i equations presenting the algebra A_i. To prove that $\mathcal{K}$ is isomorphic to $\mathbf{Mod}\,\Gamma$, we will show that (1) every Σ-algebra in $\mathcal{K}$ has a unique extension to a model of Γ and (2) for every model of Γ the Σ-reduct (i.e., the Σ-algebra obtained by forgetting the partial operations in $\Sigma^* - \Sigma$) lies in $\mathcal{K}$.

(1) Let K lie in $\mathcal{K}$. To extend K to a model of Γ, we must define, for each $i \in I$, $a \in |A_i'|_s$, and for any interpretation $f_0 \colon \{y_{i,k}\}_{k<\alpha_i} \to |K|$ such that f_0 satisfies the equations of $\mathrm{Def}(h_{i,a})$, the result $\left(h_{i,a}\right)_K(f_0)$, subject to the equations (i) and (ii) above. It is clear that f_0 satisfies the equations of $\mathrm{Def}(h_{i,a})$ iff there exists a Σ-homomorphism $f \colon A_i \to K$ with $f_0 = f \cdot p_i$. Since K lies in $\mathcal{K} = \mathcal{M}^{\perp}$, there then exists a unique Σ-homomorphism $f' \colon A_i' \to K$ with $f = f' \cdot m_i$. Put

$$(*) \qquad \left(h_{i,a}\right)_K(f_0) = f_s'(a) \qquad \text{for each } a \in \left|A_i'\right|_s.$$

Then the equations of (i) are satisfied because f' is a Σ-homomorphism, and those of (ii) are satisfied because $f = f' \cdot m_i$, thus,

$$\left(f_0\right)_{s_{i,k}}(y_{i,k}) = \left(f' \cdot m_i \cdot p_i\right)_{s_{i,k}}(y_{i,k}) = f'_{s_{i,k}}(a_{i,k}) = h_{i,a_{i,k}}(f_0).$$

Conversely, an extension of K to a model of Γ must be defined by $(*)$ (with $f_0 = f \cdot p_i$) because the map $f' \colon |A_i'| \to |K|$ given by $(*)$ is a Σ-homomorphism, due to (i), and fulfils $f = f' \cdot m_i$, due to (ii).

(2) Let K be a model of Γ. We will prove that the Σ-reduct K^0 of K lies in $\mathcal{M}^\perp = \mathcal{K}$. Let $f\colon A_i \to K^0$ be a Σ-homomorphism. We are to show that there exists a unique Σ-homomorphism $f'\colon A_i' \to K$ with $f = f' \cdot m_i$.

(2a) Existence. The map f' defined by $(*)$ above with $f_0 = f \cdot p_i$ is a Σ-homomorphism $f'\colon A_i' \to K^0$, due to (i), and it fulfils $f = f' \cdot m_i$, due to (ii).

(2b) Unicity. Let $f', f''\colon A_i' \to K^0$ be Σ-homomorphisms with $f' \cdot m_i = f'' \cdot m_i$. Since m_i is a reflection map, A_i' lies in $\mathcal{K}$, thus, by (1) we can consider A_i' as a model of Γ. Then, obviously, $f', f''\colon A_i' \to K$ are homomorphisms of $\mathbf{Mod}\,\Gamma$. It is clear that $\mathbf{Mod}\,\Gamma$ has equalizers, and the reduct-functor $\mathbf{Mod}\,\Gamma \to \mathbf{Alg}\,\Sigma$ preserves equalizers. Consequently, if $e\colon L \to A_i'$ denotes an equalizer of f' and f'' in $\mathbf{Mod}\,\Gamma$, then $e\colon L^0 \to A_i'$ is an equalizer in $\mathbf{Alg}\,\Sigma$.

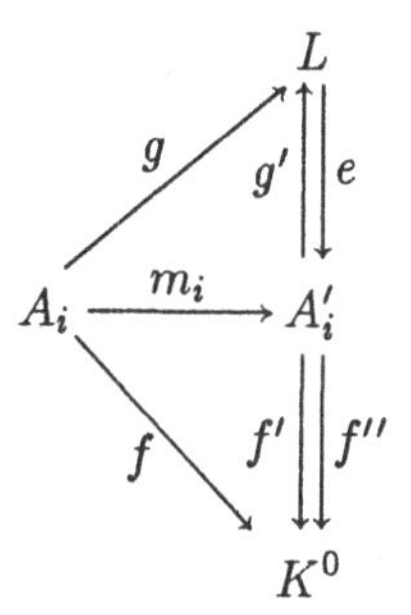

Since m_i merges f', f'', we have a Σ-homomorphism $g\colon A_i \to L^0$ with $m_i = e \cdot g$. By (2a) applied to L, there is a Σ-homomorphism $g'\colon A_i' \to L^0$ with $g = g' \cdot m_i$. Then

$$(e \cdot g') \cdot m_i = e \cdot g = m_i$$

implies $e \cdot g' = id$, since m_i is a reflection map. Consequently,

$$f' = f' \cdot e \cdot g' = f'' \cdot e \cdot g' = f''. \qquad \square$$

3.37 Remark. In the formalization of the concept of essentially algebraic theory we can proceed in a different (more complicated, but maybe more natural) way: instead of specifying a set Σ_t of "total" symbols, we specify a well-ordering $\leq$ of Σ. Then we provide, for each operation symbol σ, a set $\mathrm{Def}(\sigma)$ of operations in the signature $\{\tau \mid \tau \in \Sigma, \tau < \sigma\}$ which, again, determine the definition domain of σ. It is obvious that the concept we have introduced above is the following special case: choose any well-ordering $\leq$ in which each Σ_t-symbol precedes the first symbol of $\Sigma - \Sigma_t$.

The above theorem remains true under this generalization because (a) each locally presentable category can be described by the original (simpler) type of essentially algebraic theory and (b) the essentially algebraic theories using well-ordering lead to categories of models which are locally presentable. The latter is proved precisely as part I of the above proof.

Exercises

3.a Free Σ-algebras

(1) Free algebras of the one-sorted signature of one binary operation σ can be described as follows. A *binary tree* t labelled in a set X with $\sigma \notin X$ is a partial function from the set of all binary words into $X \cup \{\sigma\}$ such that

(a) the domain of definition of t is finite and non-empty,

(b) if $t(k_1 \dots k_{n-1} k_n)$ is defined, then $t(k_1 \dots k_{n-1}) = \sigma$.

Verify that the set of all binary trees is a Σ-algebra under the following operation assigning to a pair of trees t_0, t_1 the following tree t:

$$t(\emptyset) = \sigma, \quad t(0k_2 \dots k_n) = t_0(k_2 \dots k_n) \quad \text{and} \quad t(1k_2 \dots k_n) = t_1(k_2 \dots k_n).$$

Verify that this is a free Σ-algebra generated by X provided that each $x \in X$ is identified with the Σ-tree whose only value is $\emptyset \mapsto x$.

(2) Generalize (1) to many-sorted signatures.

3.b Epimorphisms in Alg Σ

(1) Prove that in **Alg** Σ, epimorphisms are precisely the homomorphisms surjective in each sort. (Hint: if A is a proper subalgebra of B, form a pushout P of two copies of $A \hookrightarrow B$. The two maps $B \to P$ are different because the elements of the two copies of $B \setminus A$ are never merged. Thus, for each epimorphism $e\colon A \to B$, the subalgebra $e(A)$ of B is not proper.)

(2) Prove that for each epimorphism $e\colon A \to B$ and each Σ-algebra C, given an S-sorted function $f\colon |B| \to |C|$ such that $f \cdot e\colon A \to C$ is a homomorphism, then $f\colon B \to C$ is also a homomorphism.

3.c Regular projectives. Find a dense subcategory formed by regular projectives, if possible, in the categories $\mathbf{Set}^S$, **Vec**, **Grp**, **Gra**, **Aut**, **CPO**.

3.d Finitely Generated and Finitely Presentable Algebras

(1) Prove that a vector space in **Vec** is finitely generated iff its dimension is finite. Conclude that finitely presentable and finitely generated vector spaces coincide.

(2) Find a finitely generated semigroup which is not finitely presentable. (Hint: consider generators x, y, z and equations $xy^n = xz^n$ for $n = 1, 2, 3, \dots$.)

3.e FP sketches and Varieties

(1) Construct an FP sketch describing the variety of semigroups.

(2) Construct an FP sketch describing deterministic sequential automata.

(3) Which variety of algebras is described by the FP sketch $\mathscr{S} = (\mathcal{A}, \mathbf{L}, \sigma)$, where $\mathcal{A}$ is the four-element Boolean algebra $\{0, a, \overline{a}, 1\}$, $\mathbf{L}$ consists of the two discrete diagrams $\emptyset$ and $\{a, \overline{a}\}$, $\sigma(\emptyset)$ is the empty cone with domain 1, and $\sigma(\{a, a'\})$ is the cone $a \leftarrow 0 \rightarrow a'$?

3.f Quasivarieties

(1) Let Σ be the one-sorted signature of a unary operation σ and nullary operations $\tau_0, \tau_1, \tau_2, \ldots$. Is the class of all Σ-algebras A satisfying

$$\left((\tau_{2n})_A = (\tau_{2n+1})_A \quad \text{for all } n \in \omega\right) \implies \sigma_A(x) = x$$

a (finitary) quasivariety? Is it an epireflective subcategory of $\mathbf{Alg}\,\Sigma$?

(2) Can the quasivariety of torsion-free groups be described by finitely many implications?

(3) For each many-sorted relational signature Σ find a quasivariety of algebras equivalent to $\mathbf{Rel}\,\Sigma$. (Hint: see Example 3.20(4)).

3.g Effective Equivalence Relations

(1) Verify that **Set** has effective equivalence relations.

(2) Prove that if $\mathcal{K}$ is a category with effective equivalence relations, then so is $\mathcal{K}^{\mathcal{A}}$ for each small category $\mathcal{A}$.

(3) Verify that **Pos** does not have effective equivalence relations.

3.h Absolute Coequalizers

(1) Verify that in every category a coequalizer

$$A \underset{g}{\overset{f}{\rightrightarrows}} B \xrightarrow{c} C$$

is absolute whenever there exist morphisms $\overline{f}\colon B \to A$ and $\overline{c}\colon C \to B$ with $f \cdot \overline{f} = id_B$, $c \cdot \overline{c} = id_C$, and $g \cdot \overline{f} = \overline{c} \cdot c$. (Hint: show that the given equations together with $c \cdot f = c \cdot g$ imply that c is a coequalizer of f and g. Every functor "preserves" such equations.)

(2) Prove that in $\mathbf{Set}^S$ every coequalizer of an equivalence relation is absolute.

(3) Conclude from (1) and (2) that the forgetful functor

$$U\colon \mathbf{Alg}\,\Sigma \to \mathbf{Set}^S$$

creates coequalizers of equivalence relations.

3.i Theory of $\mathcal{V}$-algebras

(1) For each variety $\mathcal{V}$ of one-sorted finitary algebras prove that the theory $\mathbf{Th}\,\mathcal{V}$ (see Definition 3.14) is isomorphic to the category whose objects are natural numbers, and whose morphisms from n to m are all m-tuples $(\tau_1, \dots, \tau_m)$ of terms in the variables $x_1, \dots, x_n$ (i.e., m-tuples of elements of $F\{x_1, \dots, x_n\}$) with composition defined by substitution as follows. Given morphisms

$$k \xrightarrow{(\tau_1,\dots,\tau_n)} n \xrightarrow{(\sigma_1,\dots,\sigma_m)} m$$

their composite is

$$(\tau_1, \dots, \tau_n) \cdot (\sigma_1, \dots, \sigma_m) = \big(\tau_1(\sigma_1, \dots, \sigma_m)), \dots, \tau_n(\sigma_1, \dots, \sigma_m)\big)$$

where $\tau_i \mapsto \tau_i(\sigma_1, \dots, \sigma_m)$ denotes the substitution of x_i by σ_i for $i = 1, \dots, m$.

(2) In an analogous way, describe $\mathbf{Th}\,\mathcal{V}$ for a variety of S-sorted finitary algebras as the category whose objects are expressions $s_1 \times \cdots \times s_n$ and morphisms from $s_1 \times \cdots \times s_m$ to $t_1 \times \cdots \times t_m$ are m-tuples $(\tau_1, \dots, \tau_m)$ where τ_i is an element of sort t_i in $F\{x_1^{s_1}, \dots, x_n^{s_n}\}$.

Historical Remarks

The father of universal algebra is Garret Birkhoff who introduced (one-sorted) algebras as sets endowed with operations, and who, inter alia, characterized equational classes of algebras as HSP-classes (Theorem 3.9) in [Birkhoff 1935]. The many-sorted approach, inspired by early computer science, was first formalized in [Birkhoff, Lipson 1970].

Categories of algebras presented by equations and implications were studied e.g. by [Mal'cev 1958], [Isbell 1964], [Linton 1966], and [Felscher 1968]. The characterization of quasivarieties in Theorem 3.22 is due to [Mal'cev 1956]. The abstract characterization of quasivarieties (Theorems 3.24 and 3.33) is due to [Isbell 1964], an abstract characterization of varieties was presented in [Lawvere 1963].

An entirely new view of algebraic categories was presented by F. W. Lawvere in his dissertation [Lawvere 1963]: an algebra is a finite-product-preserving functor, and a homomorphism is a natural transformation (Theorem 3.16). A generalization of Lawvere's algebraic theories to infinitary algebras is due to [Linton 1966].

Monads as a means of describing algebraic categories were first used by [Eilenberg, Moore 1965], and a characterization of monadic categories over **Set** was presented in [Linton 1966].

The idea of essentially algebraic category stems from [Freyd 1972], and it was studied in [Adámek, Herrlich, Rosický 1988] and [Adámek, Herrlich, Tholen 1989]. The characterization in Theorem 3.36 has not been, to our knowledge, published before. Further information about varieties of partial algebras can be obtained in [Reichel 1984].

Chapter 4

Injectivity Classes

In the present chapter we study two special classes of accessible categories: weakly locally presentable categories, which are the accessible categories with products, and locally multipresentable categories, which are the accessible categories with connected limits. Recall that locally presentable categories can be characterized by orthogonality (in the sense that they are just the small-orthogonality classes in categories $\mathbf{Set}^{\mathcal{A}}$, see Theorem 1.46). We will show that weakly locally presentable categories can be characterized by injectivity (Theorem 4.11), and locally multipresentable categories by cone-orthogonality (Theorem 4.30). And, while locally presentable categories are precisely the categories sketchable by limit-sketches, weakly locally presentable categories are sketchable by limit–epi sketches (i.e., sketches whose models are set-valued functors preserving certain limits and certain epimorphisms) (Theorem 4.13), and locally multipresentable categories are sketchable by limit–coproduct sketches (i.e., models preserve certain limits and certain products) (Theorem 4.32).

Orthogonality w.r.t. a morphism can be generalized to orthogonality (or injectivity) w.r.t. a cone as follows: an object K is orthogonal if every morphism from the domain of the cone to K has a unique factorization through a unique member of the cone. (And K is injective if every morphism from the domain of the cone to K has a factorization through some member.) Accessible categories are then described as precisely the categories of objects injective w.r.t. a set of cones in some locally presentable category (Corollary 4.18). (This is closely related to the characterization of accessible categories as cone-reflective subcategories of the categories $\mathbf{Set}^{\mathcal{A}}$, see Corollary 2.54.) Locally multipresentable categories are described as precisely the categories of objects orthogonal w.r.t. a set of cones in a locally presentable category (Theorem 4.30).

4.A Weakly Locally Presentable Categories

Injectivity

The following concept "relieves" that of orthogonality by omitting the uniqueness requirement:

4.1 Definition

(1) An object K is said to be *injective* with respect to a morphism

$$m: A \to A'$$

provided that for each morphism $f: A \to K$ there exists a morphism $f': A' \to K$ such that the triangle

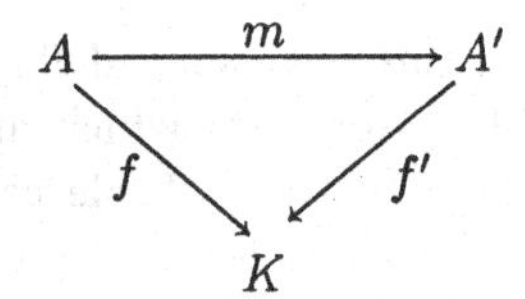

commutes.

(2) For each class $\mathcal{M}$ of morphisms in a category $\mathcal{K}$ we denote by $\mathcal{M}$-**Inj** the full subcategory of $\mathcal{K}$ of all objects injective w.r.t. each morphism in $\mathcal{M}$.

Conversely, a full subcategory of $\mathcal{K}$ is called a *(small-) injectivity class* provided that it has the form $\mathcal{M}$-**Inj** for a (small) collection $\mathcal{M}$ of morphisms in $\mathcal{K}$.

Observe that in the case where $m: A \to A'$ is a monomorphism, we can say that injectivity is the possibility of extending any morphism from the subobject A of A' to the whole A'.

4.2 Examples

(1) A poset P is injective w.r.t $\mathcal{M}$ = strong monomorphisms in the category **Pos** iff P is a complete lattice.

In fact, we have seen in Example 1.33(5) that each poset P is a strong subobject of the complete lattice $I(P)$ of ideals of P. If P is injective with respect to the embedding $P \hookrightarrow I(P)$, then this embedding is a split monomorphism, and given $h: I(P) \to P$ with $h\{x \in P \mid x \leq p\} = p$, then each subset $M \subseteq P$ has a join, viz., $h(\overline{M})$, where $\overline{M}$ is the smallest ideal containing M.

Conversely, if P is a complete lattice, then for every strong monomorphism $m\colon A \to A'$ we can extend each order-preserving function

$$f\colon A \to P$$

to A' by the following rule

$$f'(a) = \bigvee \{\, f(x) \mid x \in A,\ m(x) \leq a \,\} \qquad \text{for all } a \in A'.$$

(2) Injective distributive lattices ($\mathcal{M}$ = monomorphisms) are precisely the complete Boolean algebras (see [Banaschewski, Bruns 1968]).

(3) Injective Abelian groups ($\mathcal{M}$ = monomorphisms) are precisely the divisible groups. This is a small-injectivity class given by $\mathcal{M} = \hom(\mathbb{Z}, \mathbb{Z})$. Several other injectivity concepts are investigated in the category of Abelian groups, e.g., if $\mathcal{M}$ = ω-pure embeddings, then $\mathcal{M}$-injective groups are the algebraically compact groups (see [Fuchs 1970]).

(4) If $\mathcal{M}$ is a collection of epimorphisms, then $\mathcal{M}^{\perp} = \mathcal{M}\text{-}\mathbf{Inj}$. Thus, for example, posets form a small-injectivity class of graphs (see Example 1.33(4)).

(5) In the category **Top** of topological spaces consider the embedding m of the discrete space $A = \{0,1\}$ to the unit interval $A' = [0,1]$. The $\{m\}$-injective spaces are precisely the pathwise connected spaces (i.e., those in which each pair of points can be connected by an arc).

In contrast, the class of all connected spaces is not an injectivity class (see Exercise 4.a).

4.3 Proposition. *Every injectivity class of a category $\mathcal{K}$ is closed in $\mathcal{K}$ under products and split subobjects.*

PROOF. Given a collection $\mathcal{M}$ of morphisms in $\mathcal{K}$, we are to show that

(i) for each set X_i $(i \in I)$ of $\mathcal{M}$-injective objects the product $\prod_{i\in I} X_i$ is $\mathcal{M}$-injective,

(ii) for each split monomorphism $u\colon X \to Y$, if Y is $\mathcal{M}$-injective, then X is $\mathcal{M}$-injective.

Let $m\colon A \to A'$ be a morphism in $\mathcal{M}$. In (i), given a morphism $f\colon A \to \prod_{i\in I} X_i$, each component of f can be extended to f_i' along m (since X_i is $\mathcal{M}$-injective), and the morphism $f'\colon A' \to \prod_{i\in I} X_i$ whose components are f_i' extends f along m. In (ii), given a morphism $f\colon A \to X$, there exists $f'\colon A' \to Y$ with $f' \cdot m = u \cdot f$ (since Y is $\mathcal{M}$-injective) and we just compose f' with any morphism v such that $v \cdot u = id_X$. ☐

4.4 Remarks

(1) In a category with pushouts each orthogonality class is an injectivity class. In fact, for each morphism $m: A \to A'$ we can form the pushout of m and m, and then factorize $id_{A'}$, $id_{A'}$:

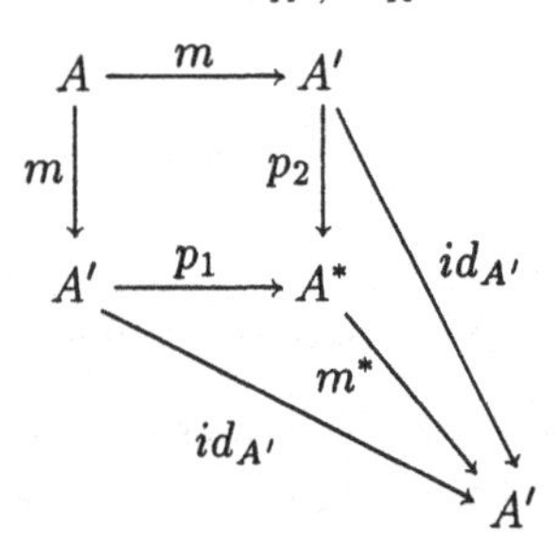

Given a class $\mathcal{M}$ of morphisms, put $\mathcal{M}^* = \mathcal{M} \cup \{m^* \mid m \in \mathcal{M}\}$. Then

$$\mathcal{M}^\perp = \mathcal{M}^*\text{-}\mathbf{Inj}.$$

(2) In a category with pushouts we thus have the following hierarchy of conditions on full, isomorphism-closed subcategories:

$$\begin{array}{ccc} \text{small-orthogonality class} & \Longrightarrow & \text{small-injectivity class} \\ \Downarrow & & \Downarrow \\ \text{orthogonality class} & \Longrightarrow & \text{injectivity class} \\ \Downarrow & & \Downarrow \\ \text{closed under limits} & \Longrightarrow & \text{closed under products} \\ & & \text{and split subobjects} \end{array}$$

4.5 Definition. Let $\mathcal{A}$ be a subcategory of $\mathcal{K}$. By a *weak reflection* of a $\mathcal{K}$-object K in $\mathcal{A}$ is meant a morphism $r: K \to K^*$ with K^* in $\mathcal{A}$, such that every morphism $f: K \to A$ with A in $\mathcal{A}$ factorizes (not necessarily uniquely) through r. If each $\mathcal{K}$-object has a weak reflection, then $\mathcal{A}$ is said to be *weakly reflective*.

Remarks

(1) The following implications

$$\text{reflective} \Longrightarrow \text{weakly reflective} \Longrightarrow \text{cone-reflective}$$

are obvious.

(2) For subcategories of a category with products we have

$$\text{cone-reflective and closed under products} \Longrightarrow \text{weakly reflective.}$$

(In fact, for each cone-reflection $(K \xrightarrow{r_i} A_i)_{i \in I}$ the morphism $(r_i): K \to \prod_{i \in I} A_i$ is a weak reflection.)

(3) Any weakly reflective subcategory $\mathcal{A}$ of $\mathcal{K}$ closed under split subobjects is an injectivity class, therefore it is closed under products. This is not true for weakly reflective subcategories in general (see Exercise 4.c(3)).

(4) For locally presentable categories we will later see that, under certain set-theoretical assumptions, every injectivity class is weakly reflective, and every class closed under products and split subobjects is an injectivity class, see Theorem 6.26.

4.6 Examples

(1) Posets with a greatest element form a weakly reflective, full subcategory of **Pos**.

(2) The full subcategory of **Pos** formed by complete lattices (cf. Example 4.2(1)) is weakly reflective in **Pos**. MacNeille completions (see Exercise 4.c(2)) are weak reflections. This subcategory is not reflective.

Analogously, complete Boolean algebras form a full, weakly reflective subcategory in the category of distributive lattices.

(3) Divisible Abelian groups form a weakly reflective subcategory in **Ab**. Divisible hulls are weak reflections. Again, this subcategory is not reflective.

(4) The cone-reflective subcategory of unary algebras with a cycle (see Example 2.52(3)) is not weakly reflective in the category of unary algebras (because it is not closed under products).

Small-injectivity Classes

We know that, for each locally presentable category $\mathcal{K}$, the small-orthogonality classes in $\mathcal{K}$ are precisely the reflective, accessibly embedded subcategories (see Theorem 1.39 and Remark 1.35). In the present section we will prove that the small-injectivity classes in $\mathcal{K}$ are precisely the weakly reflective, accessibly embedded subcategories.

4.7 Proposition. *Every small-injectivity class in an accessible category is an accessibly embedded, accessible subcategory.*

PROOF. Let $\mathcal{M}$ be a set of morphisms in an accessible category $\mathcal{K}$. There exists a regular cardinal λ such that $\mathcal{K}$ is λ-accessible and each domain and codomain of an $\mathcal{M}$-morphism is a λ-presentable object.

I. $\mathcal{M}$-**Inj** is closed under λ-directed colimits: This is quite analogous to the proof in Proposition 1.35. If $(K_t \xrightarrow{k_t} K)_{t \in T}$ is a λ-directed colimit of $\mathcal{M}$-injective objects K_t, then for every $m: A \to A'$ in $\mathcal{M}$, each morphism $f: A \to K$ has a factorization $f = k_t \cdot f_1$ for some $f_1: A \to K_t$ and, choosing $f_1': A' \to K_t$ with $f_1 = f_1' \cdot m$, we get $f' = k_t \cdot f_1'$ with $f = f' \cdot m$. Thus K is $\mathcal{M}$-injective.

II. $\mathcal{M}$-**Inj** is accessible: By Corollary 2.36 it is sufficient to observe that $\mathcal{M}$-**Inj** is clearly closed under λ-pure subobjects. □

Example. The category of complete lattices and order-preserving maps is not accessibly embedded into **Pos**, thus, it does not form a small-injectivity class (although it is an injectivity class, see Example 4.2(1)).

4.8 Theorem. *Let $\mathcal{A}$ be a full subcategory of a locally presentable category $\mathcal{K}$. Then the following conditions are equivalent:*

(i) *$\mathcal{A}$ is a small-injectivity class in $\mathcal{K}$;*

(ii) *$\mathcal{A}$ is accessible, accessibly embedded, and closed under products in $\mathcal{K}$;*

(iii) *$\mathcal{A}$ is weakly reflective and accessibly embedded in $\mathcal{K}$.*

PROOF. (i) $\Rightarrow$ (ii) follows from Propositions 4.7 and 4.3.

(ii) $\Rightarrow$ (i). Let $\mathcal{A}$ fulfil (ii). There is a regular cardinal λ such that $\mathcal{K}$ and $\mathcal{A}$ are λ-accessible categories and the inclusion of $\mathcal{A}$ into $\mathcal{K}$ preserves λ-directed colimits and λ-presentable objects (see Theorem 2.19). Let K be a λ-presentable object of $\mathcal{K}$ and let $(f_i: K \to A_i)_{i \in I}$ be a solution set for K (see Corollary 2.45). The morphism

$$f = (f_i): K \to \prod_{i \in I} A_i$$

is, clearly, a weak reflection of K in $\mathcal{A}$. There is a factorization of f through some $r: K \to K^*$, where K^* is λ-presentable in $\mathcal{A}$. It is evident that r is a weak reflection of K in $\mathcal{A}$.

We have verified that every λ-presentable object K in $\mathcal{K}$ has a weak reflection $r_K: K \to K^*$ with K^* λ-presentable in $\mathcal{A}$. We will now prove that

$$\mathcal{A} = \{\, r_K \mid K \in \mathbf{Pres}_\lambda \mathcal{K} \,\}\text{-}\mathbf{Inj}.$$

In fact, each object of $\mathcal{A}$ is, obviously, $\{r_K\}$-injective. Conversely, let A be an $\{r_K\}$-injective object of $\mathcal{K}$. Since $\mathcal{K}$ is locally λ-presentable, A is a colimit of its canonical diagram $D\colon \mathcal{D} \to \mathcal{K}$ w.r.t. $\mathbf{Pres}_\lambda \mathcal{K}$, and D is λ-filtered, see Proposition 1.22. Next, let D^* be the canonical diagram of A w.r.t. $\{ K^* \mid K \in \mathbf{Pres}_\lambda \mathcal{K} \}$. Then D^* is a subdiagram of D (since K^* is λ-presentable in $\mathcal{K}$ for each $K \in \mathbf{Pres}_\lambda \mathcal{K}$) which is cofinal (see 0.11) because for each $\mathcal{D}$-object $K \xrightarrow{f} A$ the r_K-injectivity of A implies that there is a $\mathcal{D}^*$-object $K^* \xrightarrow{f^*} A$ with $f = f^* \cdot r_K$ (i.e., such that $r_K\colon f \to f^*$ is a $\mathcal{D}$-morphism). It follows that D^* is λ-filtered (see Exercise 1.o(3)), and that A is a colimit of D^*. Therefore, A belongs to $\mathcal{A}$ because $\mathcal{A}$ is closed under λ-filtered colimits in $\mathcal{K}$ (see Remark 1.21).

(ii) $\Rightarrow$ (iii). This follows from Corollary 2.45 and Remark 4.5(2).

(iii) $\Rightarrow$ (ii). This follows from Theorem 2.53 and Remark 4.5(3), see also Observation 2.4. □

Remark. The above proof of the equivalence of conditions (i) and (ii) is independent of Theorem 2.53.

Weakly Locally Presentable Categories

Just as initiality generalizes orthogonality, we will now generalize cocompleteness and local presentability. Recall that a category is locally presentable iff it is accessible and cocomplete.

4.9 Definition

(1) By a *weak colimit* of a diagram D is meant a compatible cocone of D through which every compatible cocone of D factorizes (not necessarily uniquely).

    A category is *weakly cocomplete* if every diagram in it has a weak colimit.

(2) A category is called *weakly locally λ-presentable* provided that it is λ-accessible and weakly cocomplete.

Remark. Any weakly reflective, full subcategory $\mathcal{A}$ of a cocomplete category $\mathcal{K}$ is a weakly cocomplete: a weak colimit of a diagram $D\colon \mathcal{D} \to \mathcal{A}$ is given by a weak reflection of a colimit of D in $\mathcal{K}$.

4.10 Examples

(1) The category of non-empty sets and mappings is weakly locally finitely presentable: it has colimits of all non-empty diagrams, and any object is weakly initial.

(2) The category of divisible Abelian groups is weakly locally finitely presentable.

(3) The category of CPO's with bottom and continuous (not necessarily strict) maps is weakly locally ω_1-presentable.

4.11 Characterization Theorem. *The following conditions on a category $\mathcal{K}$ are equivalent:*

(i) *$\mathcal{K}$ is weakly locally presentable;*

(ii) *$\mathcal{K}$ is an accessible category with products;*

(iii) *$\mathcal{K}$ is equivalent to a weakly reflective, accessibly embedded subcategory of $\mathbf{Set}^{\mathcal{A}}$ for some small category $\mathcal{A}$;*

(iv) *$\mathcal{K}$ is equivalent to a small-injectivity class in a locally presentable category.*

PROOF. (i) ⇒ (iii). Let $\mathcal{K}$ be a λ-accessible category with weak colimits and put $\mathcal{A} = \mathbf{Pres}_\lambda\, \mathcal{K}$. Then the canonical functor

$$E \colon \mathcal{K} \to \mathbf{Set}^{\mathcal{A}^{op}}$$

is full, faithful, and λ-accessible (see Propositions 2.8 and 1.26). Each object of $\mathbf{Set}^{\mathcal{A}^{op}}$ is a colimit of objects from $E(\mathcal{K})$ (because a set-valued functor is a colimit of hom-functors). We will prove that if a diagram $D \colon \mathcal{D} \to E(\mathcal{K})$ has a colimit $\left(Dd \xrightarrow{c_d} C\right)_{d \in \mathcal{D}^{obj}}$ in $\mathbf{Set}^{\mathcal{A}^{op}}$, then C has a weak reflection in $E(\mathcal{K})$. In fact, let $\left(Dd \xrightarrow{\overline{c}_d} \overline{C}\right)$ be a weak colimit of D in $E(\mathcal{K})$. There exists $r \colon C \to \overline{C}$ with $r \cdot c_d = \overline{c}_d$ for all $d \in \mathcal{D}^{obj}$. Then r is a weak reflection of C in $E(\mathcal{K})$, since any morphism $f \colon C \to A$ with A in $E(\mathcal{K})$ defines a compatible cocone $\left(Dd \xrightarrow{f \cdot c_d} A\right)_{d \in \mathcal{D}^{obj}}$ of D in $E(\mathcal{K})$. The unique $\overline{f} \colon \overline{C} \to A$ with $\overline{f} \cdot \overline{c}_d = f \cdot c_d$ for all $d \in \mathcal{D}^{obj}$ fulfils $\overline{f} \cdot r = f$ because $\overline{f} \cdot r \cdot c_d = f \cdot c_d$ for all $d \in \mathcal{D}^{obj}$.

(iii) ⇒ (iv) follows from Theorem 4.8.

(iv) ⇒ (ii) follows from Propositions 4.7 and 4.3.

(ii) $\Rightarrow$ (i). Let $\mathcal{K}$ be a λ-accessible category with products. Then $\mathcal{K}$ is equivalent to a full subcategory $E(\mathcal{K})$ of $\mathbf{Set}^{\mathcal{A}^{op}}$ closed under products and λ-directed colimits, where $\mathcal{A} = \mathbf{Pres}_\lambda \mathcal{K}$ (cf. Propositions 2.8 and 1.26). Hence $E(\mathcal{K})$ is weakly reflective in $\mathbf{Set}^{\mathcal{A}^{op}}$ (by Theorem 4.8), and thus it has weak colimits (see Remark 4.9). □

Remark. The above theorem cannot be specialized to a given λ (in contrast to Theorem 1.39). In fact, there are full, weakly reflective subcategories of locally finitely presentable categories which are closed under directed colimits, but are not finitely accessible.

Consider the following embedding of graphs:

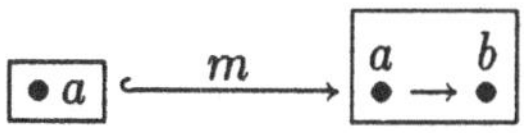

$$\boxed{\bullet\, a} \xhookrightarrow{\quad m \quad} \boxed{\overset{a}{\bullet} \to \overset{b}{\bullet}}$$

Then $\{m\}$-**Inj** is the full subcategory of **Gra** over all graphs such that any vertex is a source of an arrow. This subcategory is weakly reflective and closed under directed colimits in **Gra** (see Theorem 4.8) but is not finitely accessible (see Remark 2.59).

4.12 Definition. By a *limit–epi sketch* is meant a sketch $\mathscr{S} = (\mathcal{A}, \mathbf{L}, \mathbf{C}, \sigma)$ such that each $\mathbf{C}$-diagram is a span

$$\begin{array}{ccc} A & \xrightarrow{\ e\ } & B \\ {\scriptstyle e}\downarrow & & \\ B & & \end{array}$$

to which σ assigns the cocone

$$\begin{array}{ccc} A & \xrightarrow{\ e\ } & B \\ {\scriptstyle e}\downarrow & & \downarrow{\scriptstyle id_B} \\ B & \xrightarrow[id_B]{} & B \end{array}$$

Thus, models of a limit–epi sketch are functors turning specified cones to limits, and specified morphisms to epimorphisms.

Remark. The sketches in Examples 2.57(1) and (3) are limit–epi sketches.

4.13 Theorem. *A category is weakly locally presentable iff it is sketchable by a limit–epi sketch.*

PROOF. I. Let $\mathcal{K}$ be a λ-accessible category with weak colimits and put $\mathcal{B} = \mathbf{Pres}_\lambda \mathcal{K}$. We first prove that any λ-small diagram $D\colon \mathcal{D} \to \mathcal{B}$ has a weak colimit in $\mathcal{B}$. Let $(Dd \xrightarrow{k_d} K)_{d\in\mathcal{D}^{\mathrm{obj}}}$ be a weak colimit in $\mathcal{K}$. Since $\mathcal{K}$ is λ-accessible, K is a λ-directed colimit of a diagram D^* of objects $D_i^* \in \mathcal{B}$ and morphisms $d_{i,j}^*\colon D_i^* \to D_j^*$, say $\operatorname{colim} D^* = (D_i^* \xrightarrow{k_i^*} K)_{i\in I}$. For each $d \in \mathcal{D}^{\mathrm{obj}}$, since Dd is λ-presentable, the morphism k_d factorizes through $k_{i(d)}^*$ for some $i(d) \in I$. The number of objects d of $\mathcal{D}$ is less than λ, and I is λ-directed, thus, there is an upper bound $i \in I$ of all $i(d)$'s. Then each k_d factorizes as

$$k_d = k_i^* \cdot h_d \qquad \text{for some} \quad h_d\colon Dd \to D_i^*.$$

Given a morphism $\delta\colon d_1 \to d_2$ in $\mathcal{D}$, the equation

$$k_i^* \cdot \left(h_{d_2} \cdot D\delta\right) = k_{d_2} \cdot D\delta = k_{d_1} = k_i^* \cdot h_{d_1}$$

implies, since Dd_1 is λ-presentable, that there exists $i(\delta) \geq i$ such that $d_{i,i(\delta)}^* \cdot h_{d_2} \cdot D\delta = d_{i,i(\delta)}^* \cdot h_{d_1}$. Again, there is an upper bound $j \in I$ of all $i(\delta)$'s, and then the morphisms

$$\overline{h}_d = d_{i,j}^* \cdot h_d\colon Dd \to D_j^*$$

form a compatible cocone of D: for each $\delta\colon d_1 \to d_2$ in $\mathcal{D}$ we have

$$\begin{aligned}
\overline{h}_{d_2} \cdot D\delta &= d_{i,j}^* \cdot h_{d_2} \cdot D\delta \\
&= d_{i(\delta),j}^* \cdot d_{i,i(\delta)}^* \cdot h_{d_2} \cdot D(\delta) \\
&= d_{i(\delta),j}^* \cdot d_{i,i(\delta)}^* \cdot h_{d_1} \\
&= d_{i,j}^* \cdot h_{d_1} \\
&= \overline{h}_{d_1}.
\end{aligned}$$

The cocone $\left(Dd \xrightarrow{\overline{h}_d} D_j^*\right)$ is a weak colimit of D because the given weak colimit $(Dd \xrightarrow{k_d} K)$ factorizes through it: we have $k_d = k_j^* \cdot \overline{h}_d$ for each d. The object D_j^* is λ-presentable.

We are going to present a limit-epi sketch $\mathscr{S}^*$ for $\mathcal{K}$ by modifying the sketch $\mathscr{S}$ of Remark 2.58. In that sketch $\mathscr{S} = (\mathcal{A}, \mathbf{L}, \mathbf{C}, \sigma)$, every object of $\mathcal{A}$ of the form $A(D)$, where $D \in \mathbf{L}$, is the domain of the limit cone $\sigma(D)$. Put $\sigma(D) = \left(A(D) \xrightarrow{a_{D,d}} Dd\right)_{d\in\mathcal{D}^{\mathrm{obj}}}$, where $D\colon \mathcal{D} \to Y(\mathcal{B}^{\mathrm{op}})$ is a λ-small diagram. As proved above, every λ-small diagram in $\mathcal{B}$ has a weak colimit in $\mathcal{B}$—thus, every λ-small diagram in $Y(\mathcal{B}^{\mathrm{op}}) \cong \mathcal{B}^{\mathrm{op}}$ has a weak limit in $Y(\mathcal{B}^{op})$. Let us choose a weak limit $\left(B(D) \xrightarrow{b_{D,d}} Dd\right)_{d \in \mathcal{D}^{obj}}$ of D with $B(D)$ in $Y(\mathcal{B}^{op})$. Denote by

$$e_D : B(D) \to A(D)$$

the unique factorization (defined by $a_{D,d} \cdot e_D = b_{D,d}$). The sketch $\mathscr{S}^*$ is obtained from $\mathscr{S}$ by substituting each of the canonical diagrams of $A(D)$ (i.e., the diagrams from $\mathbf{C}$) by the span $A(D) \xleftarrow{e_D} B(D) \xrightarrow{e_D} A(D)$. That is, $\mathscr{S}^*$ is the limit–epi sketch

$$\mathscr{S}^* = (\mathcal{A}, \mathbf{L}, \mathbf{C}^*, \sigma^*)$$

where $\mathbf{C}^*$ is the set of all spans $A(D) \xleftarrow{e_D} B(D) \xrightarrow{e_D} A(D)$ for $D \in \mathbf{L}$, and $\sigma^*(D) = \sigma(D)$ for each $D \in \mathbf{L}$. To conclude the proof, we show that $\mathscr{S}$ and $\mathscr{S}^*$ have the same models:

$$\mathbf{Mod}\,\mathscr{S}^* = \mathbf{Mod}\,\mathscr{S} \qquad (\approx \mathcal{K}, \text{ see Remark 2.58}).$$

Let $G\colon \mathcal{A} \to \mathbf{Set}$ be a model of $\mathscr{S}$. For each $D \in \mathbf{L}$ we know that G preserves the canonical colimit of $A(D)$ w.r.t $Y(\mathcal{B}^{op})$. Thus, to prove that Ge_D is an epimorphism, it is sufficient to prove that each morphism $c\colon YC \to A(D)$ with C in $\mathcal{B}^{op}$ factorizes through e_D. To this end, factorize the following compatible cone $\left(YC \xrightarrow{a_{D,d} \cdot c} Dd\right)_{d \in \mathcal{D}^{obj}}$ of D through the above weak limit: there exists $\bar{c}\colon YC \to B(D)$ with

$$b_{D,d} \cdot \bar{c} = a_{D,d} \cdot c \qquad \text{for each } d \in \mathcal{D}^{obj}.$$

Then $a_{D,d} \cdot c = a_{D,d} \cdot e_D \cdot \bar{c}$ for all $d \in \mathcal{D}^{obj}$, which implies $c = e_D \cdot \bar{c}$.

Conversely, let $G\colon \mathcal{A} \to \mathbf{Set}$ be a model of $\mathscr{S}^*$. By Lemma 2.24, to prove that G is a model of $\mathscr{S}$ it is sufficient to verify that its domain restriction F to $Y(\mathcal{B}^{op}) \cong \mathcal{B}^{op}$ is a λ-directed colimit of hom-functors. This can be done analogously to the proof of Lemma 2.24: we make use of the fact that Ge_D is an epimorphism and (in the last part of the proof) we choose, for the given $a \in G(A(D))$, an element $b \in G(B(D))$ with $Ge_D(b) = a$.

II. Let $\mathscr{S} = (\mathcal{A}, \mathbf{L}, \mathbf{C}, \sigma)$ be a limit–epi sketch. The category $\mathbf{Mod}\,\mathscr{S}$ is accessible (by Theorem 2.60) and has products (in fact it is closed under products in $\mathbf{Set}^{\mathcal{A}}$) because any product of epimorphisms in $\mathbf{Set}$ is an epimorphism. Hence, $\mathbf{Mod}\,\mathscr{S}$ is locally weakly presentable by Theorem 4.11. □

Remark. Again, the theorem cannot be stated for a fixed λ (in contrast to Remark 1.52): the sketch from Remark 2.59 is a limit–epi sketch with finite limit-diagrams whose category of models is not weakly locally finitely presentable.

4.B A Characterization of Accessible Categories

In the present section we prove that accessible categories are precisely the small-cone-injectivity classes of locally presentable categories.

4.14. Recall that a *cone* is a set of morphisms with a common domain. The set can be empty (in which case the cone is given by the object which is considered as the domain).

Definition

(1) An object K is said to be *injective* with respect to a cone $(A \xrightarrow{m_i} A_i)_{i \in I}$ provided that for each morphism $f: A \to K$ there exists $i \in I$ and a morphism $f': A_i \to K$ with $f = f' \cdot m_i$.

(2) For each class $\mathcal{M}$ of cones in a category $\mathcal{K}$ we denote by $\mathcal{M}$-**Inj** the full subcategory of $\mathcal{K}$ of all objects injective w.r.t. each cone in $\mathcal{M}$.

Conversely, a full subcategory of $\mathcal{K}$ is called a *(small-) cone-injectivity class* provided that it has form $\mathcal{M}$-**Inj** for a (small) collection of cones in $\mathcal{K}$.

Remark. The injectivity of K w.r.t. the empty cone A means that there is no morphism $A \to K$.

Observe that every cone-injectivity class is closed under split subobjects.

4.15 Examples

(1) Every injectivity class is a cone-injectivity class.

(2) An Abelian group is a torsion group iff it is injective w.r.t. the following cone:

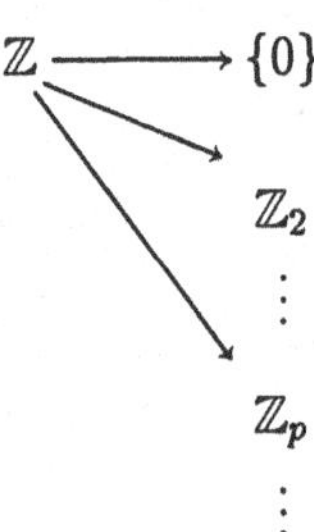

(formed by the trivial homomorphism and the canonical quotient maps of the additive group $\mathbb{Z}$ of integers to $\mathbb{Z}_p$, where p is prime). Thus, torsion groups form a small-cone-injectivity class in **Ab**.

(3) Analogously, fields form a small-cone-injectivity class in the category **Rng** of commutative unitary rings. Here we use the ring $\mathbb{Z}[x]$ of integer polynomials, which we first factor canonically onto $\mathbb{Z}_p[x]$ (p a prime), and then we take the field $\mathbb{Z}_p[x]^*$ of fractions over it, thus getting a canonical homomorphism $h_p \colon \mathbb{Z}[x] \to \mathbb{Z}_p[x]^*$. Moreover, we denote by $h_0 \colon \mathbb{Z}[x] \to \mathbb{Q}[x]^*$ the canonical embedding into the field of fractions over rational polynomials.

A ring has inverses for all non-zero elements iff it is injective to the following cone:

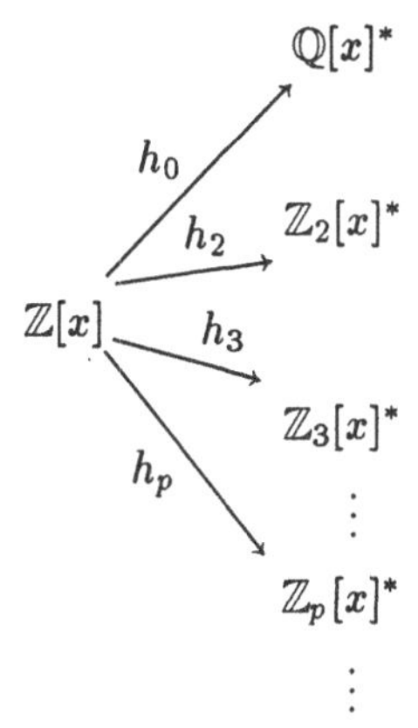

The condition $0 \neq 1$ for fields is expressed by injectivity to the empty cone with the domain $\{0\}$.

(4) Linearly ordered sets form a small-cone-injectivity class in **Pos**, presented by the following single cone:

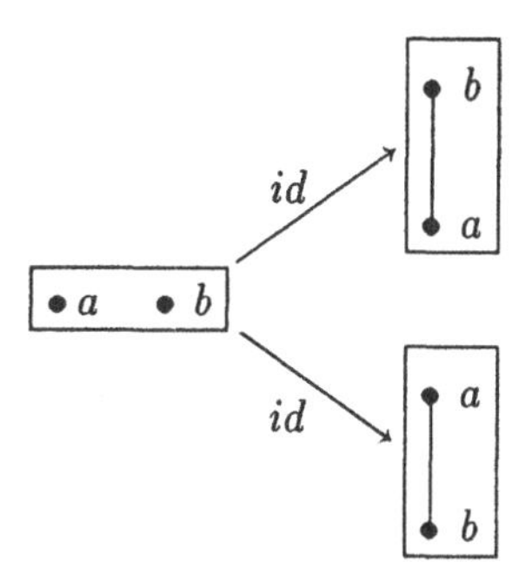

(5) Connected graphs (i.e., non-empty graphs in which each pair of nodes can be connected by a directed path) form a small-cone-injectivity class

in **Gra**, presented by the unique morphism from the empty graph to any discrete non-empty graph, and by the following cone:

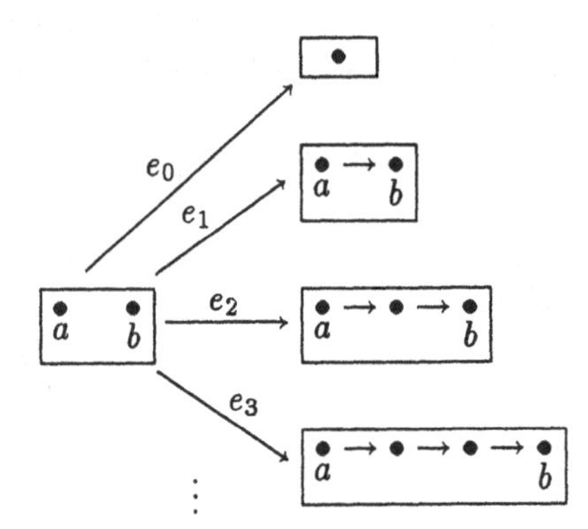

4.16 Proposition. *Every small-cone-injectivity class in an accessible category is an accessible, accessibly embedded subcategory.*

The proof is quite analogous to that of Proposition 4.7.

4.17 Theorem. *For each locally presentable category $\mathcal{K}$ the small-cone-injectivity classes in $\mathcal{K}$ are precisely the accessible, accessibly embedded subcategories of $\mathcal{K}$.*

PROOF. We know that every small-cone-injectivity class in $\mathcal{K}$ is accessible and accessibly embedded by Proposition 4.16. Conversely, assume that $\mathcal{A}$ is an accessible, accessibly embedded subcategory of a locally presentable category $\mathcal{K}$. There is a regular cardinal λ such that $\mathcal{K}$ and $\mathcal{A}$ are λ-accessible categories, and the inclusion of $\mathcal{A}$ into $\mathcal{K}$ preserves λ-directed colimits and λ-presentable objects (see Theorem 2.19). Let K be a λ-presentable object in $\mathcal{K}$, and $(f_i\colon K \to A_i)_{i\in I}$ be a solution set for K (see Corollary 2.45). For any $i \in I$ there is a factorization of f_i through some $r_{K,i}\colon K \to K_i^*$ where K_i^* is λ-presentable in $\mathcal{A}$. Denote the cone $\left(r_{K,i}\right)_{i\in I}$ by r_K. Then

$$\mathcal{A} = \{\, r_K \mid K \in \mathbf{Pres}_\lambda\, \mathcal{K}\,\}\text{-}\mathbf{Inj}\,.$$

The proof is the same as in Theorem 4.8. □

4.18 Corollary. *A category is accessible iff it is equivalent to a small-cone-injectivity class in $\mathbf{Set}^{\mathcal{A}}$ for some small category $\mathcal{A}$.*

This follows from Theorem 4.17 and the fact that each λ-accessible category $\mathcal{K}$ is equivalent to the image of the canonical functor into $\mathbf{Set}^{\mathcal{A}}$ (Proposition 2.8) for $\mathcal{A} = (\mathbf{Pres}_\lambda\, \mathcal{K})^{\mathrm{op}}$. □

4.C Locally Multipresentable Categories

The concept of orthogonality has been generalized above in two steps: first to injectivity (by giving up the uniqueness) and then to cone-injectivity (by working with cones rather than single arrows). In the present section we take the second step alone: we consider orthogonality to a cone, and we characterize the small-cone-orthogonality classes of locally presentable categories as precisely the accessible categories with connected limits, or equivalently, the accessible categories with multicolimits. (This corresponds nicely to the characterization theorem 4.11.) Accessible categories with multicolimits are called locally multipresentable categories.

Cone-orthogonality Classes

4.19 Definition

(1) An object K is said to be *orthogonal* to a cone $(A \xrightarrow{m_t} A_t)_{t\in T}$ provided that for each morphism $f: A \to K$ there exists a unique $t \in T$ such that f factorizes through m_t and that the factorization is unique. That is, there exists a unique morphism $f': A_t \to K$ such that the triangle

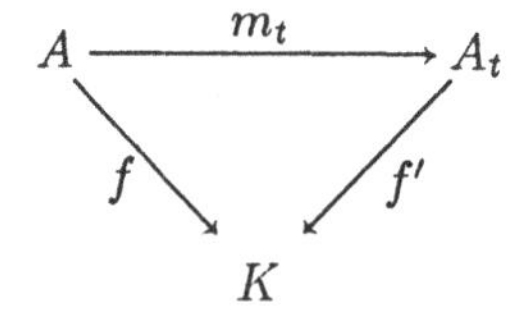

commutes.

(2) For each class $\mathcal{M}$ of cones in a category $\mathcal{K}$ we denote by $\mathcal{M}^\perp$ the full subcategory of $\mathcal{K}$ of all objects orthogonal to each cone in $\mathcal{M}$.

Conversely, a full subcategory of $\mathcal{K}$ is called a *(small-) cone-orthogonality class* provided that it has the form $\mathcal{M}^\perp$ for a (small) collection $\mathcal{M}$ of cones in $\mathcal{K}$.

4.20 Examples

(1) Every orthogonality class is, of course, a cone-orthogonality class.

(2) Fields and torsion Abelian groups are small-cone-orthogonality classes, see Examples 4.15.

(3) Let $\mathbf{Pos}^@$ denote the category of strict partially ordered sets (i.e., sets with an antireflexive, antisymmetric, transitive relation) and strict

order-preserving maps. The class of linearly ordered sets is a small-cone-orthogonality class presented by the following single cone:

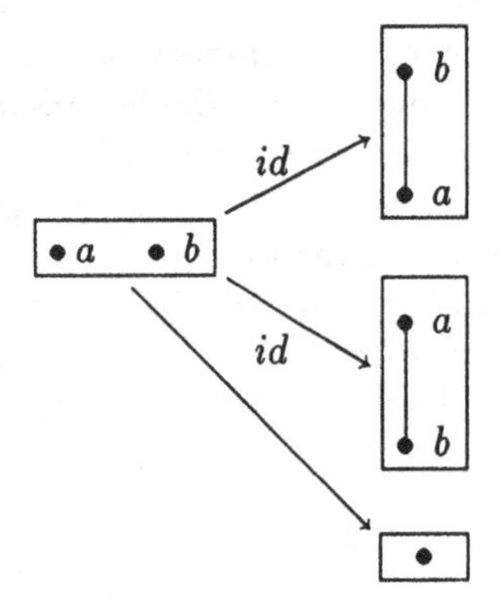

In contrast, linearly ordered sets do not form a cone-orthogonality class in **Pos**, as we can see from the following:

4.21 Remark. Let $\mathcal{K}$ be a category with a terminal object 1. Every (small-) cone-orthogonality class of $\mathcal{K}$ containing 1 is a (small-) orthogonality class.

In fact, let $\mathcal{M}$ be a collection of cones with $1 \in \mathcal{M}^{\perp}$. For each cone $(A \xrightarrow{m_t} A_t)_{t \in T}$ the unique morphism $A \to 1$ factorizes through a unique t, thus, T contains just one element.

4.22. We know that orthogonality classes are closed under limits (Observation 1.34) and injectivity classes are closed under products (Proposition 4.3). We will now show that cone-orthogonality classes are closed under *connected limits*, i.e., limits of diagrams $D \colon \mathcal{D} \to \mathcal{K}$ such that $\mathcal{D}$ is a *connected category* (which means that $\mathcal{D}$ is non-empty and is not a coproduct of two non-empty categories).

Proposition. *Every cone-orthogonality class is closed under connected limits.*

PROOF. Let $\mathcal{M}$ be a class of cones in a category $\mathcal{K}$, and let $D \colon \mathcal{D} \to \mathcal{K}$ be a connected diagram with Dd in $\mathcal{M}^{\perp}$ for each $d \in \mathcal{D}^{\text{obj}}$. If $(L \xrightarrow{l_d} Dd)_{d \in \mathcal{D}^{\text{obj}}}$ is a limit of D, we will prove that L lies in $\mathcal{M}^{\perp}$. Choose a cone $\left(A \xrightarrow{m_t} A_t\right)_{t \in T}$ in $\mathcal{M}$. For each morphism $f \colon A \to L$ and each $d \in \mathcal{D}^{\text{obj}}$ there exists a unique $t(d) \in T$ such that $l_d \cdot f$ factorizes (uniquely) through $m_{t(d)}$. We will prove that $t(d)$ is independent of d. Since $\mathcal{D}$ is a connected category, it is sufficient to prove that for each morphism $\delta \colon d \to d'$ in $\mathcal{D}$ we have

$t(d) = t(d')$. In fact, $l_{d'} \cdot f$ factorizes both through $t(d')$ and through $t(d)$ (since $l_{d'} \cdot f = D\delta \cdot (l_d \cdot f)$), and since Dd' is orthogonal to the given cone, we conclude that $t(d) = t(d')$. Put $t = t(d)$ for any d. Then for each $d \in \mathcal{D}^{\text{obj}}$ we have a unique $f'_d \colon A_t \to Dd$ with $l_d \cdot f = f'_d \cdot m_t$. The cone (f'_d) is compatible: given $\delta \colon d \to d'$ in $\mathcal{D}$ then $l_{d'} \cdot f$ factorizes through m_t both via $f'_{d'}$ and via $D\delta \cdot f'_d$, thus, $f'_{d'} = D\delta \cdot f'_d$. Let $f' \colon A_t \to L$ be the unique morphism with $f'_d = l_d \cdot f'$ for all $d \in \mathcal{D}^{\text{obj}}$. Then, obviously, $f = f' \cdot m_t$. The uniqueness of t and f' is easy to verify. □

4.23 Proposition. *In a category with pushouts each (small-) cone-orthogonality class is a (small-) cone-injectivity class.*

Remark. This is analogous to Remark 4.4(1).

PROOF. For each cone $(A \xrightarrow{m_t} A_t)_{t \in T}$ we can form the pushouts of m_s and m_t for all pairs $s, t \in T$:

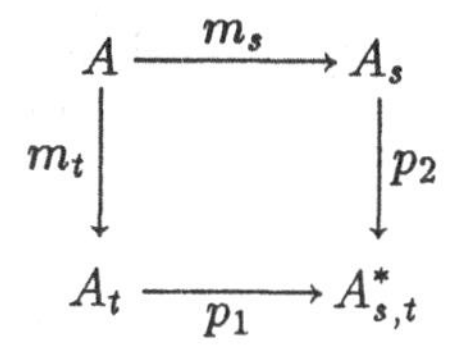

and for $s = t$ we have a unique factorization

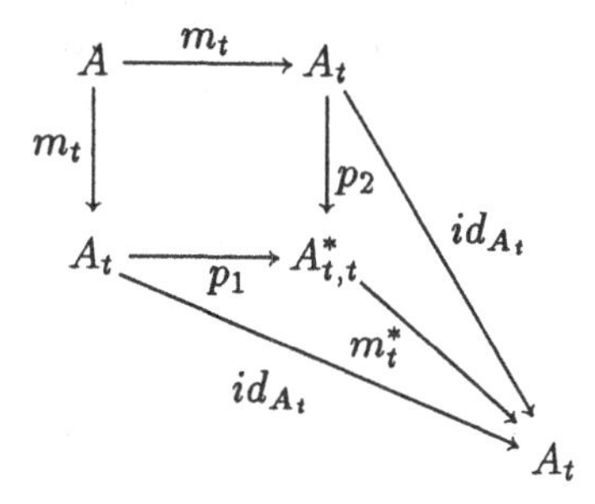

Given a class $\mathcal{M}$ of cones, let $\mathcal{M}^*$ be the class of cones extending $\mathcal{M}$ by adding to each cone $(A \xrightarrow{m_t} A_t)_{t \in T}$

(i) the empty cones with domains $A^*_{s,t}$ (for all $s, t \in T$, $s \neq t$),

(ii) the morphisms $A^*_{t,t} \xrightarrow{m^*_t} A_t$ (for all $t \in T$), considered as singleton cones.

Then

$$\mathcal{M}^{\perp} = \mathcal{M}^*\text{-}\mathbf{Inj}\,. \qquad \square$$

Locally Multipresentable Categories

4.24 Definition

(1) A functor $F\colon \mathcal{K} \to \mathcal{L}$ is called a *right multiadjoint* provided that for each object L of $\mathcal{L}$ there exists a cone $(L \xrightarrow{k_i} FK_i)_{i\in I}$, with each K_i in $\mathcal{K}$, such that for every arrow $L \xrightarrow{k} FK$ with K in $\mathcal{K}$ there exists a unique $i \in I$ for which a commutative triangle

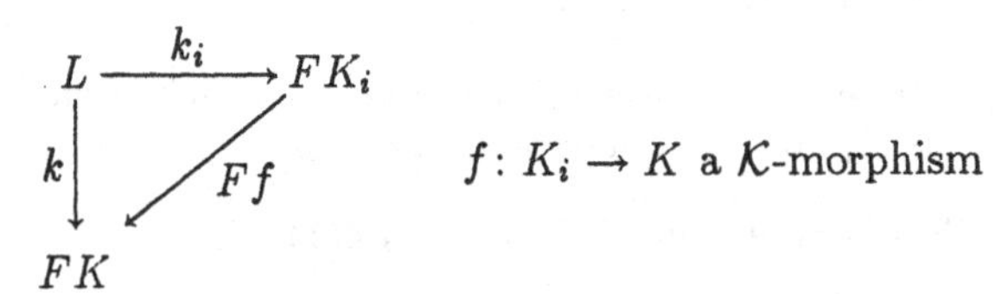

exists, and moreover, f is also unique.

(2) A subcategory $\mathcal{K}$ of a category $\mathcal{L}$ is called *multireflective* provided that the inclusion functor $\mathcal{K} \hookrightarrow \mathcal{L}$ is a right multiadjoint.

4.25 Examples

(1) Linearly ordered sets are multireflective in the category $\mathbf{Pos}^{@}$ of strictly ordered posets. Given a connected strictly ordered poset $(X, <)$ (i.e., no non-trivial equivalence relation $\sim$ on X exists such that $x < y$ implies $x \sim y$), then a multireflection of $(X, <)$ is the cone of all $id_X : (X, <) \to (X, \sqsubset)$ where $\sqsubset$ is a linear ordering extending $<$. Example: for $X = \{a, b, c\}$ with $a < b$ and $a < c$ we have the following multireflection:

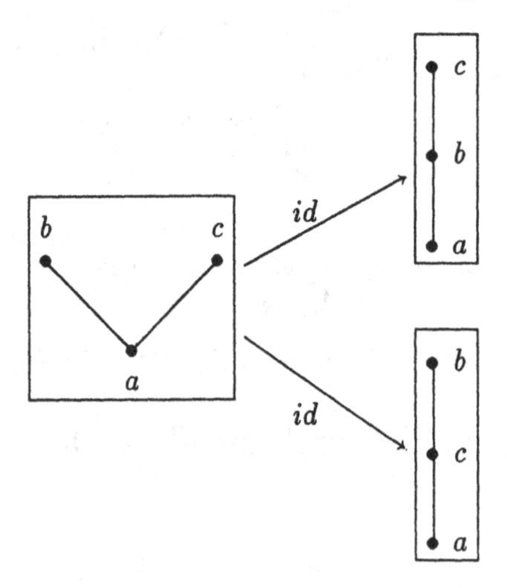

If $(X, <)$ is disconnected, a multireflection is obtained analogously, but taking into account all possible ways of gluing together elements of different components. An example is the following:

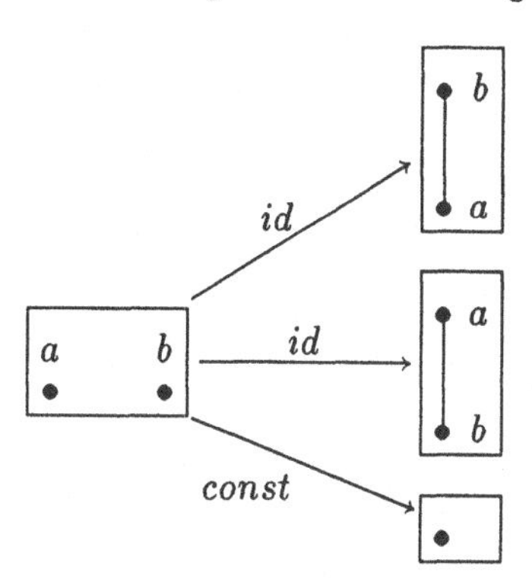

(2) The category **Fld** of fields is multireflective in the category **Rng** of commutative unitary rings. In fact, given a ring A which is not an integral domain, let $\mathrm{Id}(A)$ be the set of all maximal ideals of A. For each ideal $I \in \mathrm{Id}(A)$ we have the quotient field A/I and the canonical homomorphism $h_I : A \to A/I$. Then

$$(*) \qquad \left(A \xrightarrow{h_I} A/I\right)_{I \in \mathrm{Id}(A)}$$

is a multireflection of A in **Fld**. If A is an integral domain, a multireflection of A is obtained as follows: we extend the cone $(*)$ by the embedding

$$A \hookrightarrow \text{field of fractions of } A.$$

4.26 Theorem.

(i) *Each right multiadjoint functor preserves connected limits.*

(ii) *If $\mathcal{K}$ is a category with connected limits, then a functor $F : \mathcal{K} \to \mathcal{L}$ is a right multiadjoint iff it preserves connected limits and satisfies the solution-set condition.*

The proof, which is an easy adaptation of the corresponding result on adjoints, is left to the reader. □

4.27 Proposition. *Every small-cone-orthogonality class in a locally presentable category is multireflective.*

PROOF. Let $\mathcal{M}$ be a set of cones in a locally presentable category $\mathcal{K}$. Then $\mathcal{M}^\perp$ is closed under connected limits (see Proposition 4.22) and the embedding $\mathcal{M}^\perp \hookrightarrow \mathcal{K}$ satisfies the solution-set condition (see Propositions 4.23, 4.16 and Corollary 2.45). Thus, $\mathcal{M}^\perp$ is multireflective by Theorem 4.26. □

4.28 Definition

(1) By a *multicolimit* of a diagram $D\colon \mathcal{D} \to \mathcal{K}$ is meant a set $\left(Dd \xrightarrow{k_d^i} K^i\right)_{d\in\mathcal{D}^{\mathrm{obj}}}$ of compatible cocones of D, $i \in I$, such that for each compatible cocone $\left(Dd \xrightarrow{k_d} K\right)_{d\in\mathcal{D}^{\mathrm{obj}}}$ of D there is a unique $i \in I$ for which a factorization

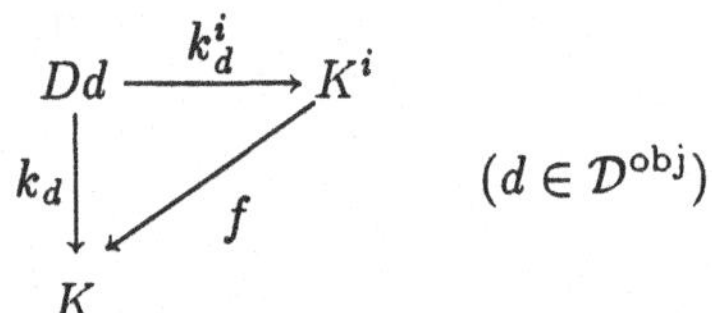

via a morphism $f\colon K^i \to K$ exists, and moreover, f is also unique.

(2) A category is called *locally λ-multipresentable* provided that it is λ-accessible and has multicolimits.

Remark. A diagram with no compatible cocone has a multicolimit, (viz., the empty set of cocones).

Small-cone-orthogonality Classes

Let us call a full subcategory of a category $\mathcal{K}$ a *λ-cone-orthogonality class* of $\mathcal{K}$ if it has the form $\mathcal{M}^\perp$ for a collection $\mathcal{M}$ of cones $(A \xrightarrow{m_t} A_t)_{t\in T}$ such that A and each A_t are λ-presentable objects of $\mathcal{K}$.

The following is a natural extension of Theorem 1.39:

4.29 Theorem. *Let $\mathcal{K}$ be a locally λ-presentable category. The following conditions on a full subcategory $\mathcal{A}$ of $\mathcal{K}$ are equivalent:*

(i) *$\mathcal{A}$ is a λ-cone-orthogonality class in $\mathcal{K}$,*

(ii) *$\mathcal{A}$ is a multireflective subcategory of $\mathcal{K}$ closed under λ-directed colimits.*

They imply that $\mathcal{A}$ is locally λ-multipresentable.

PROOF. (i) $\Rightarrow$ (ii). Let $\mathcal{A} = \mathcal{M}^{\perp}$ for a set $\mathcal{M}$ of cones with λ-presentable domains and codomains. Following Proposition 4.27, $\mathcal{A}$ is a multireflective subcategory of $\mathcal{K}$. Moreover, $\mathcal{A}$ is closed under λ-directed colimits; the proof is completely analogous to that for small-orthogonality classes (Proposition 1.35).

(ii) $\Rightarrow$ (i). This is completely analogous to the corresponding implication in Theorem 1.39 (here $\mathcal{M}$ is the collection of all multireflections of λ-presentable objects of $\mathcal{K}$ in $\mathcal{A}$).

It remains to show that $\mathcal{A}$ is multicocomplete. For each diagram

$$D: \mathcal{D} \to \mathcal{A}$$

we form a colimit $(Dd \xrightarrow{c_d} C)_{d \in \mathcal{D}^{\mathrm{obj}}}$ in $\mathcal{K}$, and we consider a multireflection $(C \xrightarrow{r_t} A_t)_{t \in T}$ of C in $\mathcal{A}$. It is easy to verify that the cocones

$$\left(Dd \xrightarrow{r_t \cdot c_d} A_t\right)_{d \in \mathcal{D}^{\mathrm{obj}}} \qquad (t \in T)$$

form a multicolimit of D in $\mathcal{A}$. □

4.30 Characterization Theorem. *The following conditions on a category $\mathcal{K}$ are equivalent:*

(i) *$\mathcal{K}$ is locally λ-multipresentable;*

(ii) *$\mathcal{K}$ is λ-accessible and has connected limits;*

(iii) *$\mathcal{K}$ is equivalent to a full, multireflective subcategory of* $\mathbf{Set}^{\mathcal{A}}$ *closed under λ-directed colimits for some small category $\mathcal{A}$;*

(iv) *$\mathcal{K}$ is equivalent to a λ-cone-orthogonality class in a locally λ-presentable category.*

The proof is completely analogous to that of Theorem 4.11.

4.31 Definition. A sketch $\mathscr{S} = (\mathcal{A}, \mathbf{L}, \mathbf{C}, \sigma)$ is called a *limit–coproduct* sketch provided that every diagram in $\mathbf{C}$ is discrete.

Example. The sketch for lists (see 2.57(2)) is a limit–coproduct sketch.

4.32 Theorem. *A category is locally multipresentable iff it is sketchable by a limit–coproduct sketch.*

Remark. In more detail, we will prove that a category is locally λ-multipresentable iff it is equivalent to $\mathbf{Mod}\,\mathscr{S}$ for a limit-coproduct sketch in which all $\mathbf{L}$-diagrams are λ-small.

PROOF. I. Let $\mathcal{K}$ be a locally λ-multipresentable category, and let $\mathcal{B} = \mathbf{Pres}_\lambda \mathcal{K}$. For each λ-small diagram D in $\mathcal{B}$ and each multicolimit of D the codomains of the multicolimit are all λ-presentable (the proof is analogous to that of Proposition 1.16), thus, we can say that $\mathcal{B}$ is closed under the formation of λ-small multicolimits.

We are going to present a limit–coproduct sketch $\mathscr{S}^*$ for $\mathcal{K}$ by modifying the sketch $\mathscr{S}$ of Remark 2.58. In that sketch $\mathscr{S} = (\mathcal{A}, \mathbf{L}, \mathbf{C}, \sigma)$, every object of $\mathcal{A}$ of the form $A(D)$, where $D \in \mathbf{L}$, is the domain of the limit cone $\sigma(D)$. Put $\sigma(D) = \big(A(D) \xrightarrow{a_{D,d}} Dd\big)_{d \in \mathcal{D}^{\mathrm{obj}}}$, where $D\colon \mathcal{D} \to Y(\mathcal{B}^{\mathrm{op}})$ is a λ-small diagram. Since $\mathcal{B}$ has λ-small multicolimits, $Y(\mathcal{B}^{\mathrm{op}})$ has λ-small multilimits, thus, there exists a multilimit

$$\big(B_i(D) \xrightarrow{b_{i,D,d}} Dd\big)_{d \in \mathcal{D}^{\mathrm{obj}}} \qquad \text{for } i \in I$$

with $B_i(D)$ in $Y(\mathcal{B}^{\mathrm{op}})$ for each $i \in I$. For each $i \in I$ denote by

$$e_{i,D}: B_i(D) \to A(D)$$

the unique factorization (defined by $a_{D,d} \cdot e_{i,D} = b_{i,D,d}$).

The sketch $\mathscr{S}^*$ is obtained from $\mathscr{S}$ by substituting each of the canonical diagrams of $A(D)$ by the discrete diagram $\{\, B_i(D) \,\}_{i \in I}$. That is, $\mathscr{S}^*$ is the limit–coproduct sketch

$$\mathscr{S}^* = (\mathcal{A}, \mathbf{L}, \mathbf{C}^*, \sigma^*)$$

where $\mathbf{C}^*$ consists of the discrete diagrams $\{\, B_i(D) \,\}_{i \in I}$ for $D \in \mathbf{L}$ with

$$\sigma^*(\{\, B_i(D) \,\}) = \big(B_i(D) \xrightarrow{e_{i,D}} A(D)\big)_{i \in I}$$

and

$$\sigma^*(D) = \sigma(D) \qquad \text{for all } D \in \mathbf{L}.$$

To conclude the proof, we show that $\mathscr{S}$ and $\mathscr{S}^*$ have the same models:

$$\mathbf{Mod}\,\mathscr{S}^* = \mathbf{Mod}\,\mathscr{S} \quad (\approx \mathcal{K}, \text{ see Remark 2.58}).$$

This follows from the observation that for each $D \in \mathbf{L}$ the discrete category of $B_i(D) \xrightarrow{e_{i,D}} A(D)$, $i \in I$, is a full, cofinal subcategory of the comma-category $Y(\mathcal{B}^{\mathrm{op}}) \downarrow A(D)$. In fact, if $i \neq j$, then $e_{i,D}$ does not factorize through $e_{j,D}$ (by the definition of multilimit). Given an arrow $YC \xrightarrow{c} A(D)$ with C in $\mathcal{B}^{\mathrm{op}}$, the compatible cone $(YC \xrightarrow{a_{D,d} \cdot c} Dd)$ of D factorizes through $b_{i,D,d}$ for a unique $i \in I$, i.e., we have $\bar{c}\colon YC \to B_i(D)$ with $b_{i,D,d} \cdot \bar{c} = a_{D,d} \cdot c$ (for each $d \in \mathcal{D}^{\mathrm{obj}}$). This implies $a_{D,d} \cdot e_{i,D} \cdot \bar{c} = a_{D,d} \cdot c$

for each $d \in \mathcal{D}^{\mathrm{obj}}$, i.e., $c = e_{i,D} \cdot \bar{c}$; thus $\bar{c}$ is the unique morphism from c to $e_{i,D}$ in the comma-category $Y(\mathcal{B}^{\mathrm{op}}) \downarrow A(D)$. Consequently, a functor preserves the canonical colimit of $A(D)$ w.r.t. $Y(\mathcal{B}^{\mathrm{op}})$ iff it preserves the coproduct $\left(B_i(D) \xrightarrow{e_{i,D}} A(D)\right)_{i \in I}$.

II. Let $\mathscr{S} = (\mathcal{S}, \mathbf{L}, \mathbf{C}, \sigma)$ be a limit–coproduct sketch. The category $\mathbf{Mod}\,\mathscr{S}$ is accessible (by Theorem 2.60) and accessibly embedded to $\mathbf{Set}^{\mathcal{A}}$ (in fact, closed under λ-directed colimits for any regular cardinal λ such that all diagrams in D are λ-small). Moreover, because connected limits commute with coproducts in $\mathbf{Set}$ (see Exercise 4.f), $\mathbf{Mod}\,\mathscr{S}$ is closed in $\mathbf{Set}^{\mathcal{A}}$ under connected limits. Therefore $\mathbf{Mod}\,\mathscr{S}$ is locally multipresentable by Theorem 4.30. □

Exercises

4.a Connected Topological Spaces. Prove that connected spaces do not form an injectivity class in **Top** (although they are closed under products and split subobjects).

(Hint: for any cardinal λ there are arbitrarily large connected spaces having all subspaces of cardinality λ discrete. Consequently, the two-point discrete space belongs to any injectivity class in **Top** containing all connected spaces.)

4.b Injective Modules. Let R be a unitary ring. Prove that an R-module A is injective w.r.t. the class $\mathcal{M}$ of all monomorphisms iff for any right ideal U in R and any R-module homomorphism $f\colon U \to A$ there is an extension of f to a R-module homomorphism $f'\colon R \to A$ (Baer's criterion). Conclude that $\mathcal{M}$-**Inj** is a small-injectivity class.

4.c Weakly Reflective Subcategories

(1) Show that all graphs containing a loop form a small-injectivity class in **Gra** closed under split subobjects. Describe all weak reflections of the graph with one vertex and no edge.

(2) For each poset P and each set $A \subseteq P$ put $A^- = \{p \in P \mid p \leq a \text{ for each } a \in A\}$ and $A^+ = \{p \in P \mid p \geq a \text{ for each } a \in A\}$. The *MacNeille completion* of P is the poset P^* of all cuts, i.e., all $A \subseteq P$ with $A = A^{+-}$, ordered by inclusion. Verify that the map $r\colon P \to P^*$ defined by $r(p) = \{p\}^-$ is a weak reflection of P in the category of complete lattices and order-preserving functions.

(3) Show that all sets with at least two elements form a weakly reflective full subcategory of **Set** which is neither closed under split subobjects nor under products.

(4) Show that if $\mathcal{A}$ is a weakly reflective subcategory of $\mathcal{B}$ and $\mathcal{B}$ is a weakly reflective subcategory of $\mathcal{C}$, then $\mathcal{A}$ is weakly reflective in $\mathcal{C}$.

4.d Injective Hulls

(1) Let $\mathcal{M}$ be a collection of morphisms in a category $\mathcal{K}$. A morphism $m\colon A \to A'$ in $\mathcal{M}$ is called *$\mathcal{M}$-essential* provided that for each morphism $f\colon A' \to A''$ we have that

$$fm \in \mathcal{M} \quad \text{implies} \quad f \in \mathcal{M}.$$

By an *$\mathcal{M}$-injective hull* of an object K is meant an $\mathcal{M}$-essential morphism from K into an $\mathcal{M}$-injective object.

Show that an $\mathcal{M}$-injective hull is essentially unique and that it forms a weak reflection in the full subcategory of $\mathcal{M}$-injective objects. That is, if $p: K \to P$ is an $\mathcal{M}$-injective hull of K, then

(a) for each $\mathcal{M}$-injective hull $p': K \to P'$ there exists an isomorphism

$$i: P \to P'$$

with $p' = ip$,

and

(b) for each morphism $f: K \to A$ such that A is $\mathcal{M}$-injective there exists a morphism $f': P \to A$ with $f = f' \cdot p$.

(2) A full subcategory $\mathcal{A}$ of a category $\mathcal{K}$ is called *stably weakly reflective* if for any K in $\mathcal{K}$ there is a weak reflection $r: K \to K^*$ to $\mathcal{A}$ such that

$$f \cdot r = r \implies f \text{ is an isomorphism}$$

for any morphism $f: K^* \to K^*$.

Prove that the following statements are equivalent for any full, isomorphism-closed subcategory $\mathcal{A}$ of a category $\mathcal{K}$:

(a) $\mathcal{A}$ is an injectivity class with injective hulls;

(b) $\mathcal{A}$ is stably weakly reflective.

(3) Consider the following morphism m in **Gra**:

$$\emptyset \longrightarrow \boxed{\begin{matrix} \bullet \to \bullet \to \bullet & \cdots \\ 0 \quad 1 \quad 2 & \end{matrix}}$$

Show that $\{m\}$-**Inj** is not stably weakly reflective in **Gra**.

4.e Projectivity. The dual concept of injectivity is that of projectivity: an object is said to be *projective* w.r.t. a morphism $m: A \to A'$ provided that for each morphism $f: K \to A'$ there exists a morphism $f': K \to A$ such that $m \cdot f' = f$.

(1) Prove that, for any unitary ring R, an R-module is projective w.r.t. $\mathcal{M}$ = monomorphisms iff it is a split subobject of a free module. Conclude that an Abelian group is projective iff it is free.

(2) Let

$$0 \to A \to B \xrightarrow{e} C \to 0$$

be an exact sequence of Abelian groups. Prove that an Abelian group G is projective w.r.t. e iff $\mathrm{Ext}(G, A) = 0$.

4.f Connected Limits. Prove that connected limits commute with coproducts in **Set**.

4.g Locally Multipresentable Categories. Prove the following generalization of Proposition 1.16 to multilimits: For any λ-small diagram $D: \mathcal{D} \to \mathcal{K}$ consisting of λ-presentable objects, its multicolimit $\left(Dd \xrightarrow{k^*_{d,i}} K^i\right)_{d \in \mathcal{D}^{\mathrm{obj}}}$, $i \in I$, has all objects K^i, $i \in I$, λ-presentable.

4.h A Limit–Coproduct Sketch for Fields. Find a limit-coproduct sketch whose category of models is equivalent to **Fld**. (Hint: proceed in the spirit of Example 1.50(2), adding to the objects a and a^2 two objects t (terminal) and u (units). The only colimit diagram is the discrete diagram with objects u and t, and the corresponding cocone has codomain a.)

Historical Remarks

Injectivity w.r.t. morphisms is a classical algebraic concept (injective modules, injective groups, etc., see e.g. [Fuchs 1970]). The categorical concept of an injectivity class was introduced by [Maranda 1964]. Limit-epi sketches are introduced by [Lair 1987]. Theorems 4.11 and 4.13 are from [Adámek, Rosický 1994a], a different, but closely related, result to Theorem 4.13 can be found in [Lair 1987].

Cone-injectivity was introduced by [John 1977] (in the dual form), and it was studied in the spirit of our treatment by H. Andréka, I. Németi, and I. Sain in the late 1970's. See [Andréka, Németi 1979a] and [Németi, Sain 1982]; in the latter paper the authors attribute the idea of using cones to E. Nelson. The equivalence of sketchable categories and small-cone-injectivity classes in locally presentable categories was proved by [Guitart, Lair 1980]. Our treatment (Corollary 4.18) follows [Adámek, Rosický 1993].

Locally multipresentable categories are due to J. Diers who introduced the concept and studied its properties in [Diers 1980a,b]; Theorem 4.32 is due to [Guitart, Lair 1980]. Further generalizations and the relationship to sketches can be found in [Ageron 1992].

Chapter 5

Categories of Models

In this chapter we study the axiomatization of classes of structures in (finitary and infinitary) first-order logic. We will show that locally presentable categories are precisely the categories of models of limit theories, and accessible categories are precisely the categories of models of basic theories. There is a substantial difference between those two cases: in the locally presentable case we can specialize to a given cardinal λ, and we see that locally λ-presentable categories are precisely the categories of models of limit theories in the λ-ary logic L_λ. Nothing like that is possible in the case of accessible categories: we will see that a finitely accessible category need not have an axiomatization in the finitary logic L_ω, and that a basic theory in L_ω need not have a finitely accessible category of models.

5.A Finitary Logic

Formulas, Models, and Satisfaction

Up to now, we have worked with many-sorted operations and many-sorted relations separately (see Example 1.2(4) and Chapter 3). We will now combine them: by a signature we will mean a set Σ of operation and relation symbols of prescribed arities. A Σ-structure A is, then, a many-sorted set together with appropriate operations and relations, and a homomorphism is an S-sorted function preserving the given operations and relations. Although infinitary logic is needed for general locally presentable categories, we start with the classical finitary (but many-sorted) first-order logic.

5.1. Σ-structures and homomorphisms. Let S be a set (of sorts). A *finitary S-sorted signature* is a set $\Sigma = \Sigma_{\mathrm{ope}} \cup \Sigma_{\mathrm{rel}}$ with Σ_{ope} and Σ_{rel} disjoint. The elements of Σ_{ope} are called *operation symbols*, and for each operation symbol $\tau \in \Sigma_{\text{ope}}$ an arity $s_1 \times \cdots \times s_n \to s$ is given ($s, s_1, \ldots, s_n \in S$); notation: $\tau\colon s_1 \times \cdots \times s_n \to s$. The elements of Σ_{rel} are called *relation symbols*, and for each relation symbol $\sigma \in \Sigma_{\text{rel}}$ an arity $s_1 \times \cdots \times s_n$ is given ($s_1, \ldots, s_n \in S$). For example, the (one-sorted) signature Σ_{OG} of ordered Abelian groups has $(\Sigma_{\text{OG}})_{\text{ope}} = \{+, -, 0\}$ of arities $s \times s \to s$, $s \to s$ and $\to s$, respectively, and $(\Sigma_{\text{OG}})_{\text{rel}} = \{\leq\}$ of arity $s \times s$.

By a Σ-*structure* A is meant an S-sorted set $|A| = (A_s)_{s\in S}$ together with operations

$$\sigma_A\colon A_{s_1} \times \cdots \times A_{s_n} \to A_s$$

for all $\sigma\colon s_1 \times \cdots \times s_n \to s$ in Σ_{ope}, and relations

$$\sigma_A \subseteq A_{s_1} \times \cdots \times A_{s_n}$$

for all σ in Σ_{rel} of arity $s_1 \times \cdots \times s_n$. A *homomorphism* from a structure A to a structure B is an S-sorted map $f\colon |A| \to |B|$ satisfying

$$f_s\big(\sigma_A(a_1, \ldots, a_n)\big) = \sigma_B\big(f_{s_1}(a_1), \ldots, f_{s_n}(a_n)\big)$$

for each operation $\sigma\colon s_1 \times \cdots \times s_n \to s$, and

$$(f_{s_1} \times \cdots \times f_{s_n})(\sigma_A) \subseteq \sigma_B$$

for each relation σ of arity $s_1 \times \cdots \times s_n$. The category of Σ-structures and homomorphisms is denoted by $\mathbf{Str}\,\Sigma$.

Remark. The category $\mathbf{Str}\,\Sigma$ has a number of properties analogous to the corresponding category $\mathbf{Alg}\,\Sigma_{\text{ope}}$ of algebras. Let us denote by

$$(-)^0\colon \mathbf{Str}\,\Sigma \to \mathbf{Alg}\,\Sigma_{\text{ope}} \qquad \text{and} \qquad |-|\colon \mathbf{Str}\,\Sigma \to \mathbf{Set}^S$$

the natural forgetful functors.

(1) $\mathbf{Str}\,\Sigma$ has (regular epi, mono)- and (epi, regular mono)-factorizations of morphisms. Given a Σ-structure A, by a *substructure* of A is meant a Σ-structure B with $|B| \subseteq |A|$ such that for each relation symbol σ of arity $s_1 \times \cdots \times s_n$ we have $\sigma_B = \sigma_A \cap (B_{s_1} \times \cdots \times B_{s_n})$. If A is a substructure of B, then the inclusion $A \hookrightarrow B$ is a regular monomorphism (and every regular monomorphism in $\mathbf{Str}\,\Sigma$ has such a representation).

For each homomorphism $f\colon A \to B$ consider the (regular epi, regular mono)-factorization

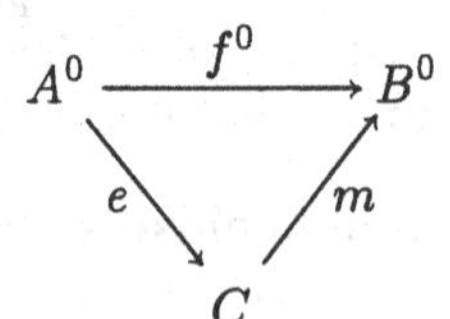

in $\mathbf{Alg}\,\Sigma_{\text{ope}}$. Then an (epi, regular mono)-factorization of f is obtained by defining the relations of C to form a substructure of B, i.e., $\sigma_C = \big(m_{s_1} \times \cdots \times m_{s_n}\big)^{-1}(\sigma_B)$ for each relation symbol σ of arity $s_1 \times \cdots \times s_n$. A (regular epi, mono)-factorization is obtained by defining $\sigma_C = (e_{s_1} \times \cdots \times e_{s_n})(\sigma_A)$ for each relation symbol σ of arity $s_1 \times \cdots \times s_n$.

(2) $\mathbf{Str}\,\Sigma$ is complete. A limit of a diagram D in $\mathbf{Str}\,\Sigma$ is obtained by forming a limit $A^0 = \lim D^0$ of the corresponding diagram D^0 of algebras in $\mathbf{Alg}\,\Sigma_{\text{ope}}$, and defining, for each $\sigma \in \Sigma_{\text{rel}}$, the relation σ_A as the largest relation on A^0 for which the limit-maps are Σ-homomorphisms.

For example, a product $A = \prod_{i \in I} A_i$ in $\mathbf{Str}\,\Sigma$ is obtained from the product $\prod_{i \in I} A_i^0$ of the corresponding algebras (with operations defined coordinate-wise) by defining σ_A, for $\sigma \in \Sigma_{\text{rel}}$ of arity $s_1 \times \cdots \times s_n$, to consist of precisely those n-tuples $\big((a_i^1)_{i \in I}, \ldots, (a_i^n)_{i \in I}\big)$ such that for each $i \in I$ we have $(a_i^1, \ldots, a_i^n) \in \sigma_A$.

(3) $\mathbf{Str}\,\Sigma$ is cocomplete. A colimit of a diagram D in $\mathbf{Str}\,\Sigma$ is obtained by forming a colimit $A^0 = \operatorname{colim} D^0$ of the corresponding diagram of Σ_{ope}-algebras, and defining, for each $\sigma \in \Sigma_{\text{rel}}$, the relation σ_A as the smallest relation on A for which the colimit-maps are Σ-homomorphisms.

For example, if $D\colon (I, \leq) \to \mathbf{Str}\,\Sigma$ is a directed diagram, then a colimit is formed on the level of sets: if $(|D_i| \xrightarrow{c_i} C)_{i \in I}$ is a colimit in $\mathbf{Set}^S$, then we define operations on C in a unique manner that makes each c_i a Σ-homomorphism (see Remark 3.4(4)), and for each $\sigma \in \Sigma_{\text{rel}}$ of arity $s_1 \times \cdots \times s_n$, we put $\sigma_C = \bigcup_{i \in I} (c_i)_{s_1} \times \cdots \times (c_i)_{s_n}(\sigma_{D_i})$.

(4) A Σ-structure A is λ-presentable in $\mathbf{Str}\,\Sigma$ iff

(a) the algebra A^0 obtained from A by forgetting the relations is λ-presentable (i.e., can be presented by less than λ generators and less than λ equations, see Theorem 3.12)

and

(b) A has finitely many edges, i.e.,

$$\operatorname{card} \bigcup_{\sigma \in \Sigma_{\text{rel}}} \sigma_A < \aleph_0.$$

(5) $\mathbf{Str}\,\Sigma$ is a locally finitely presentable category. In fact, for each Σ-structure A we have, by Remark 3.4(6), a directed diagram D of finitely presentable algebras in $\mathbf{Alg}\,\Sigma_{\text{ope}}$ whose colimit is the algebra A^0. Consider the (obviously directed) diagram D^* of all finitely presentable Σ-structures B with B^0 an object of D; then A is a colimit of D^*.

(6) The forgetful functor $|-|\colon \mathbf{Str}\,\Sigma \to \mathbf{Set}^S$ is a right adjoint because it is a composite of the forgetful functor of $\mathbf{Alg}\,\Sigma_{\mathrm{ope}}$ (which is a right adjoint by Remark 3.4) and the functor $(-)^0$ (which has a left adjoint assigning to each Σ_{ope}-algebra A the Σ-structure with $\sigma_A = \emptyset$ for each relational symbol σ). For a left adjoint $F\colon \mathbf{Set}^S \to \mathbf{Str}\,\Sigma$ of $|-|$ we call FX the *free Σ-structure* generated by the S-sorted set X.

5.2 Formulas. As in Chapter 3, we assume that a set V of standard variables is given, with V_s denumerable for each sort s. *Terms* are defined as in 3.A, using operations only, that is (a) each variable of sort s is a term of sort s, (b) for each operation $\sigma\colon s_1 \times \cdots \times s_n \to s$ and each n-tuple of terms τ_i of sort s_i $(i = 1, \dots, n)$, the expression $\sigma(\tau_1, \dots, \tau_n)$ is a term of sort s. The set $\mathrm{Var}(\tau)$ of (free) variables of a term τ is defined by

$$\mathrm{Var}(x) = x$$
$$\mathrm{Var}\big(\sigma(\tau_1 \dots \tau_n)\big) = \mathrm{Var}(\tau_1) \cup \cdots \cup \mathrm{Var}(\tau_n).$$

By an *atomic formula* is meant either an equation $\tau_1 = \tau_2$ (i.e., a pair of terms), or an expression $\sigma(\tau_1, \dots, \tau_n)$ where σ is a relation symbol of arity $s_1 \times \cdots \times s_n$ and τ_i is a term of sort s_i $(i = 1, \dots, n)$. For example, $x + y = y - x$ and $x \le y$ are atomic formulas in Σ_{OG}.

Formulas are built up from the atomic formulas $\varphi, \psi, \dots$ by the well-known logical operators:

$\neg\varphi$	(negation of φ)
$\varphi \Rightarrow \psi$	(φ implies ψ)
$\varphi \wedge \psi$	(conjunction of φ and ψ)
$\varphi \vee \psi$	(disjunction of φ and ψ)
$(\forall x)\varphi, (\exists x)\varphi$	(universal and existential quantifiers using a variable $x \in V$).

The *free variables* of a formula φ are those variables which appear non-quantified in φ. That is, the set $\mathrm{Var}\,\varphi$ of free variables is defined by the following induction:

$$\begin{aligned}
\mathrm{Var}(\tau_1 = \tau_2) &= \mathrm{Var}\,\tau_1 \cup \mathrm{Var}\,\tau_2 \\
\mathrm{Var}\,\sigma(\tau_1, \dots, \tau_n) &= \mathrm{Var}\,\tau_1 \cup \cdots \cup \mathrm{Var}\,\tau_n \\
\mathrm{Var}(\neg\varphi) &= \mathrm{Var}\,\varphi \\
\mathrm{Var}(\varphi \Rightarrow \psi) &= \mathrm{Var}(\varphi \wedge \psi) = \mathrm{Var}(\varphi \vee \psi) = \mathrm{Var}\,\varphi \cup \mathrm{Var}\,\psi \\
\mathrm{Var}(\forall x)\varphi &= \mathrm{Var}(\exists x)\varphi = \mathrm{Var}\,\varphi - \{x\}.
\end{aligned}$$

Example: $\mathrm{Var}(x+y) = \{x, y\}$ and $\mathrm{Var}((\forall x)(\exists y)(x+y) = 0) = \emptyset$. We write $\varphi(x_1, \ldots, x_n)$ to indicate that φ is a formula with $\mathrm{Var}\,\varphi \subseteq \{x_1, \ldots, x_n\}$.

5.3 The Logic L_ω. For each Σ-structure A and each formula $\varphi(x_1, \ldots, x_n)$ we are going to define the meaning of "A satisfies the formula $\varphi(x_1, \ldots, x_n)$ under the assignment $x_i \mapsto a_i$"; we use the symbol

$$A \models \varphi[a_1, \ldots, a_n].$$

Here, a_i is an element of the underlying set of A of the same sort s_i as x_i $(i = 1, \ldots, n)$. We know from Proposition 3.2 how to define the assignment of values to terms $\tau(x_1, \ldots, x_n)$: we have $\tau(a_1, \ldots, a_n) = f^{\#}(\tau)$ for $f\colon \{x_1, \ldots, x_n\} \to A$ given by $f_{s_i}(x_i) = a_i$. For atomic formulas we write

$$A \models (\tau_1 = \tau_2)[a_1, \ldots, a_n]$$

iff $\tau_1(a_1, \ldots, a_n) = \tau_2(a_1, \ldots, a_n)$ in A and

$$A \models \sigma(\tau_1, \ldots, \tau_m)[a_1, \ldots, a_n]$$

iff σ_A contains $\big(\tau_1(a_1, \ldots, a_n), \ldots, \tau_m(a_1, \ldots, a_n)\big)$. The logical operators have their usual meaning:

$$
\begin{array}{ll}
A \models \neg\varphi[a_1, \ldots, a_n] & \text{iff it is not true that } A \models \varphi[a_1, \ldots, a_m] \\
A \models (\varphi \Rightarrow \psi)[a_1, \ldots, a_n] & \text{iff } A \models \neg\varphi[a_1, \ldots, a_n] \text{ or } A \models \psi[a_1, \ldots, a_n] \\
A \models (\varphi \wedge \psi)[a_1, \ldots, a_n] & \text{iff } A \models \varphi[a_1, \ldots, a_n] \text{ and } A \models \psi[a_1, \ldots, a_n] \\
A \models (\varphi \vee \psi)[a_1, \ldots, a_n] & \text{iff } A \models \varphi[a_1, \ldots, a_n] \text{ or } A \models \psi[a_1, \ldots, a_n] \\
A \models \big((\forall x)\varphi\big)[a_1, \ldots, a_n] & \text{iff } \mathrm{Var}(\varphi) \subseteq \{x, x_1, \ldots, x_n\} \text{ and} \\
 & \quad A \models \varphi[a, a_1, \ldots, a_n] \\
 & \quad \text{for any } a \text{ in } X \text{ of the same sort as } x \\
A \models \big((\exists x)\varphi\big)[a_1, \ldots, a_n] & \text{iff } \mathrm{Var}(\varphi) \subseteq \{x, x_1, \ldots, x_n\} \text{ and} \\
 & \quad A \models \varphi[a, a_1, \ldots, a_n] \\
 & \quad \text{for some } a \text{ in } X \text{ of the same sort as } x.
\end{array}
$$

The collection of all formulas together with the satisfaction relation $\models$ is called the *finitary logic* L_ω.*

A formula without free variables is called a *sentence*. Thus

$$(\forall x)(\exists y)(x+y) = 0$$

*Some authors use the symbol $L_{\omega\omega}$ to stress the fact that both conjunctions and quantifications are performed over finitely many variables only.

is a sentence of Σ_{OG} (satisfied by each group).

Two formulas $\varphi(x_1, \ldots, x_n)$ and $\psi(x_1, \ldots, x_n)$ are called *equivalent* if for any Σ-structure A and for any elements $a_1, \ldots, a_n$ we have

$$A \models \varphi[a_1, \ldots, a_n] \text{ iff } A \models \psi[a_1, \ldots, a_n].$$

For example, the formulas $(\forall x)\varphi$ and $\neg(\exists x)\neg\varphi$ are equivalent (thus, we can introduce $\forall$ via $\exists$ and $\neg$). The formula $(\varphi \Rightarrow \psi) \wedge (\psi \Rightarrow \varphi)$ is denoted by $\varphi \Leftrightarrow \psi$. We write $(\forall x_1 \ldots x_n)\varphi$ instead of $(\forall x_1) \ldots (\forall x_n)\varphi$; analogously for $\exists$. We also use the symbol $\exists!$ for the unique existence: $(\exists! x)\varphi$ denotes

$$(\exists x)\, \varphi(x) \wedge (\forall x, y)\big(\varphi(x) \wedge \varphi(y) \Rightarrow (x = y)\big).$$

A set of sentences is called a *theory*. For each theory T we denote by $\mathbf{Mod}\, T$ the full subcategory of $\mathbf{Str}\, \Sigma$ consisting of all *models* of T, i.e., structures satisfying each sentence in T. In the case where T consists of a single formula φ we write just $\mathbf{Mod}\, \varphi$. Any finite theory $T = \{\varphi_1, \ldots, \varphi_n\}$ can be reduced to this simple case:

$$\mathbf{Mod}\, T = \mathbf{Mod} \bigwedge_{i=1}^{n} \varphi_i.$$

5.4 Examples

(1) Any set of equations is a theory (such theories are called *equational*). The corresponding categories of models are finitary varieties.

(2) Any set of implications in the sense of Section 3.B is a theory whose category of models is a finitary quasivariety.

(3) The theory of partially ordered sets is presented by the following sentences in the signature Σ of one-sorted binary relation $\leq$:

$$(\forall x)(x \leq x)$$
$$(\forall x, y)(x \leq y \wedge y \leq x \Rightarrow x = y)$$
$$(\forall x, y, z)(x \leq y \wedge y \leq z \Rightarrow x \leq z).$$

By adding the following sentence

$$(\forall x, y)(x \leq y \vee y \leq x)$$

we get the theory of linearly ordered sets.

(4) The theory of ordered Abelian groups is presented by adding the following sentences in Σ_{OG} to the theory of linearly ordered sets:

$$(\forall x, y, z)((x+y)+z = x+(y+z))$$
$$(\forall x, y)(x+y = y+x)$$
$$(\forall x)(x-x=0)$$
$$(\forall x)(x+0=x)$$
$$(\forall x, y, z)(x \leq y \Rightarrow x+z \leq y+z).$$

Limit Theories and Locally Finitely Presentable Categories

5.5 Remark: Presentation Formula. Each finitely presentable Σ-structure A allows the construction of a formula π_A with the following property: the comma-category $A \downarrow \mathbf{Str}\,\Sigma$ is precisely the category of models of π_A. We will now show how such a formula can be constructed.

The algebra A^0 is finitely presentable. Thus, without loss of generality, we can consider A^0 as a quotient algebra of the free algebra on n generators modulo a congruence generated by m equations for some natural numbers n and m. More precisely, there exist variables $x_1, \dots, x_n$ in V of sorts $s_1, \dots, s_n$, respectively, and equations $\tau_i = \tau_i'$ $(i = 1, \dots, m)$ between terms of the free algebra $T_{\Sigma_{\mathrm{ope}}}\{x_1, \dots, x_n\}$ such that

$$A^0 = T_{\Sigma_{\mathrm{ope}}}\{x_1, \dots, x_n\}/{\sim}$$

for the congruence $\sim$ generated by $\tau_i = \tau_i'$ $(i = 1, \dots, m)$. Next, the relational part of A consists of finitely many edges, say,

$$\sigma_1([\varrho_{1,1}], \dots, [\varrho_{1,p_1}]), \quad \dots, \quad \sigma_k([\varrho_{k,1}], \dots, [\varrho_{k,p_k}])$$

where $\sigma_1, \dots, \sigma_k$ are symbols in Σ_{rel}, and $\varrho_{i,j}$ are terms in $T_{\Sigma_{\mathrm{ope}}}\{x_1, \dots, x_n\}$. Here $[\,]$ denotes the congruence class of the above congruence $\sim$.

The *presentation formula* of A expressing the above presentation is the following formula

$$(\pi_A) \qquad (\tau_1 = \tau_1') \wedge \dots \wedge (\tau_m = \tau_m') \wedge \\ \wedge\, \sigma_1(\varrho_{1,1}, \dots, \varrho_{1,p_1}) \wedge \dots \wedge \sigma_k(\varrho_{k,1}, \dots, \varrho_{k,p_k}).$$

It has the following properties.

(i) For each Σ-structure B and each Σ-homomorphism $f\colon A \to B$ we have

$$B \models \pi_A[b_1, \dots, b_n]$$

where $b_i = f_{s_i}([x_i])$.

(ii) For each Σ-structure B and each n-tuple $b_1, \ldots, b_n$ of elements of B with
$$B \models \pi_A[b_1, \ldots, b_n]$$
there is a unique homomorphism $f\colon A \to B$ with $f_{s_i}([x_i]) = b_i$.

In fact, (i) is obvious. For (ii), the function $f_0\colon \{x_1, \ldots, x_n\} \to B$ defined by $(f_0)_{s_i}(x_i) = b_i$ can be extended to a unique Σ_{ope}-homomorphism

$$f_0^{\#}: T_{\Sigma_{\text{ope}}}\{x_1, \ldots, x_n\} \to B.$$

For each i we have $(f_0^{\#})_{t_i}(\tau_i) = (f_0^{\#})_{t_i}(\tau_i')$, where t_i is the sort of τ_i, τ_i'. Thus $f_0^{\#}$ factorizes through A^0 and the resulting Σ_{ope}-homomorphism

$$f\colon A^0 \to B^0$$

is, obviously, a Σ-homomorphism due to the σ_i's in π_A.

5.6 Remark: Orthogonality Formula. For each Σ-homomorphism

$$h\colon A \to B \qquad A, B \text{ finitely presentable}$$

we will construct a formula π_h whose models are the Σ-structures orthogonal to h (i.e., Σ-structures C such that each morphism from A to C uniquely factorizes through h). Let $\pi_A(x_1, \ldots, x_n)$ and $\pi_B(y_1, \ldots, y_m)$ be presentation formulas for A and B, respectively. For each x_i there exists a term τ_i in $T_{\Sigma_{\text{ope}}}\{y_1, \ldots, y_m\}$ such that $h_{s_i}([x_i]) = [\tau_i]$. Define the following formula π_h:

$$(\pi_h) \qquad (\forall x_1, \ldots, x_n)(\pi_A(x_1, \ldots, x_n) \Rightarrow (\exists! y_1, \ldots, y_m)\big(\pi_B \wedge (x_1 = \tau_1) \wedge \cdots \wedge (x_n = \tau_n)\big).$$

The formula π_h has the above-mentioned property:

$$\{h\}^{\perp} = \mathbf{Mod}\, \pi_h.$$

In fact, let C be a Σ-structure satisfying π_h. Then for each homomorphism $f\colon A \to C$ we have

$$C \models \pi_A\big[f_{s_1}[x_1], \ldots, f_{s_n}[x_n]\big],$$

thus, there is a unique m-tuple $b_1, \ldots, b_m$ in $|B|$ satisfying both π_B and $x_i = \tau_i$. In other words, there is a unique Σ-homomorphism $f'\colon B \to C$ such that

$$f_{s_i}([x_i]) = f'_{s_i}([\tau_i]) = f'_{s_i} \cdot h_{s_i}([x_i]), \qquad i = 1, \ldots, n;$$

equivalently, $f = f' \cdot h$. Conversely, any Σ-structure orthogonal to h is, obviously, a model of π_h.

5.7 Definition. A *limit sentence* is a sentence of the following type:

$$(\forall x_1, \dots, x_n)\big(\varphi(x_1, \dots, x_n) \Rightarrow (\exists! y_1, \dots y_m)\psi(x_1, \dots, x_n, y_1, \dots, y_m)\big)$$

where φ and ψ are conjunctions of atomic formulas. A set of limit sentences is called a *limit theory*.

Remark. For each limit sentence σ the subcategory $\mathbf{Mod}\,\sigma$ is closed in $\mathbf{Str}\,\Sigma$ under limits. In fact, closedness under products follows from the coordinate-wise satisfaction of formulas, and closedness under equalizers from the fact that we use only "unique existence" (i.e., $\exists!$).

5.8 Examples

(1) Any implication (see Section 3.B) is a limit sentence.

(2) The "orthogonality formula" above is a limit sentence.

(3) We know that the theory of small categories is given by an essentially algebraic theory. This can be formalized as a limit theory if we replace the partial composition-operation $x \cdot y = z$ by a ternary relation. Thus, let Σ_{cat} be the one-sorted signature of

$$\begin{aligned} (\Sigma_{\text{cat}})_{\text{ope}} &= \{\text{dom}, \text{codom}\} && \text{both of arity } s \to s \\ (\Sigma_{\text{cat}})_{\text{rel}} &= \{\text{comp}\} && \text{of arity } s \times s \times s. \end{aligned}$$

Small categories are given by the following limit theory:

$$\begin{aligned} &(\forall x, y, z)\big(\text{comp}(x, y, z) \Rightarrow \text{dom}(x) = \text{codom}(y)\big) \\ &(\forall x, y)\big(\text{dom}(x) = \text{codom}(y) \Rightarrow (\exists! z)\text{comp}(x, y, z)\big) \\ &(\forall x)\big(\text{comp}(x, \text{dom}(x), x) \wedge \text{comp}(\text{codom}(x), x, x)\big) \\ &(\forall x_1, \dots, x_6)\big(\text{comp}(x_1, x_2, x_3) \wedge \text{comp}(x_3, x_4, x_5) \\ &\qquad\qquad \Leftrightarrow \text{comp}(x_2, x_4, x_6) \wedge \text{comp}(x_1, x_6, x_5)\big) \end{aligned}$$

(where the last axiom is a conjunction of two limit sentences).

(4) Analogously, any finitary essentially algebraic category (see 2.31) can be axiomatized by a limit theory:

5.9 Characterization Theorem. *A category is locally finitely presentable iff it is equivalent to a category of models of a limit theory in L_ω.*

PROOF. Necessity follows from Theorem 1.46 and Example 1.41: any locally finitely presentable category $\mathcal{K}$ is equivalent to an orthogonality class in some $\mathbf{Rel}\,\Sigma$, given by a set of homomorphisms $h_i\colon A_i \to B_i$, $i \in I$, with A_i, B_i finitely presentable. Then $\mathcal{K}$ is axiomatizable by the limit theory $\{\pi_{h_i} \mid i \in I\}$ where π_{h_i} is the orthogonality formula of Remark 5.6 above.

Conversely, let T be a limit theory in L_ω. We will prove that $\mathbf{Mod}\,T$ is closed under limits and directed colimits in $\mathbf{Str}\,\Sigma$. It follows that $\mathbf{Mod}\,T$ is finitely presentable by the reflection theorem, 2.46, and Remark 5.1(5).

I. Limits. Let D be a diagram with a limit $(A \xrightarrow{a_d} Dd)_{d \in \mathcal{D}^{\mathrm{obj}}}$ in $\mathbf{Str}\,\Sigma$ and with each Dd a model of T. From the description of limits in Remarks 3.4(1) and 5.1(2) it is easy to see that given a conjunction $\varphi(x_1, \ldots, x_n)$ of atomic formulas and an interpretation $f\colon \{x_i\} \to |A|$ of variables in A, then

$$A \models \varphi[f] \qquad \text{iff} \qquad Dd \models \varphi[a_d \cdot f] \quad \text{for each } d \in \mathcal{D}^{\mathrm{obj}}.$$

Now consider any formula

$$\varrho = (\forall x_1, \ldots, x_n)\big(\varphi(x_1, \ldots, x_n) \Rightarrow (\exists! y_1, \ldots y_m)\psi(x_1, \ldots, x_n, y_1, \ldots, y_m)\big)$$

of T. To prove that A is a model of ϱ, let $f\colon \{x_i\} \to |A|$ be an interpretation of the variables $\{x_i\}$ in ϱ with $A \models \varphi[f]$. For each d we have $Dd \models \varphi[a_d \cdot f]$, and since Dd is a model of ϱ, there exists a unique interpretation $g_d\colon \{y_j\} \to |Dd|$ of the variables $\{y_j\}$ of ϱ in Dd with $Dd \models \psi[a_d \cdot f, g_d]$. For each morphism $\delta\colon d \to d'$ of $\mathcal{D}$ we have $D\delta \cdot g_d = g_{d'}$ (in fact, $g_{d'}$ is unique and, since $D\delta\colon Dd \to Dd'$ is a Σ-homomorphism, it is easy to see that $Dd' \models \psi[a_{d'} \cdot f, D\delta \cdot g_d]$), thus, the cone (g_d) is compatible for

$$|{-}| \cdot D\colon \mathcal{D} \to \mathbf{Set}^S.$$

This implies (since $|{-}|$ preserves limits) that there exists a unique interpretation $g\colon \{y_j\} \to A$ with $g_d = a_d \cdot g$ for each $d \in \mathcal{D}^{\mathrm{obj}}$. Now

$$Dd \models \psi[a_d \cdot f, a_d \cdot g] \qquad \text{for all } d \in \mathcal{D}^{\mathrm{obj}}$$

implies

$$A \models \psi[f, g],$$

and g is obviously unique with the last property. Therefore, A is a model of ϱ for each $\varrho \in T$.

II. Directed colimits. Let $(D_t \xrightarrow{a_t} A)_{t \in T}$ be a directed colimit in $\mathbf{Str}\,\Sigma$. From the description of directed colimits in Remarks 3.4(4) and 5.1(3) it is easy to see that given a conjunction $\varphi(x_1, \ldots, x_n)$ of atomic formulas and an interpretation $f\colon \{x_i\} \to |A|$ of variables in A, then $A \models \varphi[f]$ iff $D_t \models \varphi[f']$ for some $t \in T$ and for an essentially unique interpretation $f'\colon \{x_i\} \to |D_t|$ with $f = a_t \cdot f'$. It follows easily that if each D_t is a model of a formula ϱ as above, then A is a model of ϱ too. □

Universal Horn Theories

Quasivarieties of universal algebras, as studied in Section 3.B, have a direct generalization to classes of models of Horn theories:

5.10 Definition. By a *universal Horn theory* is meant a theory consisting of sentences of the following form:

$$(\forall x_1, \ldots, x_n)(\varphi_1 \wedge \cdots \wedge \varphi_m \Rightarrow \varphi)$$

where $\varphi_1, \ldots, \varphi_m$, and φ are atomic formulas.

5.11 Examples

(1) The theory of partially ordered sets (Example 5.4(3)) is a universal Horn theory.

(2) Any set of implications is a universal Horn theory.

(3) In contrast, the theory of linearly ordered sets (or ordered groups) is not a universal Horn theory because of the linearity axiom.

5.12. We now generalize the characterization of finitary quasivarieties as SP-classes closed under directed colimits (see Theorem 3.22):

Theorem. *Categories of models of universal Horn theories of signature Σ are precisely the full subcategories of* $\mathbf{Str}\,\Sigma$ *closed under products, substructures, and directed colimits.*

PROOF. I. Let $\mathcal{K}$ be a full subcategory of $\mathbf{Str}\,\Sigma$ closed under products, substructures, and directed colimits. Since by Remark 5.1 $\mathbf{Str}\,\Sigma$ is complete, wellpowered, and has (epi, strong mono)-factorizations of morphisms, it follows that $\mathcal{K}$ is an epireflective subcategory of $\mathbf{Str}\,\Sigma$, see 0.8. Then $\mathcal{K}$, being closed under directed colimits, is the orthogonality class

$$\mathcal{K} = \{\, c_A \mid A \in \mathbf{Pres}_\omega\, \mathbf{Str}\,\Sigma \,\}$$

of reflections $c_A : A \to A^*$ of finitely presentable Σ-structures A in $\mathcal{K}$ (see Theorem and Remark 1.39). Each c_A is surjective in every sort, thus, we can find presentation formulas (see Remark 5.5) for A and A^* using the same variables $x_1, \ldots, x_n$. The orthogonality to c_A is then axiomatized by the formula

$$(\forall x_1, \ldots, x_n)(\pi_A \Rightarrow \pi_{A^*})$$

and, since both π_A and π_{A^*} are conjunctions of atomic formulas, that sentence can, obviously, be substituted by a set of universal Horn sentences.

II. If a structure A satisfies a universal Horn sentence, then every substructure A' satisfies that sentence too. This follows from the fact that for each atomic formula the satisfaction under assignment of values in A' is interpreted in A and A' in the same way. Thus, categories of models of universal Horn theories are closed under substructures. The argument for products and directed colimits is analogous, due to the description of limits and directed colimits in Remarks 3.4 and 5.1. □

5.13 Remarks

(1) We see that categories of models of universal Horn theories are precisely the full, epireflective subcategories of $\mathbf{Str}\,\Sigma$ closed under directed colimits.

(2) Let T be a universal Horn theory. Regular epimorphisms in the category $\mathbf{Mod}\,T$ are surjective (the verification is the same as in Remark 3.4(2)). In contrast to algebraic theories, the converse is not true. Example: the morphism $id_X(X, \emptyset) \to (X, X \times X)$ for $X \neq \emptyset$ is surjective but not a regular epimorphism in $\mathbf{Gra}$.

(3) For each universal Horn theory, the category $\mathbf{Mod}\,T$ has a generator $\mathcal{P}$ formed by regular projectives (see Remark 3.4(5)); e.g., the set of reflections of free Σ-structures generated by finitely presentable sets (see Remark 5.1(6)).

Example of a locally finitely presentable category which is not equivalent to any category of models of a universal Horn theory. Let Σ be a (one-sorted) signature of two unary relation symbols α, β and one binary relation symbol ϱ. Let $\mathcal{K}$ be the category of models of the following sentence:

$$(\forall x)\big(\alpha(x) \wedge \beta(x) \Rightarrow (\exists! y)\varrho(x,y)\big).$$

The category $\mathcal{K}$ is locally finitely presentable by Theorem 5.9. We will show that $\mathcal{K}$ does not have a generator formed by regular projectives. Let A be the structure in $\mathcal{K}$ of one element with all relations empty. Let $B \in \mathcal{K}$ have two elements a, b and $\alpha_B = \{a\}$, $\beta_B = \{b\}$, $\varrho_B = \emptyset$. We can describe a coequalizer

$$A \underset{v}{\overset{u}{\rightrightarrows}} B \xrightarrow{w} C$$

in $\mathcal{K}$, where $u(a) = a$, $v(a) = b$ as follows: $|C| = \{c, d\}$, $\alpha_C = \beta_C = \{c\}$, $\varrho_C = \{(c, d)\}$, and $w(a) = w(b) = c$. Since w is not surjective, for any regular projective object P of $\mathcal{K}$ we have $\alpha_P \cup \beta_P \neq \emptyset$. Therefore, no morphism from P to A exists.

Characterization of Purity

Recall from Definition 2.27 that a morphism $f\colon A \to B$ is called *ω-pure* provided that in each commutative square

$$\begin{array}{ccc} A' & \xrightarrow{f'} & B' \\ {\scriptstyle u}\downarrow & & \downarrow{\scriptstyle v} \\ A & \xrightarrow[f]{} & B \end{array}$$

with A' and B' finitely presentable, the morphism u factorizes through f'. We will characterize ω-pure morphisms in $\mathbf{Str}\,\Sigma$ by means of preservation of the satisfaction of certain formulas. We can suppose, without loss of generality, that A is a substructure of B and f is the inclusion map, see Proposition 2.31 and Remark 5.1(1).

5.14 Definition. A formula is called *positive-primitive* if it has the following form

$$(\exists\, y_1, \dots, y_m)\psi(x_1, \dots, x_n, y_1, \dots, y_m),$$

where ψ is a conjunction of atomic formulas.

5.15 Proposition. *A substructure A of a structure B (of finitary signature) is ω-pure iff for each positive-primitive formula $\varphi(x)$ and each assignment a of values in A we have*

$$A \models \varphi[a] \qquad \textit{iff} \qquad B \models \varphi[a].$$

Remark. Since A is a substructure of B, it is obvious that $A \models \varphi[a]$ always implies $B \models \varphi[a]$. Thus, it is only the reverse implication that matters.

PROOF. I. Let A be an ω-pure substructure of B, and let

$$\varphi(x_1, \dots, x_n) = (\exists\, y_1, \dots, y_m)\, \psi(x_1, \dots, x_n, y_1, \dots, y_m)$$

be a formula as in Definition 5.14. There exists a finitely presentable Σ-structure Q for which ψ is a presentation formula. In fact, suppose that

$$\psi = (\tau_1 = \tau_1') \wedge \dots \wedge (\tau_k = \tau_k') \wedge \bigwedge_{i=1}^{m} \sigma_i(\varrho_{i,1}, \dots, \varrho_{i,r(i)})$$

and let $\overline{Q}$ be the quotient algebra of the free algebra of signature Σ_{ope} over the variables $\{x_1, \dots, x_n, y_1, \dots, y_m\}$ modulo the smallest congruence $\sim$

with $\tau_1 \sim \tau_1', \ldots, \tau_k \sim \tau_k'$. Then Q is the Σ-structure obtained from $\overline{Q}$ by defining the relations as follows:

$$\sigma_Q = \{ ([\varrho_{i,1}], \ldots, [\varrho_{i,r(i)}]) \mid i = 1, \ldots, m \text{ with } \sigma_i = \sigma \} \qquad \text{for } \sigma \in \Sigma_{\text{rel}}.$$

Then we have a canonical interpretation $b\colon \{x_1, \ldots, x_n, y_1, \ldots, y_m\} \to |Q|$ of variables in Q such that homomorphisms $f\colon Q \to B$ precisely correspond to assignments $f \cdot b\colon \{x_1, \ldots, x_n, y_1, \ldots, y_m\} \to |B|$ with $B \models \psi[f \cdot b]$. Denote by $g\colon P \to Q$ the unique homomorphism extending $x_i \mapsto b(x_i)$.

For each assignment $a\colon \{x_1, \ldots, x_n\} \to |A|$ such that $B \models \varphi[a]$ we want to prove that $A \models \varphi[a]$. There exists an extension to an assignment $\overline{a}\colon \{x_1, \ldots, x_n, y_1, \ldots, y_m\} \to |B|$ with $B \models \psi[\overline{a}]$. This means that there exists a homomorphism $f\colon Q \to B$ with $\overline{a} = f \cdot b$. Let P be the free Σ_{ope}-algebra over $\{x_1, \ldots, x_n\}$ and let $a^*\colon P \to A$ be the unique extension of a to a homomorphism $a^*\colon P \to A$. The following square

$$\begin{array}{ccc} P & \xrightarrow{g} & Q \\ {\scriptstyle a^*}\downarrow & & \downarrow{\scriptstyle f} \\ A & \hookrightarrow & B \end{array}$$

commutes. Since A is an ω-pure substructure of B, and since both P and Q are finitely presentable Σ-structures, there exists a homomorphism $h\colon Q \to A$ with $a^* = h \cdot g$. Then $\psi = \pi_Q$ implies $A \models \psi[h \cdot b]$, thus, $A \models \varphi[a]$.

II. Let A be a substructure of B satisfying the above condition. Let

$$\begin{array}{ccc} P & \xrightarrow{g} & Q \\ {\scriptstyle u}\downarrow & & \downarrow{\scriptstyle v} \\ A & \hookrightarrow & B \end{array}$$

be a commutative square in $\mathbf{Str}\,\Sigma$ with P and Q both finitely presentable. Consider the orthogonality formula of g (see Remark 5.7) and denote by ψ the following subformula of π_g:

$$\psi = \pi_P(y_1, \ldots, y_m) \wedge (x_1 = \tau_1) \wedge \cdots \wedge (x_n = \tau_n).$$

The assignment $a\colon x_i \mapsto u(x_i)$ has the property that $B \models \psi^*[a]$ for

$$\psi^* = (\exists\, y_1, \ldots, y_m)\, \psi(x_1, \ldots, x_n, y_1, \ldots, y_m).$$

Since ψ^* is a positive-primitive formula, we know that $A \models \psi^*[a]$, which implies, due to the part π_Q of ψ, that u factorizes through g. $\square$

Axiomatizability

We now present some basic properties of categories axiomatizable in the finitary logic L_ω: each such category is closed under ultraproducts (see below) in $\mathbf{Str}\,\Sigma$, consequently, if it is closed under finite limits, then it is closed under all limits (in fact, then it is locally presentable). Moreover, directed colimits in each axiomatizable category are "standard", i.e., formed as in $\mathbf{Str}\,\Sigma$. We use these results to show that a logical characterization of finitely accessible categories (in the logic L_ω) is not possible: there exist finitely accessible categories which are not axiomatizable in the logic L_ω.

5.16 Definition. A class of Σ-structures is called *axiomatizable* (in the logic L_ω) provided that it is equivalent to $\mathbf{Mod}\,T$ for some theory T of Σ-structures (in L_ω).

5.17 Remark. Recall that a collection $\mathcal{U}$ of subsets of a set I is called an *ultrafilter* provided that

(1) $\mathcal{U}$ is closed under super-sets, i.e., $U \in \mathcal{U} \Rightarrow U' \in \mathcal{U}$ for each $U \subseteq U'$;

(2) $\mathcal{U}$ is closed under finite intersections, i.e., $U, U' \in \mathcal{U} \Rightarrow (U \cap U') \in \mathcal{U}$;

(3) for each set $U \subseteq I$, either $U \in \mathcal{U}$ or $I - U \in \mathcal{U}$.

We will use the following fact (equivalent to the axiom of choice, see [Jech 1978]): given a collection $\mathcal{U}_0$ of subsets of I such that any two members of $\mathcal{U}_0$ have a non-empty intersection, there exists an ultrafilter containing $\mathcal{U}_0$.

For each collection A_i ($i \in I$) of Σ-structures and each ultrafilter $\mathcal{U}$ on the index set I we denote by

$$\prod\nolimits_{\mathcal{U}} A_i$$

the *ultraproduct* of the structures A_i modulo $\mathcal{U}$, which is a colimit of the directed diagram of all products $\prod_{i \in U} A_i$, where U ranges through $\mathcal{U}$, and all the canonical maps

$$\prod_{i \in U} A_i \to \prod_{j \in U'} A'_j,$$

where $U' \subseteq U$ are members of $\mathcal{U}$.

A more concrete description: $\prod_{\mathcal{U}} A_i$ is the quotient of the cartesian product $\prod_{i \in I} A_i$ modulo the equivalence $\sim$ defined by

$$(x_i)_{i \in I} \sim (y_i)_{i \in I} \quad \text{iff } \mathcal{U} \text{ contains } \{\, i \in I \mid x_i = y_i \,\}.$$

Thus, the elements of sort s in the ultraproduct $A = \prod_{\mathcal{U}} A_i$ are the equivalence classes $[x_i]_{i\in I}$ of elements $(x_i)_{i\in I} \in \prod_{i\in I}(A_i)_s$. Operations are defined coordinate-wise: given $\tau \in \Sigma_{\text{ope}}$, $\tau\colon s_1 \times \cdots \times s_n \to s$,

$$\tau_A([a_{1,i}], \ldots, [a_{n,i}]) = [\tau_{A_i}(a_{1,i}, \ldots, a_{n,i})].$$

Relations are determined as follows: given $\sigma \in \Sigma_{\text{rel}}$ of arity $s_1 \times \cdots \times s_n$, then

$$([a_{1,i}], \ldots, [a_{n,i}]) \in \sigma_A \quad \text{iff } \mathcal{U} \text{ contains } \{\, i \in I \mid (a_{1,i}, \ldots, a_{n,i}) \in \sigma_{A_i} \,\}.$$

5.18 Proposition. *Every axiomatizable class in L_ω is closed under ultraproducts (i.e., it contains the ultraproduct $\prod_{\mathcal{U}} A_i$ whenever it contains each A_i).*

PROOF. By induction on the complexity of a formula $\varphi(x_1, \ldots, x_n)$ we will prove that given elements $[a_{1,i}]$, ..., $[a_{n,i}]$ in the ultraproduct $\prod_{\mathcal{U}} A_i$ of corresponding sorts, then the following holds:

$$\prod\nolimits_{\mathcal{U}} A_i \models \varphi[[a_{1,i}], \ldots, [a_{n,i}]] \qquad \text{iff}$$
$$\mathcal{U} \text{ contains } \{i \in I \mid A_i \models \varphi[a_{1,i}, \ldots, a_{n,i}]\}.$$

This proves the proposition: if each A_i is a model of φ, then so is $\prod_{\mathcal{U}} A_i$.

(1) Let φ be an atomic formula. Then the statement follows immediately from the above description of $\prod_{\mathcal{U}} A_i$.

(2) The cases $\varphi \wedge \varphi'$ and $\varphi \vee \varphi'$ are clear.

(3) For $\neg\varphi$, we have $\prod_{\mathcal{U}} A_i \models \neg\varphi[[a_{1,i}], \ldots, [a_{n,i}]]$ iff it is not true that $\prod_{\mathcal{U}} A_i \models \varphi[[a_{1,i}], \ldots, [a_{n,i}]]$, and this holds iff $\mathcal{U}$ does not contain

$$\{i \in I \mid A_i \models \varphi[[a_{1,i}], \ldots, [a_{n,i}]]\}.$$

The last holds iff $\mathcal{U}$ contains the complement of that subset of I, which is the set $\{i \in I \mid A_i \models \neg\varphi[a_{1,i}, \ldots, a_{n,i}]\}$.

(4) The case $((\exists x)\varphi)$. Suppose that $\varphi(x, x_1, \ldots, x_n)$ is a formula with

$$\prod\nolimits_{\mathcal{U}} A_i \models (\exists x)\varphi(x, x_1, \ldots, x_n)[[a_{1,i}], \ldots, [a_{n,i}]].$$

Then we will verify that $\mathcal{U}$ contains

$$U = \{\, i \in I \mid A_i \models (\exists x)\varphi(x, x_1, \ldots, x_n)[a_{i,1}, \ldots, a_{i,n}] \,\}.$$

By the definition of $\exists$, there exists an element $[b_i]$ of $\prod_{\mathcal{U}} A_i$ for which $\prod_{\mathcal{U}} A_i \models \varphi\big[[b_i], [a_{i,1}], \ldots, [a_{i,n}]\big]$. This last is equivalent to $\mathcal{U}$ containing

$$\{\, i \in I \mid A_i \models \varphi(b_i, a_{i,1}, \ldots, a_{i,n}) \,\}$$

which is a subset of U. Thus $U \in \mathcal{U}$.

Conversely, if $U \in \mathcal{U}$, we will verify that there is an element $[b_i]$ with $\prod_{\mathcal{U}} A_i \models \varphi\big[[b_i], [a_{i,1}], \ldots, [a_{i,n}]\big]$. For each $i \in U$ we can choose $b_i \in A_i$ with $A_i \models \varphi[b_i, a_{i,1}, \ldots, a_{i,n}]$. We further choose an arbitrary $b_j \in A_j$ for any $j \in I - U$, and we obtain an element $[b_i]_{i \in I}$ of $\prod_{\mathcal{U}} A_i$. By the induction hypothesis, $\prod_{\mathcal{U}} A_i \models \varphi\big[[b_i], [a_{i,1}], \ldots, [a_{i,n}]\big]$.

(5) The case $(\forall x)\varphi$ follows from (4) and (3). □

5.19 Corollary (Compactness Theorem). *A theory T in L_ω has a model iff every finite theory $T' \subseteq T$ has a model.*

PROOF. Let I denote the set of all finite subsets of T, and for each $i \in I$ put

$$\uparrow i = \{\, j \in I \mid i \subseteq j \,\}.$$

These sets are pairwise non-disjoint, thus, we can choose an ultrafilter $\mathcal{U}$ containing each $\uparrow i$.

If for each $i \in I$ there exists a model A_i of the theory i, then we form the ultraproduct $\prod_{\mathcal{U}} A_i$. Observe that for every sentence σ in T we have

$$\uparrow i \subseteq \{\, j \in I \mid A_j \models \sigma \,\} \qquad \text{whenever } \sigma \in i.$$

Consequently, $\mathcal{U}$ contains $\{\, j \in I \mid A_j \models \sigma \,\}$, therefore, $\prod_{\mathcal{U}} A_i$ is a model of every $\sigma \in T$. □

5.20 Theorem. *Every axiomatizable class of Σ-structures in L_ω, closed under finite limits in* **Str** Σ, *is closed under limits and directed colimits in* **Str** Σ.

Remark. We will prove a somewhat stronger result: every axiomatizable class $\mathcal{A}$ closed in **Str** Σ under equalizers is closed under

(1) directed limits,*

(2) directed colimits.

*Directed limits, the dual concept to directed colimits, are limits of diagrams over down-directed posets. (That is, posets $(I, \leq)$ such that each pair of elements of I has a lower bound.)

The theorem then follows: if $\mathcal{A}$ is, moreover, closed under finite products, than it is closed under products (because every product $\prod_{i\in I} A_i$ is a directed limit of all products $\prod_{j\in J} A_j$ with $J \subseteq I$ finite) and thus closed under limits.

PROOF. For simplicity of notation, we assume that Σ is a one-sorted signature. The reader will have no difficulty in verifying that the proof "works" for many-sorted signatures too. Let $\mathcal{A}$ be as a full subcategory of $\mathbf{Str}\,\Sigma$ closed under equalizers and ultraproducts (see Proposition 5.18).

I. Let us prove that $\mathcal{A}$ is closed under directed limits. That is, if

$$D\colon (I, \leq) \to \mathcal{A}$$

is a down-directed diagram, then $\mathcal{A}$ contains the limit-object of D in $\mathbf{Str}\,\Sigma$. By Remark 5.1, this limit-object is the substructure L of $\prod_{i\in I} D_i$ over all I-tuples $(x_i)_{i\in I}$, $x_i \in D_i$, satisfying

$$D_{j,i}(x_j) = x_i \quad \text{for all } j \leq i \text{ in } I$$

where $D_{j,i}$ are the connecting maps of the diagram D.

Since $(I, \leq)$ is down-directed, the down-sets $\downarrow i = \{j \in I \mid j \leq i\}$ have non-empty pairwise intersections. Thus, there exists an ultrafilter $\mathcal{U}$ on I containing $\downarrow i$ for each $i \in I$. Given $i_0 \in I$, form the ultrapower of D_{i_0} modulo $\mathcal{U}$,

$$D^*_{i_0} = \prod\nolimits_{\mathcal{U}} D_{i_0},$$

which is a quotient of $D^I_{i_0}$ lying in $\mathcal{A}$. Thus, the ultraproducts

$$A = \prod\nolimits_{\mathcal{U}} D_i \qquad \text{and} \qquad A^* = \prod\nolimits_{\mathcal{U}} D^*_i$$

both lie in $\mathcal{A}$. Define homomorphisms $f_1, f_2\colon A \to A^*$ by

$$f_1([x_i]_{i\in I}) = [[x_i, x_i, x_i, \ldots]]_{i\in I}$$

and

$$f_2([x_i]_{i\in I}) = [[D_{j,i}(x_j)]_{j\leq i}]_{i\in I}.$$

(More precisely, $[D_{j,i}(x_j)]_{j\leq i}$ denotes the equivalence class of any $(a_j)_{j\in I} \in D^I_i$ with $a_j = D_{j,i}(x_j)$ for each $j \leq i$.) The morphism f_1 is clearly well-defined; to verify the same about f_2, we must show that if $[x_i] = [x'_i]$, then there exists $U \in \mathcal{U}$ such that for each $i \in U$, $[D_{j,i}(x_j)]_{j\leq i} = [D_{j,i}(x'_j)]_{j\leq i}$.

In fact, the set $U = \{i \in I \mid x_i = x'_i\}$ lies in $\mathcal{U}$, and for each $i \in U$ we know that $U \cap \downarrow i \in \mathcal{U}$ and $D_{j,i}(x_j) = D_{j,i}(x'_j)$, for all $j \in U \cap \downarrow i$.

By hypothesis, $\mathcal{A}$ contains the equalizer of f_1, f_2, which is the substructure E of $\prod_{\mathcal{U}} D_i$ over all $[x_i]_{i\in I}$ for which there exists $U \in \mathcal{U}$ such that

$$(1) \qquad i \in U \implies \mathcal{U} \text{ contains } U_i \stackrel{\text{def}}{=} \{\, j \in I \mid j \leq i, D_{j,i}(x_j) = x_i \,\}.$$

We shall prove that E is isomorphic to L; hence, L lies in $\mathcal{A}$.

Define a morphism $h\colon L \to E$ by

$$h(x_i)_{i\in I} = [x_i]_{i\in I}.$$

This is well-defined since for each $(x_i)_{i\in I}$ in L we have $D_{j,i}(x_j) = x_i$ for all $j \leq i$, thus, $[x_i]_{i\in I}$ lies in E: put $U = I$ in the above condition, then $U_i = \downarrow i \in \mathcal{U}$. The morphism h is one-to-one. In fact, if two elements $(x_i)_{i\in I}$ and $(y_i)_{i\in I}$ of L are merged by h, then there exists $V \in \mathcal{U}$ with $x_i = y_i$ for each $i \in V$. For any $j \in I$ choose $i \in V$ with $i \leq j$ (which is possible, since $V \cap \downarrow j \in \mathcal{U}$) to conclude that

$$x_j = D_{i,j}(x_i) = D_{i,j}(y_i) = y_j;$$

thus, $(x_i)_{i\in I} = (y_i)_{i\in I}$. The morphism h is also surjective: given $[x_i]_{i\in I}$ in E, we have U as in (1) above, and we define $(y_i)_{i\in I}$ by

$$(2) \qquad y_i = D_{j,i}(x_j) \qquad \text{for any } j \in U,\, j \leq i.$$

(This is independent of j: given $j', j'' \in U$, there exists $j \in U_{j'} \cap U_{j''}$ and we have $D_{j',i}(x_{j'}) = D_{j',i}D_{j,j'}(x_j) = D_{j,i}(x_j)$; analogously for $D_{j'',i}(x_{j''})$.) It is obvious that $(y_i) \in L$ and $h(y_i) = [x_i]$: for $i \in U$, we have $x_i = y_i$. Consequently, h is a bijection. To see that h is an isomorphism, it remains to show that for each n-ary relation symbol σ of Σ the preimage of σ_E under h^n is contained in σ_L. This follows from the definition of the ultraproduct $A = \prod_{\mathcal{U}} D_i$: any n-tuple in σ_E has the form $\big([x_i^1]_{i\in I}, \ldots, [x_i^n]_{i\in I}\big)$ where the n-tuple $\big((x_i^1)_{i\in I}, \ldots, (x_i^n)_{i\in I}\big)$ lies in $\sigma_{\prod D_i}$; that is, for each $i \in I$ we have $(x_i^1, \ldots, x_i^n) \in \sigma_{D_i}$. For every $t = 1, \ldots, n$ let $(y_i^t) \in L$ be the above-constructed I-tuple mapped by h to (x_i^t), see (2). Since j in (2) can be chosen independently of $t = 1, \ldots, n$ we can easily see that $(y_i^1, \ldots, y_i^n) \in \sigma_{D_i}$ for each $i \in I$. Consequently, the n-tuple $\big((y_i^1), \ldots, (y_i^n)\big)$ lies in σ_L, and it is mapped by h^n to the given n-tuple $\big([x_i^1], \ldots, [x_i^n]\big)$.

II. Let us prove that $\mathcal{A}$ is closed under directed colimits. That is, if $D\colon (I, \leq) \to \mathcal{A}$ is an (up-)directed diagram, then $\mathcal{A}$ contains the colimit object of D in $\mathbf{Str}\,\Sigma$. This colimit object, as described in Remark 5.1, is the naturally defined structure C on the quotient set of $\bigcup_{i\in I} |D_i| \times \{i\}$ under the equivalence $\sim$ defined by

$$(x,i) \sim (x',i') \qquad \text{iff} \qquad D_{i,j}(x) = D_{i',j}(x') \quad \text{for some } j \geq i,\, j \geq i'.$$

Since $(I, \leq)$ is directed, there exists an ultrafilter $\mathcal{U}$ on I containing each of the sets $\uparrow i = \{j \in I \mid j \geq i\}$ for $i \in I$. Define, as above, $D^*_{i_0} = \prod_{\mathcal{U}} D_{i_0}$ and $A = \prod_{\mathcal{U}} D_i$, $A^* = \prod_{\mathcal{U}} D^*_i$. Further define $f_1, f_2 \colon A \to A^*$ by

$$f_1([x_i]_{i\in I}) = [[x_i, x_i, x_i, \dots]]_{i\in I},$$
$$f_2([x_i]_{i\in I}) = [[D_{i,j}(x_i)_{i\leq j}]_{i\in I}.$$

Then $\mathcal{A}$ contains the equalizer of f_1, f_2, which is the substructure E of $\mathcal{A}$ over all $[x_i]_{i\in I}$ for which there exists $U \in \mathcal{U}$ such that

$$i \in U \implies \mathcal{U} \text{ contains } U_i \stackrel{\text{def}}{=} \{\, j \in I \mid j \geq i,\, D_{i,j}(x_i) = x_j \,\}.$$

The morphism

$$h\colon C \to E, \qquad h([x,i]) = [D_{i,j}(x)]_{j\geq i}$$

is one-to-one. In fact, given (x,i), (x',i') with $[D_{j,i}(x)]_{j\geq i} = [D_{j,i'}(x')]_{j\geq i'}$, there exists $U \in \mathcal{U}$ such that for each $j \in U \cap (\uparrow i) \cap (\uparrow i')$ $(\in \mathcal{U})$ we have $D_{i,j}(x) = D_{i',j}(x')$, thus, $[x,i] = [x',i']$. The morphism h is also surjective: for each $[x_i]_{i\in I} \in E$ we choose U as above, then given any $i_0 \in U$ we have $h([x_{i_0}, i_0]) = [x_i]_{i\in I}$; in fact, $U_{i_0} \in \mathcal{U}$ and for each $j \in U_{i_0}$ we know that $x_j = D_{i_0,j}(x_{i_0})$. To see that h is an isomorphism, it remains to show that for each n-ary relation symbol $\sigma \in \Sigma$ the pre-image of σ_E under h^n is contained in σ_C. Any n-tuple $([x^1_i], \dots, [x^n_i])$ in σ_E can be chosen so that $(x^1_i, \dots, x^n_i) \in \sigma_{D_i}$ for each $i \in I$ (see Remark 5.1(3)). For every $t = 1, \dots, n$ we have $U^t \in \mathcal{U}$ corresponding to $[x^t_i] \in E$, and when choosing $i_0 \in \bigcap_{t=1}^n U^t$, we get $h([x^t_{i_0}, i_0]) = [x^t_i]_{i\in I}$ for $t = 1, \dots, n$. Since C is a quotient of $\coprod_{i\in I} D_i$ and $(x^1_{i_0}, \dots, x^n_{i_0}) \in \sigma_{D_{i_0}}$, we conclude that $([x^1_{i_0}, i_0], \dots, [x^n_{i_0}, i_0]) \in \sigma_C$. □

5.21 Corollary. *Every axiomatizable class of Σ-structures in L_ω closed under finite limits in* **Str** Σ *is a locally finitely presentable category (as a full subcategory of* **Str** Σ*).*

In fact, this follows from Theorems 5.20 and 2.48; the subcategory is reflective in **Str** Σ.

5.22 Example of a finitely accessible category which is not axiomatizable in the logic L_ω. Let $\mathcal{A}$ be the full subcategory of **Alg** Σ where Σ is the one-sorted signature of operations σ (unary) and α (nullary), given by the condition $\sigma^n(\alpha) = \sigma^{n-1}(\alpha)$ for some $n \geq 2$. $\mathcal{A}$ is finitely accessible. However, $\mathcal{A}$ is not axiomatizable in L_ω because it is closed under finite limits in $\mathbf{Alg}\,\Sigma$, but is not closed under infinite products. A stronger property of $\mathcal{A}$ is shown in Exercise 5g.

5.23 Theorem. *Directed colimits in axiomatizable categories are "standard", i.e., the inclusion* $\mathbf{Mod}\,T \hookrightarrow \mathbf{Str}\,\Sigma$ *preserves directed colimits for each theory* T *in* L_ω.

PROOF. Let $D\colon I \to \mathbf{Mod}\,T$ be a directed diagram with colimits

$$\left(D_i \xrightarrow{a_i} A\right)_{i\in I} \qquad \text{in } \mathbf{Mod}\,T$$

and

$$\left(D_i \xrightarrow{\overline{a}_i} \overline{A}\right)_{i\in I} \qquad \text{in } \mathbf{Str}\,\Sigma.$$

We will first prove that the homomorphism $h\colon \overline{A} \to A$ defined by $h \cdot a_i = \overline{a}_i$ $(i \in I)$ is ω-pure, and then we will conclude that h is an isomorphism.

Denote by $\overline{\Sigma}$ the extension of Σ by a nullary operation symbol

$$c_a \quad \text{of sort } s$$

for each element a of $\overline{A}$ of sort s. We consider $\overline{A}$ as a $\overline{\Sigma}$-structure. We can characterize ω-pure Σ-homomorphisms $f\colon \overline{A} \to B$ into models B of T via an extension of T: let $\overline{T}$ be the theory in signature $\overline{\Sigma}$ whose axioms are

(1) all axioms of T,

(2) all positive-primitive sentences (Definition 5.14) of signature $\overline{\Sigma}$ which are satisfied by A,

(3) all negated positive-primitive sentences of signature $\overline{\Sigma}$ which are satisfied by $\overline{A}$.

Let B be a model of $\overline{T}$. We obtain (i) a model B_0 of T by forgetting the constants c_a and (ii) a Σ-homomorphism $f\colon \overline{A} \to B_0$ assigning to each a in $|\overline{A}|$ the interpretation of the constant c_a in B. By Proposition 5.15, the homomorphism f is ω-pure. Conversely, given a model B_0 of T and an ω-pure Σ-homomorphism $f\colon \overline{A} \to B_0$, we obtain a model B of $\overline{T}$ by interpreting each c_a as the value of f.

To prove that h is an ω-pure homomorphism, it is sufficient to show that the theory $\overline{T}$ has a model. In fact, we then get an ω-pure homomorphism $f\colon \overline{A} \to B_0$ for B_0 in $\mathbf{Mod}\,T$. The cone $f \cdot \overline{a}_i$ is compatible with D, thus, it factorizes as $f \cdot \overline{a}_i = f' \cdot a_i$ $(i \in I)$ for some $f'\colon A \to B_0$. It follows that

$f = f' \cdot h$. The purity of f thus implies the purity of h, see Remark 2.28. We are going to use the compactness theorem, 5.19: let $T' \subseteq \overline{T}$ be finite. We will prove that there exists $i_0 \in I$ such that D_{i_0} is a model of T' for a "suitable" definition of the constants c_a. We proceed by induction on the size of T': the case $T' = \emptyset$ is clear, and for the induction step it is sufficient to prove that for each sentence φ of one of the above types (1)–(3) there exists $i_0 \in D$ such that each D_{i_0} is a model of φ. The case $\varphi \in T$ is clear. Let us turn to the case that φ is a positive-primitive sentence

$$\varphi = (\exists\, x_1 \ldots x_n)(\psi_1 \wedge \cdots \wedge \psi_k)$$

satisfied by $\overline{A}$, where the $\psi_j(x_1, \ldots, x_n)$ are atomic formulas ($j = 1, \ldots, k$). Each ψ_j is either an equation $\alpha = \beta$ for two terms of signature $\overline{\Sigma}$, or $\psi_j = \sigma(\alpha_1, \ldots, \alpha_n)$ for a relation symbol $\sigma \in \Sigma$ and for terms $\alpha_1, \ldots, \alpha_n$ of signature $\overline{\Sigma}$. A term α of signature $\overline{\Sigma}$ can be "translated" into a term α^* of signature Σ by choosing, for each occurrence of c_a in α, a variable x_a of the same sort (with $x_a \neq x_1, \ldots, x_n$ and $x_a \neq x_b$ if $a \neq b$) and by changing all occurrences of c_a to x_a. We denote by ψ_j^* the corresponding formula of signature Σ: if ψ_j is $\alpha = \beta$, then ψ_j^* is $\alpha^* = \beta^*$; if ψ_j is $\sigma(\alpha_1, \ldots, \alpha_n)$, then ψ_j^* is $\sigma(\alpha_1^*, \ldots, \alpha_n^*)$. Suppose that all elements of $|\overline{A}|$ for which we substituted c_a for x_a in $\psi_1, \ldots, \psi_k$ are $a_1, \ldots, a_m$, and denote

$$\varphi^* = (\exists\, x_{a_1} \ldots x_{a_m})(\exists\, x_1 \ldots x_m)(\psi_1^* \wedge \cdots \wedge \psi_k^*).$$

Since $\overline{A}$, considered as a $\overline{\Sigma}$-structure, satisfies φ, it is clear that as a Σ-structure it satisfies φ^*. Using the fact that $\overline{A}$ is a directed colimit of D, we will show that there exists $i_0 \in I$ such that D_{i_0} satisfies φ^*. It is then obvious that D_{i_0} can be turned into a model of φ as follows: interpret the constants $c_{a_1}, \ldots, c_{a_m}$ by the corresponding interpretation of the variables $x_{a_1}, \ldots, x_{a_m}$ of φ^*, and let c_a be arbitrary for $a \neq a_1, \ldots, a_m$.

(a) First, suppose $k = 1$. If ψ_1 is $\alpha = \beta$ for terms α, β, then since $\overline{A}$ satisfies φ, we have an interpretation $x_j \mapsto b_j$ ($j = 1, \ldots, n$) of variables in $|\overline{A}|$ such that $\alpha^*[b_j] = \beta^*[b_j]$ in $\overline{A}$. Since $\overline{A}$ is a directed colimit of D, there exists $i_0 \in I$ and an interpretation $x_j \mapsto b_j'$ of variables in $|D_{i_0}|$ such that $\alpha^*[b_j'] = \beta^*[b_j']$ in D_{i_0} and $\overline{a}_{i_0}(b_j') = b_j$ for $j = 1, \ldots, n$. Then D_{i_0} is a model of φ^*. Analogously with ψ_1 of the form $\sigma(\alpha_1, \ldots, \alpha_n)$.

(b) Let k be arbitrary. Then we first treat each ψ_t ($t = 1, \ldots, k$) separately: we obtain $i_t \in I$ and an interpretation of the variables $x_1, \ldots, x_n$ in D_{i_t} which makes D_{i_t} a model of $(\exists\, x_1 \ldots x_n)\psi_t$. Let i^* be an upper bound of $i_1, \ldots, i_k$ in I. For each of the variables x_n we have at most k different interpretations of x_n in D_{i^*} obtained by mapping the interpretation in D_{i_t} via the connecting maps. Since c_{i^*} maps each of these interpretations to a_n, it follows that there exists $i^+ \geq i^*$ such that the connecting map $D_{i^*} \to D_{i+}$ merges all the different interpretations of x_u to one interpretation (for any $u = 1, \ldots, n$). Thus, D_{i_0} is a model of φ.

The case where φ is a negated positive-primitive formula is quite analogous.

Consequently, by the compactness theorem, $\overline{T}$ has a model, which proves that h is an ω-pure morphism. To prove that h is an isomorphism, we proceed analogously to the proof of Proposition 2.31. The sets $\uparrow i = \{\, i \in I \mid i \leq j \,\}$ are pairwise non-disjoint for $i \in I$. Consequently, there exists an ultrafilter $\mathcal{U}$ on the set I which contains each of them. Let us form the ultrapower

$$A^* = \prod\nolimits_{\mathcal{U}} A_i \qquad A_i = A$$

of the object A. Then we can find homomorphisms

$$p, q \colon A \to A^*$$

such that h is an equalizer of p and q. (This is completely analogous to the proof of Proposition 2.31.) Since A^* is a model of T by Proposition 5.18, the equations

$$p \cdot a_i = p \cdot h \cdot \overline{a}_i = q \cdot h \cdot \overline{a}_i = q \cdot a_i \qquad (i \in I)$$

imply $p = q$. Thus, h is an isomorphism. □

5.B Infinitary Logic

We have seen above that locally finitely presentable categories exactly correspond to finitary limit theories. It is no surprise that for locally λ-presentable categories infinitary logic is needed. But then, again, there is an exact correspondence between locally λ-presentable categories and λ-ary limit theories. In case of accessible categories, we have seen that finitary logic is insufficient. In infinitary logic we introduce the concept of a basic theory and prove an exact correspondence between accessible categories and basic theories.

Formulas, Models, and Satisfaction

5.24 Infinitary Signature. As in the finitary case, we work with a mixed S-sorted signature Σ, where S is an (arbitrary) set of sorts. That is, Σ is a disjoint union of the set Σ_{ope} of operation symbols σ each having an arity

$$\sigma \colon \prod_{i \in I} s_i \to s \qquad (s_i, s \in S)$$

and the set Σ_{rel} of relation symbols σ of arities $\prod_{i\in I} s_i$ $(s_i \in S)$. Analogously to the finitary case we work with category $\mathbf{Str}\,\Sigma$ of Σ-structures and homomorphisms. The signature Σ is called *λ-ary* if λ is a regular cardinal with $\operatorname{card} I < \lambda$ for all the index sets I in the arities of operation and relation symbols of Σ. We assume that a standard many-sorted set V of variables is given with $\operatorname{card} V_s = \lambda$ for each sort s.

Terms are defined analogously to the finitary case: each term is given together with its sort, and

(a) every variable in V_s is a term of sort s,

(b) every expression $\sigma(\tau_i)_{i\in I}$ where $\sigma\colon \prod_{i\in I} s_i \to s$ in an operation symbol and τ_i is a term of sort s_i (for each $i \in I$) is a term of sort s.

Atomic formulas are, analogously to the finitary case, either equations $\tau_1 = \tau_2$ between terms, or expressions $\sigma(\tau_i)_{i\in I}$ where σ is a relation symbol of arity $\prod_{i\in I} s_i$, and τ_i is a term of sort s_i $(i \in I)$.

Formulas are built up from the atomic formulas by applying the following logical operators finitely many times:

$\neg\varphi$	(negation)
$\varphi \Rightarrow \psi$	(implication)
$\bigvee_{j\in J} \varphi_j$	(disjunction indexed by an arbitrary set J of cardinality less than λ)
$\bigwedge_{j\in J} \varphi_j$	(conjunction indexed by an arbitrary set J of cardinality less than λ)
$(\forall X)\varphi$	(universal quantification over an arbitrary set X of less than λ variables)
$(\exists X)\varphi$	(existential quantification over an arbitrary set X of less than λ variables)

Observe that the finite number of applications of these operators means, in particular, that although both $\forall$ and $\exists$ are used with an infinite number of variables, they can be interchanged finitely many times only within one formula.

When no danger of misunderstanding is possible, we write $(\forall x_1, x_2, \dots)$ instead of $(\forall\{x_1, x_2, \dots\})$ and analogously for $\exists$. We use the symbol $\exists!$ of unique existence analogously to the finitary case.

Disjunction and conjunction can also be performed over the empty index set: we write *true* instead of $\bigwedge_{j\in\emptyset} \varphi_j$ and *false* instead of $\bigvee_{j\in\emptyset} \varphi_j$.

5.25 Free Variables. The S-sorted set $\mathrm{Var}\,\varphi$ of *free variables* of a formula φ is defined by induction:

$$\begin{aligned}
\mathrm{Var}\,x &= x\\
\mathrm{Var}\,\sigma(\tau_1,\ldots,\tau_n) &= \bigcup_{i=1}^{n} \mathrm{Var}\,\tau_i\\
\mathrm{Var}\,\neg\varphi &= \mathrm{Var}\,\varphi\\
\mathrm{Var}\,\varphi \Rightarrow \psi &= \mathrm{Var}\,\varphi \cup \mathrm{Var}\,\psi\\
\mathrm{Var}\,\bigvee_{j\in J} \varphi_j &= \bigcup_{j\in J} \mathrm{Var}\,\varphi_j\\
\mathrm{Var}\,\bigwedge_{j\in J} \varphi_j &= \bigcup_{j\in J} \mathrm{Var}\,\varphi_j\\
\mathrm{Var}\,(\forall X)\varphi &= \mathrm{Var}\,\varphi - X\\
\mathrm{Var}\,(\exists X)\varphi &= \mathrm{Var}\,\varphi - X
\end{aligned}$$

Example. The free variables of the formula

$$\bigwedge_{j\in\omega} (x_{j+1} < x_j)$$

are x_0, x_1, x_2, ... , whereas the formula

$$(\exists x_0, x_1, x_2, \ldots) \bigwedge_{j\in\omega} (x_{j+1} < x_j)$$

has no free variables.

Analogously to the finitary case, where $\varphi(x_1,\ldots,x_n)$ was used to indicate $\mathrm{Var}\,\varphi \subseteq \{x_1,\ldots,x_n\}$, we write

$$\varphi(x_i)_{i\in I}$$

to indicate that $\mathrm{Var}\,\varphi \subseteq \{x_i \mid i \in I\}$. For the sake of brevity, we sometimes also abbreviate $\{x_i \mid i \in I\}$ to x and write

$$\varphi(x) \qquad (x \text{ is a mapping from the index set } I \text{ to } V).$$

Analogously, we use

$$(\forall x)\varphi(x), \qquad \text{or just} \quad (\forall x)\varphi,$$

for the formula $(\forall\{x_i \mid i \in I\})\varphi$ with $x\colon I \to V$ given by the x_i.

5.26 Logics L_λ and L_∞. For each regular cardinal λ and each λ-ary signature Σ we define a logic L_λ; then L_∞ is the union of all these logics. First, let us turn to the concept of satisfaction.

Let $\varphi(x_i)_{i\in I}$ be a formula of L_λ in variables x_i of sort s_i ($i \in I$). Given a Σ-structure A and elements a_i of sorts s_i in $|A|$ ($i \in I$), we define the satisfaction symbol

$$A \models \varphi[a_i]_{i\in I}$$

(to read: A satisfies φ under the assignment $x_i \mapsto a_i$) by an induction quite analogous to the finitary case:

$A \models (\tau_1 = \tau_2)[a_i]_{i\in I}$	iff $\tau_1(a_i) = \tau_2(a_i)$ $(\tau_1, \tau_2 \text{ terms})$
$A \models \sigma(\tau_1 \ldots \tau_j \ldots)[a_i]_{i\in I}$	iff $(\tau_1(a_i), \ldots, \tau_j(a_i), \ldots) \in \sigma_A$ (τ_j terms, $\sigma \in \Sigma_{\mathrm{rel}}$)
$A \models \neg\varphi[a_i]_{i\in I}$	iff it is not true that $A \models \varphi[a_i]_{i\in I}$
$A \models (\varphi \Rightarrow \psi)[a_i]_{i\in I}$	iff $A \models \neg\varphi[a_i]_{i\in I}$ or $A \models \psi[a_i]_{i\in I}$
$A \models \bigwedge_{j\in J} \varphi_j[a_i]_{i\in I}$	iff $A \models \varphi_j[a_i]_{i\in I}$ for each $j \in J$
$A \models \bigvee_{j\in J} \varphi_j[a_i]_{i\in I}$	iff $A \models \varphi_j[a_i]_{i\in I}$ for some $j \in J$
$A \models ((\forall X)\varphi)[a_i]_{i\in I}$	iff $\operatorname{Var}\varphi \subseteq X \cup \{x_i \mid i \in I\}$ and given arbitrary elements b_x in $|A|$ of sort s_x for $x \in X$, then $A \models \varphi[b_x, a_i]_{x\in X, i\in I}$
$A \models ((\exists X)\varphi)[a_i]_{i\in I}$	iff $\operatorname{Var}\varphi \subseteq X \cup \{x_i \mid i \in I\}$ and there exist elements b_x in $|A|$ of sort s_x for $x \in X$ with $A \models \varphi[b_x, a_i]_{x\in X, i\in I}$.

If the assignment $i \mapsto a_i$ is formalized as a mapping $a\colon I \to |A|$ from the index set I to the underlying set of A (with $a(i)$ of sort s_i), then we simply write

$$A \models \varphi[a].$$

The *logic* L_λ is the collection of all formulas of L_λ with the above satisfaction relation $\models$. The notion of sentence is defined precisely as in the finitary logic L_ω.

When no cardinal restrictions are considered, we work with the logic L_∞: here the standard variables form a proper class in each sort, and the formulas and satisfaction form a union of the logics L_λ for all regular cardinals λ.

Analogously to L_ω, a set T of sentences in L_λ (or L_∞) is called a *theory*, and the class of all *models* of T (i.e., all Σ-structures satisfying every sentence in T) is denoted by $\mathbf{Mod}\,T$. We do not consider large theories T (even in L_∞); thus every theory in L_∞ is actually a theory in L_λ for some λ. Observe that in the logic L_∞ (which has no size restriction on conjunction nor quantification) we always have $\mathbf{Mod}\, T = \mathbf{Mod}\, \varphi$ for $\varphi = \bigwedge T$. A class of Σ-structures is called *axiomatizable in* L_λ (or L_∞) if it has the form $\mathbf{Mod}\, T$ for some theory T in L_λ (or L_∞, resp.).

5.27 Examples

(1) Finite sets are axiomatizable in the logic L_{ω_1} (but not in L_ω). In fact, consider the following sentence in the trivial signature (one sort, no symbol):

$$(\forall\, x_0, x_1, x_2, \ldots) \bigvee_{\substack{i,j \in \omega \\ i \neq j}} (x_i = x_j).$$

More generally: all sets of cardinality less than λ form an axiomatizable class in L_{λ^+} (where λ^+ is the cardinal successor of λ).

(2) Torsion groups are axiomatizable in L_{ω_1}: extend the theory of Abelian groups (see Example 3.1) by the following sentence:

$$(\forall x) \bigvee_{n \in \omega} (x^n = 1).$$

(3) Analogously to the finitary case, every infinitary quasivariety is axiomatizable in L_∞ by a ***universal Horn theory***, which is a theory of sentences $(\forall x)(\bigwedge_{j \in J} \varphi_j \Rightarrow \varphi)$ with φ and φ_j $(j \in J)$ atomic formulas. More precisely: a class of structures is closed under products, subobjects, and λ-directed colimits iff it is axiomatizable by a universal Horn theory in L_λ. The proof is quite analogous to that of Theorem 5.12.

(4) σ-complete semilattices are axiomatizable in L_{ω_1}: extend the theory of partially ordered sets (see Example 5.4(2)) by the following sentence:

$$(*) \qquad (\forall X)\,(\exists y)\left[\left(\bigwedge_{x \in X} x \leq y\right) \wedge (\forall z)\left(\left(\bigwedge_{x \in X} x \leq z\right) \Rightarrow (y \leq z)\right)\right]$$

where X is countable.

(5) Well-ordered sets are axiomatizable in L_{ω_1}: extend the theory of linearly ordered sets (Example 5.4(3)) by the sentence

$$(\forall \{x_i\}_{i \in \omega})\left(\neg \bigwedge_{i \in \omega} ((x_{i+1} \leq x_i) \wedge \neg(x_i = x_{i+1}))\right)$$

Limit Theories and Locally Presentable Categories

5.28 Presentation and Orthogonality Formulas. Analogously to the finitary case (see Remark 5.1(4)), for each λ-ary signature Σ a Σ-structure A is λ-presentable iff the underlying algebra A^0 is λ-presentable and $\operatorname{card} \bigcup_{\sigma \in \Sigma_{\mathrm{rel}}} \sigma_A < \lambda$. We can assume, without loss of generality, that A^0 is the quotient algebra of the free algebra generated by less than λ standard variables x_i ($i \in I$) modulo a congruence generated by less than λ equations $\tau_j(x_i)_{i\in I} = \tau'_j(x_i)_{i\in I}$ for $j \in J$. The relations of A have less than λ edges, so we can list them as $\sigma_k(\varrho_{k,l})_{l\in I_k}$ for $k \in K$ (where $\operatorname{card} K < \lambda$). We define a sentence in L_λ as follows:

$$(\pi_A) \qquad \bigwedge_{j\in J} \left(\tau_j(x_i) = \tau'_j(x_i)\right) \wedge \bigwedge_{k\in K} \sigma_k\left(\varrho_{k,l}\right)_{l\in I_k}.$$

Models of π_A form the comma-category

$$A \downarrow \mathbf{Str}\,\Sigma,$$

in the same sense as explained in detail in L_ω (see Remark 5.5).

Let $h\colon A \to B$ be a homomorphism in $\mathbf{Str}\,\Sigma$ with both A and B λ-presentable. Analogously to Remark 5.6 we choose presentations of A (in variables $\{x_i \mid i \in I\}$) and of B (in variables $\{y_j \mid j \in J\}$); for each $i \in I$ we choose a term ϱ_i whose congruence class in B is the image of the congruence class of x_i under h, and we get the following formula of L_λ:

$$(\pi_h) \qquad (\underset{i\in I}{\forall})\left[\pi_A(x_i) \implies (\underset{j\in J}{\exists!})\left(\pi_B(y_j) \wedge \bigwedge_{i\in I}(x_i = \varrho_i(y_j))\right)\right].$$

Models of π_h are just the Σ-structures orthogonal to h; this is completely analogous to the case $\lambda = \omega$ in Remark 5.6.

5.29 Definition. By a *limit theory* in L_λ is meant a theory each sentence of which has the form

$$(\forall x)(\varphi(x) \Rightarrow (\exists! y)\psi(x, y))$$

where $\varphi(x)$ and $\psi(x, y)$ are conjunctions of atomic formulas.

5.30 Characterization Theorem. *Let λ be a regular cardinal. Locally λ-presentable categories are precisely the categories equivalent to categories of models of limit theories in L_λ.*

The proof is analogous to that of Theorem 5.9.

Basic Theories and Accessible Categories

5.31 Definition. A formula is called

(1) *positive-primitive* if it has the form $(\exists y)\psi(x,y)$, where $\psi(x,y)$ is a conjunction of atomic formulas,

(2) *positive-existential* if it is a disjunction of positive-primitive formulas,

(3) *basic* if it has the form

$$(\forall x)\big(\varphi(x) \Rightarrow \psi(x)\big)$$

where $\varphi(x)$ and $\psi(x)$ are positive-existential formulas.

Remark. Every positive-primitive formula is positive-existential. The formulas *true* and *false* are positive-existential. Thus, if $\varphi(x)$ is a positive-existential formula, then $(\forall x)\varphi(x)$ is a basic sentence (since it is equivalent to $(\forall x)\big(\text{true} \Rightarrow \varphi(x)\big)$) and also $(\forall x)\neg\varphi(x)$ is a basic sentence (equivalent to $(\forall x)\big(\varphi(x) \Rightarrow \text{false}\big)$).

A *basic theory* is a set of basic sentences.

5.32 Examples

(1) Every universal Horn sentence is basic.

(2) Every limit sentence

$$(\forall x)\big(\varphi(x) \Rightarrow (\exists! y)\psi(x,y)\big)$$

can be substituted by a pair of basic sentences, viz,

$$(\forall x)\big(\varphi(x) \Rightarrow (\exists y)\psi(x,y)\big)$$
$$(\forall x,y,z)\Big(\big(\varphi(x) \wedge \psi(x,y) \wedge \psi(x,z)\big) \Rightarrow (y = z)\Big),$$

where $y = \{y_i\}_{i\in I}$, $z = \{z_i\}_{i\in I}$, and the sentence $y = z$ abbreviates $\bigwedge_{i\in I}(y_i = z_i)$.

Consequently, every limit theory can be considered to be basic.

(3) The theory of linearly ordered sets (Example 5.4(3)) is basic. Observe that, since the category of linearly ordered sets does not have products, no axiomatization by a limit theory exists.

(4) The category of sets and one-to-one functions has the following basic axiomatization. Let Σ be the one-sorted signature of one binary relation symbol σ, and consider the sentences

$$(\forall x, y)\big((\sigma(x,y)) \vee (x = y)\big)$$

and

$$(\forall x, y)\neg\big(\sigma(x,y) \wedge (x = y)\big).$$

The category of models of this theory (i.e., of sets X with the binary relation $\neq$) is isomorphic to the category of sets and one-to-one functions.

(5) The theory of fields is basic: it is sufficient to add the following basic sentences to the equational theory of unitary rings:

$$(\forall x)\big((x = 0) \vee (\exists y)(xy = 1)\big)$$

and

$$\neg(0 = 1).$$

5.33 Remark: Injectivity Formula. We know that, up to equivalence, accessible categories are precisely the small-cone injectivity classes of structures (see Theorem 4.17 and Example 1.41). We will now express small-cone injectivity by a basic sentence. Since we work in the general logic L_∞ rather than some specific logic L_λ, we can replace the (rather clumsy) construction of the presentation formula π_A by the much easier concept of a positive diagram:

Let A be a Σ-structure, and let x be a one-to-one function assigning to each element a of A_s a variable x_a of sort s. The conjunction of all atomic formulas $\varphi(x)$ which are satisfied by A under the assignment $x_a \mapsto a$ is called the *positive diagram* of A and is denoted by

$$(\pi_A^+) \qquad \bigwedge_{\substack{A \models \varphi[a] \\ \varphi \text{ atomic}}} \varphi(x_a)$$

Like the (usually much smaller) presentation formula π_A of Remark 5.5, π_A^+ axiomatizes the comma-category of A in $\mathbf{Str}\,\Sigma$:

$$f\colon A \to B \quad \text{is a homomorphism iff} \quad B \models (\pi_A^+)[f].$$

Next, let $\varrho = (A \xrightarrow{h_t} A'_t)_{t\in T}$ be a cone in $\mathbf{Str}\,\Sigma$. Let x be an assignment of variables to elements of $|A|$ as above and let, analogously, y_t be an assignment of variables to elements of A_t for $t \in T$. The *injectivity formula* for the cone ϱ is the following sentence of L_∞:

$$(\pi_\varrho^+) \qquad (\forall x)\left[(\pi_A^+)(x) \Rightarrow \bigvee_{t\in T} (\exists y_t)\left((\pi_{A_t}^+)(y_t) \wedge (x = y_t \cdot h_t)\right)\right]$$

where $x = y_t \cdot h_t$ is an abbreviation of the conjunction of the formulas $x_s(a) = (y_t \cdot h_t)_s(a)$ over all elements a of $|A|$ of sort s.

It is obvious that cone-injectivity to ϱ is equivalent to the satisfaction of π_ϱ^+. Shortly:

$$\{\varrho\} - \mathbf{Inj} = \mathbf{Mod}\,\pi_\varrho^+.$$

5.34 Proposition. *Let Σ be a λ-ary signature. A substructure A of a Σ-structure B is λ-pure iff for each positive-primitive formula $\varphi(x)$ of L_λ and each assignment a of values in A we have*

$$A \models \varphi[a] \qquad \textit{iff} \qquad B \models \varphi[a].$$

PROOF: analogous to that of Proposition 5.15. □

5.35 Characterization Theorem. *Accessible categories are precisely the categories equivalent to the categories of models of basic theories.*

PROOF. By Remark 5.33, cone-injectivity can be expressed by a basic theory, thus, every accessible category is equivalent to $\mathbf{Mod}\,T$ for a basic theory T (see Theorem 4.17 and Example 1.41).

Conversely, let T be a basic theory in L_λ. Then it is easy to verify that $\mathbf{Mod}\,T$ is closed under λ-directed colimits in $\mathbf{Str}\,\Sigma$ (this is analogous to part II of the proof of Theorem 5.9). Let us show that $\mathbf{Mod}\,T$ is closed under λ-pure subjects; then it is accessible by Corollary 2.36. Let A be a λ-pure substructure of a model B of T. Let us consider a sentence

$$(\forall x)(\varphi(x) \Rightarrow \psi(x))$$

in T. Since this sentence is basic, we have $\varphi = \bigvee_{i\in I} \varphi_i$ and $\psi = \bigvee_{j\in J} \psi_j$ with all φ_i and ψ_j positive-primitive. For each assignment a with $A \models \varphi[a]$ we have $B \models \varphi[a]$. Since B is a model of T, we conclude that $B \models \psi_j[a]$ for some j. By Proposition 5.34 we have $A \models \psi_j[a]$, thus, $A \models \psi[a]$. □

Remark. We do not claim that, given a cardinal λ, than λ-accessible categories are precisely the categories equivalent to the categories of models of basic theories in L_λ! In fact, even for $\lambda = \aleph_0$ both directions are false. We have seen a finitely accessible category which is not axiomatizable in L_ω in Example 5.22 and Exercise 5g. Conversely:

5.36 Example of a basic theory in L_ω which does not have a finitely accessible category of models. Recall from Remark 2.59 that the category of all graphs satisfying

$$(\forall x)(\exists y)(\sigma(x,y))$$

(which is precisely $\mathbf{Mod}\,\mathscr{S}$ for the sketch $\mathscr{S}$ of Example 2.57(3)) is not finitely accessible. The sentence is basic by Remark 5.31.

Elementary Embeddings

In this subsection we will present an important type of accessible category: the category of all Σ-structures and all elementary embeddings. Later we will use this result for a characterization of axiomatizable categories (Theorem 5.44).

5.37 Definition. Let λ be a regular cardinal, and let Σ be a λ-ary signature. A homomorphism $h\colon A \to B$ in $\mathbf{Str}\,\Sigma$ is called *λ-elementary* provided that for each formula φ of L_λ and each interpretation a of variables in A we have

$$A \models \varphi[a] \qquad \text{iff} \qquad B \models \varphi[h \cdot a].$$

Every λ-elementary homomorphism is λ-pure (see Proposition 5.34), thus, h is an embedding. It is customary to speak about λ-elementary substructures rather than homomorphisms.

5.38 Examples

(1) In the trivial signature Σ of one sort and no symbol we have $\mathbf{Str}\,\Sigma = \mathbf{Set}$, and a subset $A \subseteq B$ is ω-elementary iff either A is infinite, or $A = B$. (In contrast, every non-empty subset is ω-pure!)

(2) Consider the extension of the theory of linear ordering by the following sentences:

$$(\forall x, y)(\exists z) \quad (x < z < y)$$
$$(\forall x)(\exists z) \quad (x < z)$$
$$(\forall x)(\exists z) \quad (z < x)$$

(where $x < y$ abbreviates $(x \leq y) \wedge (x \neq y)$). In the resulting theory of dense linear orderings without endpoints, every ω-pure embedding is ω-elementary.

(3) Every isomorphism is λ-elementary for each λ (and vice versa).

Remark. A composite of two λ-elementary embeddings is λ-elementary. If f and $f \cdot g$ are λ-elementary, then so is g.

If $\lambda \leq \lambda'$ are regular cardinals, then every λ'-elementary embedding is λ-elementary.

5.39 Proposition. *Let Σ be a λ-ary signature (for a regular cardinal λ). Each class of Σ-structures axiomatizable by a theory in L_λ is closed in* $\mathbf{Str}\,\Sigma$ *under*

(i) *λ-elementary substructures,*

(ii) *λ-directed colimits of λ-elementary embeddings.*

Remark. In (ii), moreover, the colimit maps will also be proved to be λ-elementary embeddings.

PROOF. Let T be a theory in L_λ. Then $\mathbf{Mod}\,T$ is closed under λ-elementary substructures. Let A be a λ-elementary substructure of B, then for each $\varphi \in T$ we have $A \models \varphi$ iff $B \models \varphi$; thus, A is a model of T iff B is a model of T.

Let $D\colon (I, \leq) \to \mathbf{Mod}(T)$ be a λ-directed diagram with connecting maps $d_{i,j}$ that are λ-elementary (for all $i \leq j$ in I). If $(D_i \xrightarrow{d_i} C)_{i \in I}$ is a colimit of D in $\mathbf{Str}\,\Sigma$, we will prove that each d_i is a λ-elementary embedding—it follows that C is a model of T (by the above argument). We will prove by induction on the complexity of a formula $\varphi(x)$ that for each interpretation a of variables in D_i we have

$$D_i \models \varphi[a] \quad \text{iff} \quad C \models \varphi[d_i \cdot a].$$

(1) If φ is an atomic formula, then this follows from the fact that colimit cocones of a λ-directed diagram of regular monomorphisms are formed by regular monomorphisms (see Proposition 1.62), thus, $d_i\colon D_i \to C$ is an embedding of a substructure.

(2) The passage from φ to $\neg\varphi$ is evident.

(3) The cases of conjunctions and disjunctions are also evident.

(4) For the existential quantifier consider a formula $\varphi(x, y)$ with the above property. Then $(\exists y)\varphi(x, y)$ also has that property. In fact, it is obvious that

$$D_i \models \big((\exists y)\varphi(x, y)\big)[a] \quad \text{implies} \quad C \models \big((\exists y)\varphi(x, y)\big)[d_i \cdot a].$$

Conversely, whenever $C \models \big((\exists y)\varphi(x, y)\big)[d_i \cdot a]$, then there exists an interpretation b of the y-variables in C with $C \models \varphi[d_i \cdot a, b]$. Since the number of variables in y is smaller than λ, and C is a λ-directed colimit of its substructures D_i, there clearly exists $j \geq i$ in I and an interpretation b' of the variables of y in D_j such that

$$b = d_j \cdot b' \qquad \text{and} \qquad D_j \models \varphi[d_{i,j} \cdot a, b'].$$

Thus, $D_j \models ((\exists y)\varphi(x,y))[d_{i,j} \cdot a]$, which implies $D_i \models ((\exists y)\varphi(x,y))[a]$, since $d_{i,j}$ is λ-elementary. □

5.40 Definition. Let λ be a regular cardinal, and let T be a theory of Σ-structures in the logic L_λ. We denote by $\mathbf{Elem}_\lambda\,\Sigma$ the category of all Σ-structures and all λ-elementary embeddings, and by $\mathbf{Elem}_\lambda\,T$ the full subcategory whose objects are all T-models.

These are categories, indeed, by Remark 5.38. By Proposition 5.39, $\mathbf{Elem}_\lambda\,T$ is closed in $\mathbf{Str}\,\Sigma$ under λ-directed colimits and subobjects. We will prove that this category is accessible; it need not be λ-accessible, however:

Example. If **Set** is considered as the category of models of the empty theory $T = \emptyset$ with a trivial signature (of no symbol), then $\mathbf{Elem}_\omega\,\emptyset$ has as morphisms all identity morphisms and all monomorphisms $f\colon X \to Y$ with X infinite. This category is ω_1-accessible, but not finitely accessible.

5.41 Downward Löwenheim–Skolem Theorem. *For each λ-ary signature Σ there exists a regular cardinal $\overline{\lambda} \geq \lambda$ such that every Σ-structure A has the following property: each subset of $|A|$ of power less than $\overline{\lambda}$ is contained in a λ-elementary substructure of A of power less than $\overline{\lambda}$.*

Proof. Let A be a Σ-structure. For each subset X of $|A|$ we construct a λ-elementary substructure X_λ of A, and then we will show that there exists a regular cardinal $\overline{\lambda}$ such that if X has power less than $\overline{\lambda}$, then X_λ also has power less than $\overline{\lambda}$. We define a chain X_i $(i \leq \lambda)$ of many-sorted subsets of $|A|$ by the following transfinite induction:

First step: $X_0 = X$.

Isolated step: given X_i, consider the set of all triples (φ, Z, a) where

φ is a formula of L_λ,

Z is a subset of $\operatorname{Var}\varphi$, and

$a\colon Z \to |A|$ is a mapping for which there exists an extension $\overline{a}\colon \operatorname{Var}\varphi \to |A|$ with $A \models \varphi[\overline{a}]$.

For each such triple we choose an extension $\overline{a}\colon \operatorname{Var}\varphi \to |A|$ with $A \models \varphi[\overline{a}]$, and we denote by $\operatorname{im}\overline{a}$ the many-sorted image of $\overline{a}$. Then we define

$$X_{i+1} = X_i \cup \bigcup_{(\varphi,Z,a)} \operatorname{im}\overline{a}$$

(where (φ, Z, a) ranges through the above set).

Limit step: $X_i = \bigcup_{j<i} X_j$ for each limit ordinal i.

I. X_λ is closed under the operations in A. In fact, let $\sigma\colon \prod_{j\in J} s_j \to s$ be an element of Σ_{ope}. Let $\sigma(a_j) = a$ for some a_j in X_λ $(j \in J)$; since $\operatorname{card} J < \lambda$, there exists $i < \lambda$ with a_j in X_i for each $j \in J$. Choose pairwise distinct standard variables x_j (of sort s_j) and x (of sort s) and consider the triple

$$\big(\sigma(x_j) = x,\ \{x_j\}_{j\in J},\ b_0\big)$$

where $(b_0)_{s_j}(x_j) = a_j$ for $j \in J$. This is one of the triples considered in the isolated step from X_i to X_{i+1} because we can extend b_0 to b by $b(x) = a$, and then $A \models \varphi[b]$. Moreover, the extension of b_0 is unique in this case, thus, X_{i+1} necessarily contains $b(\operatorname{Var}(\sigma(x_j) = x))$, hence $a = b(x) \in X_{i+1} \subseteq X_\lambda$.

II. The substructure $\overline{A}$ of A induced by X_λ is λ-elementary. That is, for each formula φ of L_λ and each assignment $a\colon \operatorname{Var}\varphi \to |\overline{A}|$ we have

$$A \models \varphi[a] \qquad \text{iff} \qquad \overline{A} \models \varphi[a].$$

This is true for atomic formulas φ since $\overline{A}$ is a substructure of A. We extend this to non-atomic formulas by induction on the complexity of φ. The cases of negation, disjunction, and conjunction are clear, thus, it is sufficient to consider the existential quantification: $\varphi(X) = (\exists Y)\psi(X,Y)$. If $\overline{A} \models \varphi[a]$ then, obviously, $A \models \varphi[a]$. For the converse, we use the fact that given $a\colon X \to \overline{A}$ with $A \models \varphi[a]$, there exists $i < \lambda$ with $a(X) \subseteq X_i$ (since $\operatorname{card} X < \lambda$), and then we use the triple (ψ, X, a) in the induction step: from $A \models (\exists Y)\psi(X,Y)[a]$ we know that a can be extended as required, thus, for some extension $\overline{a}\colon X + Y \to X_{i+1}$ of a we have $A \models \psi[\overline{a}]$. By the induction hypothesis, this implies $\overline{A} \models \psi[\overline{a}]$, thus, $\overline{A} \models \varphi[a]$.

III. The definition of $\overline{\lambda}$. Choose a regular cardinal α with

$$\alpha \geq \max(\lambda, \operatorname{card}\Sigma, \operatorname{card} S),$$

and denote by $\overline{\lambda}$ the cardinal successor of $\alpha^{<\lambda}$ (see Notation 2.9). We will prove that for each subset X of $|A|$ we have that

$$\#X < \overline{\lambda} \quad \text{implies} \quad \#|\overline{A}| < \overline{\lambda}.$$

Since $\overline{\lambda}$ is independent of A, this will finish the proof.

We prove by induction on $i \leq \lambda$ that $\#X < \overline{\lambda}$ implies $\#X_i < \overline{\lambda}$. The first step is clear. In the isolated step, it is sufficient to prove that the number of triples (φ, Z, a) considered there is smaller than $\overline{\lambda}$. For this, it is sufficient to show that the number of all formulas in L_λ is smaller than $\overline{\lambda}$: for each formula φ we have less than λ free variables, thus less than $\overline{\lambda}$ subsets Z of $\operatorname{Var}\varphi$, and for each Z we have

$$(\#X_i)^{\operatorname{card} Z} \leq \left(\alpha^{<\lambda}\right)^{<\lambda} = \alpha^{<\lambda},$$

see Lemma 2.10. The set of all formulas in L_λ can be expressed as a union $F_\lambda = \bigcup_{i<\lambda} F_i$ where

F_0 is the set of all atomic formulas; since $\alpha \geq \max(\lambda, \operatorname{card}\Sigma, \operatorname{card} S)$, we clearly have $\operatorname{card} F_0 \leq \alpha^{<\lambda}$;

F_{i+1} are formulas obtained from those in F_0 by the logical operations of L_λ; if $\operatorname{card} F_i \leq \alpha^{<\lambda}$ then $\operatorname{card} F_{i+1} = (\operatorname{card} F_i)^{<\lambda} \leq (\alpha^{<\lambda})^{<\lambda} = \alpha^{<\lambda}$;

$F_j = \bigcup_{i<j} F_i$ for each limit ordinal $i < j$ $(\leq \lambda)$; if $\operatorname{card} F_i \leq \alpha^{<\lambda}$ for each $i < j$, then $\operatorname{card} F_j \leq \alpha^{<\lambda}$.

Consequently, $\operatorname{card} F_\lambda \leq \alpha^{<\lambda}$, as required.

The limit step is clear: $\#X_i = \#\bigcup_{j<i} X_j \leq \alpha^{<\lambda}$. □

5.42 Theorem. *For each theory T in L_λ, the category* $\mathbf{Elem}_\lambda\, T$ *is accessible.*

PROOF. I. We first prove the accessibility of the category $\mathbf{Elem}_\lambda\,\Sigma$ of Σ-structures and λ-elementary embeddings for any relational signature Σ. By the proof of the Löwenheim–Skolem theorem, 5.41, there exists a regular cardinal α such that for the cardinal successor $\overline{\lambda}$ of $\alpha^{<\lambda}$ we have: for each Σ-structure A, every set $X \subseteq |A|$ of power less than $\overline{\lambda}$ is contained in a λ-elementary substructure of A of power less than $\overline{\lambda}$. It follows that $\mathbf{Elem}_\lambda\,\Sigma$ is $\overline{\lambda}$-accessible. In fact, the existence of $\overline{\lambda}$-directed colimits follows from Proposition 5.39. Any Σ-structure A of power less than $\overline{\lambda}$ is $\overline{\lambda}$-presentable in $\mathbf{Rel}\,\Sigma$: for each n-ary symbol $\sigma \in \Sigma$, since $n < \lambda$, we have by Remark 2.10

$$\operatorname{card}\sigma_A \leq \left(\#|A|\right)^n \leq \left(\alpha^{<\lambda}\right)^n = \alpha^{<\lambda} < \overline{\lambda},$$

thus A is $\overline{\lambda}$-presentable by Example 1.14(2). Since every Σ-structure B is a $\overline{\lambda}$-directed union of its substructures of power less than $\overline{\lambda}$, it follows that B is a $\overline{\lambda}$-directed union of its λ-elementary substructures of power less than $\overline{\lambda}$. Consequently, it follows from Remark 5.38 that B is a $\overline{\lambda}$-directed colimit in $\mathbf{Elem}_\lambda \Sigma$ of objects which are $\overline{\lambda}$-presentable (in $\mathbf{Rel}\,\Sigma$, thus, in $\mathbf{Elem}_\lambda \Sigma$).

II. Let T be an L_λ-theory (of a λ-ary signature Σ). Without loss of generality we may assume that Σ contains relation symbols only. Indeed, an operation symbol f of arity $\prod_{i\in I} s_i \to s$ can be substituted by a relation symbol ϱ_f of arity $\prod_{i\in I} s_i \times s$ together with addition of the formula

$$(\forall\{x_i\}_{i\in I})(\exists! y)\varrho_f(x_i, y)$$

to T.

By Proposition 5.39 the full subcategory $\mathbf{Elem}_\lambda T$ of $\mathbf{Elem}_\lambda \Sigma$ is closed under λ-directed colimits and subobjects. Thus, $\mathbf{Elem}_\lambda T$ is accessible by Remark 2.36. □

5.43 Remark: Elementary Diagram. For each structure A of a λ-ary signature Σ we construct a formula π^*_A characterizing λ-elementary embeddings of A. This is quite analogous to the positive diagram in Remark 5.33. We choose pairwise distinct variables x_a for elements a of $|A|$ (with x_a and a having the same sort), and then we take the conjunction of all formulas $\varphi(x_a)$ in L_λ such that A satisfies $\varphi[a]$ (i.e., φ where the interpretation of x_a is a):

$$(\pi^*_A) \qquad \bigwedge_{A\models\varphi[a]} \varphi(x_a)$$

For each Σ-structure B we have

$$f\colon A \to B \text{ is a } \lambda\text{-elementary embedding iff } B \models \pi^*_A[f].$$

5.44 Theorem. *A class $\mathcal{K}$ of Σ-structures is axiomatizable (in L_∞) iff there exists a regular cardinal λ such that $\mathcal{K}$, as a full subcategory of $\mathbf{Elem}_\lambda \Sigma$, is accessible and accessibly embedded.*

PROOF. Every category of models, as a full subcategory of $\mathbf{Elem}_\lambda \Sigma$, is accessible (see Theorem 5.42) and accessibly embedded (see Proposition 5.39).

Conversely, let $\mathcal{K}$ be accessible and accessibly embedded in $\mathbf{Elem}_\lambda \Sigma$. There exists a regular cardinal λ' such that $\mathcal{K}$ is λ'-accessible and closed under λ'-directed colimits in $\mathbf{Elem}_\lambda \Sigma$. Let $\mathcal{A} = \mathbf{Pres}_{\lambda'} \mathbf{Elem}_\lambda \Sigma$ be a set of representative λ'-presentable objects in $\mathbf{Elem}_\lambda \Sigma$. For each $A \in \mathcal{A}$ we form the cone

$$\left(A \xrightarrow{h_i} A_i\right)_{i\in I}$$

of all morphisms into objects of $\mathbf{Pres}_{\lambda'}\mathcal{K}$. Injectivity w.r.t. this cone is described by the following sentence:

$$(\varphi_A) \qquad (\forall x)\Big(\pi_A^+(x) \implies \bigvee_{i\in I}(\exists y_i)(\pi_{A_i}^*(y_i) \wedge (x = y_i \cdot h_i))\Big).$$

More precisely, given a Σ-structure B, then

$$B \models \varphi_A \quad \text{iff} \quad \begin{array}{l}\text{for each } \Sigma\text{-homomorphism } f\colon A \to B \\ \text{there exists a } \lambda\text{-elementary embedding} \\ f'\colon A_i \to B \ (i \in I) \text{ with } f = f' \cdot h_i.\end{array}$$

It follows that

$$\mathcal{K} = \mathbf{Mod}\{\, \varphi_A \mid A \in \mathcal{A}\,\}.$$

In fact, let B be an object of $\mathcal{K}$. We express B as a λ'-directed colimit of objects in $\mathbf{Pres}_{\lambda'}\mathcal{K}$. Since $\mathcal{K}$ is closed under λ'-directed colimits in $\mathbf{Elem}_\lambda\,\Sigma$, it is obvious that B is injective w.r.t. the above cones, thus, B is a model of $\{\varphi_A\}_{A\in\mathcal{A}}$. Conversely, let B be a model of $\{\varphi_A\}_{A\in\mathcal{A}}$. We express B as a canonical λ'-directed colimit of objects in $\mathcal{A}$. Then, since B is injective w.r.t. the above cones, the canonical diagram of B w.r.t. $\mathbf{Pres}_{\lambda'}\mathcal{K}$ is a cofinal subdiagram. By 0.11, B is a λ'-directed colimit of objects in $\mathbf{Pres}_{\lambda'}\mathcal{K}$, thus, B lies in $\mathcal{K}$. □

Exercises

5.a Directed Limits

(1) Prove that a limit of a down-directed diagram $D\colon (I,\leq) \to \mathbf{Set}$ can be described as the subset of $\prod_{i\in I} D_i$ of all I-tuples $(x_i)_{i\in I}$ such that

$$i \leq j \text{ in } I \quad \text{implies} \quad d_{i,j}(x_i) = x_j .$$

(2) Describe directed limits in $\mathbf{Set}^S$.

(3) Prove that for each S-sorted signature Σ the natural forgetful functor $U\colon \mathbf{Str}\,\Sigma \to \mathbf{Set}^S$ preserves limits. Use this fact to describe directed limits in $\mathbf{Str}\,\Sigma$ as substructures of $\prod_{i\in I} D_i$, analogously to (1) above.

5.b Forgetful Functors.

Let Σ be a λ-ary signature.

(1) Prove that the natural forgetful functor from $\mathbf{Str}\,\Sigma$ to $\mathbf{Rel}\,\Sigma_{\mathrm{rel}}$ is monadic and λ-accessible.

(2) Prove that the natural forgetful functor from $\mathbf{Str}\,\Sigma$ to $\mathbf{Alg}\,\Sigma_{\mathrm{ope}}$ is λ-accessible and preserves limits and colimits.

5.c Presentation Formulas

(1) In the theory of ordered Abelian groups (Example 5.4(4)) write down a presentation formula for the usual group of integers.

(2) In the theory of partially ordered sets (Example 5.4(3)) write down a presentation formula for the set of all subsets of $\{0,1,2\}$ ordered by inclusion.

(3) Prove that if Σ is a λ-ary signature, then any conjunction $\bigwedge_{i\in I} \varphi_i$ of less than λ atomic formulas is the presentation formula of some λ-presentable Σ-structure A.

5.d Essentially Algebraic Theories

(1) Translate the essentially algebraic theory of posets (see Example 3.35(4)) into a limit theory.

(2) Find a limit theory axiomatizing torsion-free Abelian groups.

5.e Regular Theories

Regular Theories are theories using sentences of the form

$$(\forall x)\big(\varphi(x) \Rightarrow \psi(x)\big)$$

where $\varphi(x)$ and $\psi(x)$ are positive-primitive formulas. Prove that a category is weakly locally presentable iff it can be axiomatized by a regular theory. (Hint: for a regular theory T the category $\mathbf{Mod}\,T$ is closed under products in **Str** Σ. Conversely, each weakly locally presentable category has the form $\{m_i\}_{i\in I}$-**Inj** for some homomorphisms m_i ($i \in I$) in **Str** Σ, and the injectivity to m_i can be expressed by a regular sentence.)

5.f Disjunctive Theories are theories using sentences of the form

$$(\forall x)\Big(\varphi(x) \implies (\exists! y) \bigvee_{i\in I} \psi_i(x,y)\Big) \wedge (\forall x,y)\neg \bigvee_{\substack{i,j\in I \\ i\neq j}} \psi_i(x,y) \wedge \psi_j(x,y)$$

where $\varphi(x)$ and $\psi_i(x,y)$ are positive-primitive formulas. Prove that a category is locally multipresentable iff it can be axiomatized by a disjunctive theory. (Hint: analogous to the hint for 5.e.)

5.g Axiomatizability. The category $\mathcal{A}$ of Example 5.22 cannot be axiomatized in any signature Σ^*, i.e., given a full embedding $E\colon \mathcal{A} \to \mathbf{Str}\,\Sigma^*$, then $E(\mathcal{A})$ is not axiomatizable in L_ω. Prove this by assuming the contrary and deriving contradiction as follows:

(1) Let $A_n \in \mathcal{A}$ be the algebra on $\{0, 1, \ldots, n\}$ with $\alpha = 0$, $\sigma(n) = n$, and $\sigma(i) = i+1$ for $i < n$. Let $\mathcal{U}$ be an ultrafilter on ω containing no finite set. Show that there exists $B \in \mathcal{A}$ with $EB = \prod_{\mathcal{U}} EA_n$. (Hint: $E(\mathcal{A})$ is closed under ultraproducts.)

(2) Show that for each $n \in \omega$ there is an algebra $B_n \in \mathcal{A}$ such that $EB_n = (EA_n)^{\mathcal{U}}$ is the ultrapower of ω copies of EA_n, and verify that $(\sigma_{B_n})^n(\alpha) \neq (\sigma_{B_n})^{n-1}(\alpha)$ for all $n \geq 2$. (Hint: the map $d\colon EA_n \to (EA_n)^{\mathcal{U}}$ with $d(x) = [x, x, x, \ldots]$ is a monomorphism, thus, there is a monomorphism from A_n to B_n, and $(\sigma_{A_n})^n(\alpha) \neq (\sigma_{A_n})^{n-1}(\alpha)$.)

(3) Show that for each $k \in \omega$ there is a homomorphism from B to B_k. (Hint: the map $h\colon \prod_{\mathcal{U}} EA_n \to (EA_k)^{\mathcal{U}}$ defined by

$$h([x_n]) = [0, 0, \ldots, 0, x_k, x_{k+1}, \ldots]$$

is a homomorphism.)

(4) Prove that for each $k \geq 2$ we have $\sigma_B^k(\alpha) \neq \sigma_B^{k-1}(\alpha)$, in contradiction with $B \in \mathcal{A}$. (Hint: combine (2) and (3)).

Historical Remarks

Categories of models of first-order theories were studied already by [Mal'cev 1958] and [Freyd 1965]. Our treatment of first-order logic is standard, see e.g. [Chang, Keisler 1973] and [Dickmann 1975]. Limit theories in finitary logic were introduced by [Coste 1979] who proved that they characterize locally finitely presentable categories, see Theorem 5.9; the generalization to infinitary logic and Theorem 5.30 was observed in [Rosický 1981a].

Theorem 5.20 is due to [Volger 1979] and Theorem 5.23 to [Richter 1971] (but our proof of the latter is new).

The importance of finitary basic theories was first recognized, in connection with topos theory, by A. Joyal and G. Reyes, see [Reyes 1974] and [Makkai, Reyes 1977]. The characterization of accessible and axiomatizable categories in infinitary logic (Theorems 5.35 and 5.44) is due to [Rosický 1981b and 1983]; Theorems 5.35 and 5.42 were independently proved by [Makkai, Paré 1989].

Disjunctive theories in connection with locally multipresentable categories were treated in [Johnstone 1979], and regular theories are due to [Makkai 1982].

Chapter 6

Vopěnka's Principle

In the present chapter we will prove that, assuming a large-cardinal axiom of set theory called Vopěnka's principle, the structure of locally presentable categories becomes much more transparent, e.g.,

(1) a category is locally presentable iff it is complete and has a dense subcategory (Theorem 6.14);

(2) each full subcategory of a locally presentable category closed under limits (or colimits) is reflective (or coreflective, resp.) (Theorems 6.22, 6.28);

(3) each full embedding of an accessible category into a locally presentable category is accessible (Theorem 6.9);

(4) every orthogonality class in a locally presentable category is a small-orthogonality class (Corollary 6.24);

(5) every subfunctor of an accessible functor is accessible (Corollary 6.31).

In each instance we discuss the reverse implication, i.e., that the property under study implies Vopěnka's principle.

Vopěnka's principle states that no locally presentable category has a large, discrete, full subcategory. We call it a large-cardinal principle because, on the one hand, it implies the existence of measurable cardinals, and on the other hand, its consistency follows from the existence of huge cardinals. We explain this in detail in the Appendix in which all results on large cardinals needed in Chapter 6 are presented. Any reader who refutes the existence of measurable cardinals must refute Vopěnka's principle. However, the present chapter brings a message even to these readers: Do not try to construct counterexamples to the above statements (1)–(5). For example, one might ask whether a given subcategory of **Rel** Σ is reflective. The first thing to check is closedness under limits. Well, if the category *is* closed under limits and is defined "constructively" (without a large-cardinal resource), then it is reflective, simply because Vopěnka's principle implies its reflectivity.

6.A Vopěnka's Principle

In this section we introduce Vopěnka's principle, prove some equivalent formulations, and show that, under the assumption that it is true, the following hold:

(i) A category has a full embedding into a locally presentable category iff it has a dense subcategory.

(ii) Every full subcategory of a locally presentable category is cone-reflective.

(iii) Every full subcategory of a locally presentable category is co-wellpowered.

Since, moreover, each of these three statements implies that Vopěnka's principle is true, they can be considered as reformulations of that principle. Other (quite surprising) statements of this kind will be proved in subsequent sections.

6.1. A class of objects of a category is called *rigid* if it admits no morphisms except the identity morphisms (i.e., if the corresponding full subcategory is discrete).

Definition. We say that *Vopěnka's principle* holds provided that no locally presentable category has a large rigid class of objects.

6.2 Remarks

(1) The following statements are equivalent to Vopěnka's principle:

  (a) There exists no large rigid class of graphs. (In fact, this is equivalent because each locally presentable category can be fully embedded into the category **Gra**, see Theorem 2.65.)

  (b) No accessible category has a large rigid class of objects (see Proposition 2.8).

(c) No category of models of a finitary first-order theory, where morphisms are elementary embeddings, has a large rigid class. (Here the equivalence follows from Theorem 5.42 and (b).)

The last formulation is used in model theory, see e.g. [Jech 1978].

(2) It is trivial to construct, for each cardinal α, a rigid set of α objects in a locally presentable category. In fact, let $\Sigma = \{ \sigma_i \}_{i<\alpha}$ be the one-sorted signature of α unary relations. The following Σ-structures A_i, $i < \alpha$, form a rigid set in $\mathbf{Rel}\,\Sigma$: the underlying set of A_i is $\{ 0 \}$, the relation σ_j is $\{ 0 \}$ for $i = j$ and $\emptyset$ for $i \neq j$. Thus, it is the transition from sets to classes which makes Vopěnka's principle a large-cardinal principle.

6.3. Let **Ord** be the ordered class of all ordinals, considered as a category.

Lemma. *The following statements are equivalent:*

(i) *Vopěnka's principle;*

(ii) *The category* **Ord** *of ordinals cannot be fully embedded into the category* **Gra** *of graphs;*

(iii) *Given objects A_i ($i \in \mathrm{Ord}$) of an accessible category, there exists a morphism from A_i to A_j for some ordinals $i < j$.*

PROOF. (i) $\Rightarrow$ (iii). Assuming that an accessible category $\mathcal{K}$ has objects A_i with $\hom(A_i, A_j) = \emptyset$ for all ordinals $i < j$, we will show that Vopěnka's principle fails. Since $\mathcal{K}$ can be fully embedded into **Gra** (see Theorem 2.65), we can assume, without loss of generality, that $\mathcal{K} = \mathbf{Gra}$. For the given class $A_i = (X_i, \alpha_i)$, $i \in \mathrm{Ord}$, of graphs we first choose a large class $C \subseteq \mathrm{Ord}$ such that

$$i < j \quad \text{implies} \quad \operatorname{card} X_i < \operatorname{card} X_j \qquad \text{for } i, j \in C.$$

Next, for each $i \in C$ we choose a rigid binary relation β_i on the set X_i (which exists by Lemma 2.64), and we denote by γ_i the binary relation which is the complement of the diagonal relation on X_i (i.e., the relation $\neq$ on X_i). We obtain a relational structure

$$\overline{A}_i = (X_i, \alpha_i, \beta_i, \gamma_i)$$

of the one-sorted signature Σ which consists of three binary relation symbols. The negation of Vopěnka's principle is derived by showing that the class $(\overline{A}_i)_{i\in C}$ is rigid in **Rel** Σ. In fact, let $f\colon \overline{A}_i \to \overline{A}_j$ be a Σ-homomorphism. Then $f\colon A_i \to A_j$ is a graph homomorphism, thus, by our original assumption, $i \geq j$. The relations γ_i, γ_j guarantee that f is one-to-one. Since $i, j \in C$ imply $\operatorname{card} X_i \geq \operatorname{card} X_j$, it follows that $\operatorname{card} X_i = \operatorname{card} X_j$; thus, $i = j$. The relation β_i guarantees that $f = id_{X_i}$.

(iii) $\Rightarrow$ (i). This is evident.

(i) $\Rightarrow$ (ii) Suppose that $F\colon \mathbf{Ord} \to \mathbf{Gra}$ is a full embedding; put $F(i) = (X_i, \alpha_i)$. There exists a large class $C \subseteq \mathrm{Ord}$ such that $i \in C$ implies that

$$\sum_{\substack{j\in C\\ j<i}} \operatorname{card} X_j < \operatorname{card} X_i \qquad \text{for } i, j \in C.$$

For each $i \in C$ choose an element $\beta_i \in X_i$ not lying in $F(j \to i)(X_j)$ for any $j < i$, $j \in C$. For the one-sorted signature Σ of one binary relation and one nullary operation symbol we have the Σ-structures

$$A_i = \big(X_i, \alpha_i, \beta_i\big) \qquad i \in C$$

which form a rigid class in **Str** Σ. Thus, Vopěnka's principle does not hold.

(ii) $\Rightarrow$ (i). By Remark 2.65 **Gra** has a full embedding into $\mathbf{Gra}_0$ where $\mathbf{Gra}_0$ is the full subcategory of **Gra** consisting of connected graphs. Assume that Vopěnka's principle does not hold, then $\mathbf{Gra}_0$ has a large rigid class, say, A_i, $i \in \mathrm{Ord}$. Then **Ord** has a full embedding into **Gra** which assigns to each cardinal i the coproduct $\coprod_{k\leq i} A_k$. In fact, every homomorphism

$$f\colon \coprod_{k\leq i} A_k \to \coprod_{l\leq j} A_l$$

preserves the connected components, thus, for each $k \leq i$ there exists $l \leq j$ such that the domain–codomain restriction of f is a homomorphism from A_k to A_l; consequently, $k = l$ and $f/_{A_k} = id_{A_k}$. This implies that $i \leq j$ and that f is the coproduct injection. □

6.4. We know from Proposition 1.26 that every category which has a (small) dense subcategory is equivalent to a full subcategory of $\mathbf{Set}^{\mathcal{A}^{op}}$. We now study the following reverse statement: every full subcategory of a locally presentable category has a dense subcategory. We first introduce a name for the categories we will investigate:

Definition. A category $\mathcal{K}$ is called *bounded* if it has a dense subcategory.

6.5 Examples

(1) Every locally presentable category (in fact, every accessible category) is bounded, see Proposition 2.8.

(2) Every small category is bounded.

(3) A discrete category is bounded iff it is small. (The same holds for any ordered class.)

(4) Any axiomatizable class of Σ-structures forms a bounded category.

In fact, let T be a theory of L_∞ and $\overline{\lambda}$ a cardinal as in Theorem 5.41. Then models A of T with $\#|A| < \overline{\lambda}$ form a dense subcategory in $\mathbf{Mod}\,T$ (see Theorem 5.41).

(5) The category **Top** of topological spaces is not bounded, see Example 1.24(7).

6.6 Theorem. *Assuming Vopěnka's principle, a category is fully embeddable into a locally presentable category iff it is bounded.*

Remark. This statement is equivalent to Vopěnka's principle, since, asuming the negation of Vopěnka's principle, it is clear that there exist non-bounded full subcategories of locally presentable categories (since a large discrete category is not bounded).

PROOF. Every bounded category is equivalent to a full subcategory of $\mathbf{Set}^{\mathcal{A}^{\mathrm{op}}}$, see Proposition 1.26. Thus, it is isomorphic to a full subcategory of a locally presentable category by Exercise 2.i(1).

Conversely, let us prove that every full subcategory $\mathcal{K}$ of **Gra** is bounded. It will follow then from Theorem 2.65 that every full subcategory of a locally presentable category is bounded. For each regular cardinal n let $\mathcal{K}_n$ be the (essentially small) full subcategory of $\mathcal{K}$ over all $\mathcal{K}$-graphs which are n-presentable in **Gra**. It is sufficient to find n such that $\mathcal{K}_n$ is dense in $\mathcal{K}$.

Assume that, on the contrary, for each n there exists a graph $A_n = (X_n, \alpha_n)$ in $\mathcal{K}$ which is not a canonical colimit of $\mathcal{K}_n$-objects. Let $\overline{A}_n = (\overline{X}_n, \overline{\alpha}_n)$ be the colimit of the canonical diagram of A_n w.r.t. $\mathcal{K}_n$ in **Gra**, and let $a_n : \overline{A}_n \to A_n$ be the induced homomorphism. Then a_n is not an isomorphism for any n. We will derive a contradiction.

I. There exists n_0 such that a_n is surjective for each $n \geq n_0$. In fact, assuming the contrary, let C be a class of cardinals such that (1) a_n is not surjective for any $n \in C$, and (2) A_n is m-presentable in **Gra** for any $n, m \in C$, $n < m$. For each $n \in C$ choose $x_n \in X_n - a_n(\overline{X}_n)$ and consider the objects $A'_n = (X_n, \alpha_n, \{x_n\})$ of $\mathbf{Rel}\,\Sigma$, where Σ consists of one binary and one unary relation symbol. By Lemma 6.3(iii) there exists a Σ-homomorphism $f\colon A'_n \to A'_m$ for some $n < m$ in C. Then $f\colon A_n \to A_m$ is a graph homomorphism with $f(x_n) = x_m$. However, since A_n is m-presentable in **Gra**, the morphism $f\colon A_n \to A_m$ is an object of the canonical diagram of A_m w.r.t. $\mathcal{K}_m$. Hence, we have the corresponding colimit map $\overline{f}\colon A_n \to \overline{A}_m$ with $f = a_m \cdot \overline{f}$. This contradicts $f(x_n) = x_m \notin a_m(\overline{X}_m)$.

II. There exists $n_1 \geq n_0$ such that $(a_n \times a_n)(\overline{\alpha}_n) = \alpha_n$ for each $n \geq n_1$. The proof is quite analogous to I except that we now use Σ with two binary relations, and we put $A'_n = (X_n, \alpha_n, \alpha_n - (a_n \times a_n)(\overline{\alpha}_n))$.

III. There exists $n \geq n_1$ such that a_n is an isomorphism (in contradiction to our choice of a_n). To prove this, it is sufficient to show that a_n is one-to-one. Assume that, on the contrary, for each $n \geq n_1$ there exist distinct $x_n, x'_n \in \overline{X}_n$ with $a_n(x_n) = a_n(x'_n)$. Since $\overline{A}_n$ is the colimit of the canonical diagram of A_n w.r.t. $\mathcal{K}_n$, there exist homomorphisms

$$\overline{h}_n\colon H_n \to \overline{A}_n \text{ with } H_n \text{ in } \mathcal{K}_n \text{ and } x_n = \overline{h}_n(y_n) \quad \text{for some } y_n,$$

$$\overline{h}'_n\colon H'_n \to \overline{A}_n \text{ with } H'_n \text{ in } \mathcal{K}_n \text{ and } x'_n = \overline{h}'_n(y'_n) \quad \text{for some } y'_n.$$

Put $H_n = (Y_n, \beta_n)$ and $H'_n = (Y'_n, \beta'_n)$, and define $h_n = a_n \cdot \overline{h}_n$ and $h'_n = a_n \cdot \overline{h}'_n$. We assume, for simplicity of notation, that the sets X_n, Y_n, Y'_n are pairwise disjoint.

Consider the one-sorted signature Σ of three binary and five unary relation symbols. For each n we have the Σ-structure

$$A^*_n = (X_n \cup Y_n \cup Y'_n, \alpha_n \cup \beta_n \cup \beta'_n, h_n, h'_n, X_n, Y_n, Y'_n, \{y_n\}, \{y'_n\}).$$

Let C be a large class of cardinals $n \geq n_1$ such that if $n < m$ in C, then the four objects $A_n, \overline{A}_n, H_n$, and H'_n are m-presentable in **Gra**. By Lemma 6.3(iii) there exists a Σ-homomorphism from A^*_n to A^*_m for some $n < m$ in C. Due to the first three unary relations, this homomorphism has the form $f \cup p \cup p'$, where (due to the first binary relation) $f\colon A_n \to A_m$, $p\colon H_n \to H_m$, and $p'\colon H'_n \to H'_m$ are graph homomorphisms. Due to the latter two binary relations, we have

$$f \cdot h_n = h_m \cdot p \quad \text{and} \quad f \cdot h'_n = h'_m \cdot p'.$$

Since A_n is m-presentable in **Gra**, we have, as in I, the corresponding morphism $\overline{f}\colon A_n \to \overline{A}_m$ with $f = a_m \cdot \overline{f}$. Besides, h_n is a morphism of the canonical diagram from the object $f \cdot h_n$ $(= h_m \cdot p)$ to the object f;

analogously, p is a morphism of that diagram (from $h_m \cdot p$ to h_m). Since the colimit cocone of the canonical diagram is compatible, we get

$$\overline{f} \cdot h_n = \overline{h}_m \cdot p$$

where $\overline{h}_m : H_m \to \overline{A}_m$ corresponds to h_m. Analogously, $\overline{f} \cdot h'_n = \overline{h}'_m \cdot p'$. Finally, the last two unary relations imply $p(y_n) = y_m$ and $p'(y'_n) = y'_m$, thus,

$$\begin{aligned}\overline{f}(a_n(x_n)) &= \overline{f} \cdot a_n \cdot \overline{h}_n(y_n) \\ &= \overline{f} \cdot h_n(y_n) \\ &= \overline{h}_m \cdot p(y_n) \\ &= \overline{h}_m(y_m) \\ &= x_m.\end{aligned}$$

Analogously, $\overline{f}(a_n(x'_n)) = x'_m$. Since $a_n(x_n) = a_n(x'_n)$, this contradicts $x_m \neq x'_m$. □

6.7. Recall from Theorem 2.53 that every accessible, accessibly embedded subcategory of an accessible category $\mathcal{K}$ is cone-reflective in $\mathcal{K}$. Vopěnka's principle implies the following surprising "generalization":

Corollary. *Assuming Vopěnka's principle, every full subcategory of an accessible category is cone-reflective.*

PROOF. Let $\mathcal{A}$ be a full subcategory of an accessible category $\mathcal{K}$. For each object K of $\mathcal{K}$ the comma-category $K \downarrow \mathcal{K}$ is accessible, see Corollary 2.44; thus, it is bounded. Consequently, the full subcategory $K \downarrow \mathcal{A}$ is bounded. □

Remark. The statement is equivalent to Vopěnka's principle, since, assuming the negation of Vopěnka's principle, a large, discrete, full subcategory of a locally presentable category evidently fails to be cone-reflective.

6.8 Corollary. *Assuming Vopěnka's principle, each bounded category is co-wellpowered.*

PROOF. Let $\mathcal{K}$ be a bounded category. By Proposition 1.26 we can assume that $\mathcal{K}$ is a full subcategory of some $\mathbf{Set}^{\mathcal{A}^{op}}$. For each object K of $\mathcal{K}$ the comma-category $K \downarrow \mathbf{Set}^{\mathcal{A}^{op}}$ is bounded (in fact, locally presentable, see Proposition 1.57), thus, the full subcategory $K \downarrow \mathcal{K}$ is also bounded. Consequently, the full subcategory of $K \downarrow \mathcal{K}$ consisting of all epimorphisms is also bounded. This clearly implies that $\mathcal{K}$ is co-wellpowered. □

Remarks

(1) The statement is equivalent to Vopěnka's principle. In fact, assuming the negation of Vopěnka's principle, there exists a full embedding $F\colon \mathbf{Ord} \to \mathbf{Gra}$ (see Lemma 6.3). For each ordinal i we clearly have $\hom(F_i, G_1) = \hom(F_i, G_2) = \emptyset$, where $F_i = F(i)$ and $\{G_1, G_2\}$ is the following dense subcategory of **Gra**:

$$G_1 \;\boxed{\bullet} \qquad\qquad G_2 \;\boxed{\bullet \to \bullet}$$

(In fact, if $\hom(F_i, G_2) \neq \emptyset$ then for each $j > i$ such that the graph F_j has more than one edge there exist at least two homomorphisms from F_i to F_j, which is a contradiction. And $\hom(F_i, G_1) \neq \emptyset$ clearly implies $\hom(F_i, G_2) \neq \emptyset$.) The full subcategory of **Gra** consisting of G_1, G_2 and F_i ($i \in \mathrm{Ord}$) is bounded, but not co-wellpowered.

(2) In contrast, the statement

(*) each accessible category is co-wellpowered

is not equivalent to Vopěnka's principle because it follows from the weaker assumption that there exist arbitrarily large compact cardinals, see [Makkai, Paré 1989]. In the Appendix we will show that the statement (*) implies the existence of arbitrarily large measurable cardinals (Example A.19).

6.B A Characterization of Locally Presentable Categories

This section is devoted to a number of surprising properties of locally presentable and accessible categories which hold under Vopěnka's principle, e.g.:

(1) locally presentable categories are precisely the complete and bounded categories;

(2) each full embedding between accessible categories is accessible;

(3) each bounded category has all objects presentable.

It is, in particular, the first of these statements which is quite fascinating: in the world in which Vopěnka's principle holds one can define locally presentable categories without mentioning the concept of presentable object! (This is explained by (3), of course.)

We begin with a basic, although somewhat technical, generalization of (2):

6.9 Theorem. *Assuming Vopěnka's principle, for each full embedding $F: \mathcal{A} \to \mathcal{K}$, where $\mathcal{K}$ is an accessible category, there exists a regular cardinal λ such that F preserves λ-directed colimits.*

Remark. The statement is equivalent to Vopěnka's principle, as Example 6.12 below demonstrates.

PROOF. Since each accessible category can be accessibly embedded into **Gra** (see Theorem 2.65) it is sufficient to prove that given a full subcategory $\mathcal{K}$ of **Gra**, the embedding $\mathcal{K} \to$ **Gra** preserves λ-directed colimits for some λ. Suppose that, on the contrary, for each regular cardinal n there exists an n-directed diagram D_n in $\mathcal{K}$ which has a colimit $A_n = \operatorname{colim} D_n$ in $\mathcal{K}$, and this is not a colimit of D_n in **Gra**. That is, for $\overline{A}_n = \operatorname{colim} D_n$ in **Gra** the induced homomorphism

$$a_n : \overline{A}_n \to A_n$$

is not an isomorphism. We will derive a contradiction.

I. There exists n_0 such that a_n is one-to-one for each $n \geq n_0$. In fact, assuming the contrary, there clearly exists a large class C of cardinals such that a_n is not one-to-one for any $n \in C$, and if $n < m$ in C, then $\overline{A}_n$ is m-presentable in **Gra**. Since the category of graphs and one-to-one homomorphisms is accessible (see Example 2.3(6)), it follows from Lemma 6.3 that there exists a one-to-one homomorphism $f: \overline{A}_n \to \overline{A}_m$ for some $n, m \in C$, $n < m$. The object $\overline{A}_n$ is m-presentable, thus f factorizes through one of the colimit maps $\overline{d}: D \to \overline{A}_m$ of D_m, say, $f = \overline{d} \cdot f'$. Since a_n is, obviously, a reflection of $\overline{A}_n$ in $\mathcal{K}$, and D lies in $\mathcal{K}$, it follows that f' factorizes through a_n, thus f factorizes through a_n. Now f is one-to-one, which implies that a_n is one-to-one, a contradiction.

II. Some a_n, $n \geq n_0$, is an isomorphism (in contradiction to our choice of a_n). To prove this, put

$$A_n = (X_n, \alpha_n), \quad \overline{A}_n = (\overline{X}_n, \overline{\alpha}_n), \quad \text{and} \quad \alpha_n^* = (a_n \times a_n)(\overline{\alpha}_n) \subseteq X_n \times X_n.$$

For the one-sorted signature Σ of two binary and two unary relation symbols consider the objects

$$A_n^* = \big(X_n, \alpha_n^*, \alpha_n - \alpha_n^*, a_n(\overline{X}_n), X_n - a_n(\overline{X}_n)\big).$$

Let C be a large class of cardinals $n \geq n_0$ such that if $n < m$ in C, then $\overline{A}_n$ is m-presentable in **Gra**. By Lemma 6.3 and Example 2.3(6) there exists a one-to-one homomorphism $f\colon A_n^* \to A_m^*$ for some $n, m \in C$, $n < m$. We will prove that a_n is an isomorphism. Due to the first unary relation we have $f \cdot a_n(\overline{X}_n) \subseteq a_m(\overline{X}_m)$, thus, there exists a unique mapping $\overline{f}\colon \overline{X}_n \to \overline{X}_m$ with

$$f \cdot a_n = a_m \cdot \overline{f}.$$

(Recall that a_m is one-to-one.) The first binary relation guarantees, then, that $\overline{f}\colon \overline{A}_n \to \overline{A}_m$ is a homomorphism in **Gra**. Since $\overline{A}_n$ is m-presentable, $\overline{f}$ factorizes through some of the colimit maps $\overline{d}\colon D \to \overline{A}_m$ of D_m in **Gra**, say, $\overline{f} = \overline{d} \cdot f'$.

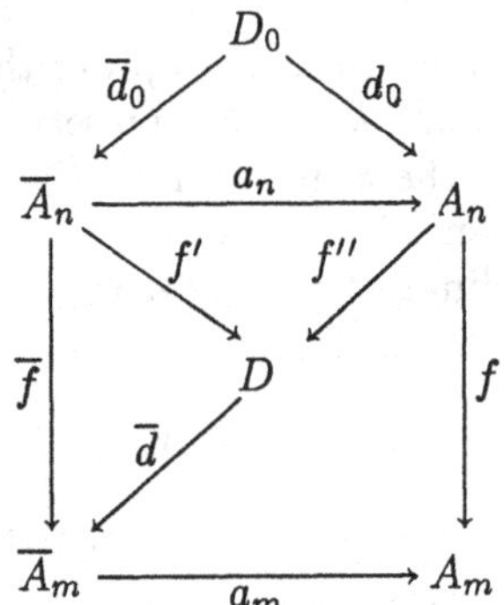

Since A_n is a reflection of $\overline{A}_n$ in $\mathcal{K}$, we have $f''\colon A_n \to D$ with $f' = f'' \cdot a_n$. We will prove that

$$f = a_m \cdot \overline{d} \cdot f''.$$

In fact, let $d_0\colon D_0 \to A_n$ be an arbitrary colimit map of D_n in $\mathcal{K}$; then we have $d_0 = a_n \cdot \overline{d}_0$ for the corresponding colimit map $\overline{d}_0$ in **Gra**, thus,

$$\begin{aligned} f \cdot d_0 &= f \cdot a_n \cdot \overline{d}_0 \\ &= a_m \cdot \overline{f} \cdot \overline{d}_0 \\ &= a_m \cdot \overline{d} \cdot f'' \cdot a_n \cdot \overline{d}_0 \\ &= \big(a_m \cdot \overline{d} \cdot f''\big) \cdot d_0. \end{aligned}$$

This implies that a_n is an isomorphism. In fact, a_n is surjective because, using the last relation, we conclude $f\big(X_n - a_n(\overline{X}_n)\big) \subseteq X_m - a_m(\overline{X}_m)$; however, $f(X_n) \subseteq a_m(\overline{X}_m)$. Thus, a_n is a bijective homomorphism. Finally, $(a_n \times a_n)(\overline{\alpha}_n) = \alpha_n$ because, using the second binary relation, we conclude $(f \times f)(\alpha_n - \alpha_n^*) \subseteq \alpha_m - \alpha_m^*$; however,

$$(f \times f)(\alpha_n) \subseteq (a_m \times a_m)(\overline{\alpha}_m) = \alpha_m^*. \qquad \square$$

6.10 Corollary. *Assuming Vopěnka's principle, each full, accessible subcategory of an accessible category is accessibly embedded.*

Remark. The statement is equivalent to Vopěnka's principle as Example 6.13 below demonstrates.

6.11 Corollary. *Assuming Vopěnka's principle, each object of a bounded category is presentable.*

PROOF. Let $\mathcal{K}$ be a category with a dense subcategory $\mathcal{A}$. The canonical functor $E\colon \mathcal{K} \to \mathbf{Set}^{\mathcal{A}^{\mathrm{op}}}$ of Proposition 1.26 preserves λ-directed colimits for some regular cardinal λ, see Theorem 6.9. Since each object EK, $K \in \mathcal{K}^{\mathrm{obj}}$ is μ-presentable in $\mathbf{Set}^{\mathcal{A}^{\mathrm{op}}}$ for some $\mu \geq \lambda$ (by Remark 1.20), K is μ-presentable in $\mathcal{K}$. □

Remark. The statement is equivalent to Vopěnka's principle, as the following example demonstrates.

6.12 Example constructed under the assumption of the negation of Vopěnka's principle: A full, epireflective, bounded subcategory $\mathcal{A}$ of the category **Gra** which has some non-presentable object, and whose embedding $\mathcal{A} \hookrightarrow \mathbf{Gra}$ does not preserve λ-directed colimits for any regular cardinal λ.

Let $(B_i)_{i \in \mathrm{Ord}}$ be a rigid class in **Gra**, and let $\mathcal{A}$ be the full subcategory of **Gra** consisting of the terminal object $T = \big(\{t\}, \{(t,t)\}\big)$ and all those graphs A such that

$$\hom(B_i, A) = \emptyset \qquad \text{for all } i \in \mathrm{Ord}.$$

(a) $\mathcal{A}$ is epireflective since it is, obviously, closed under products and subobjects in **Gra**.

(b) $\mathcal{A}$ is bounded as it contains the graphs G_1, G_2 of Remark 6.8(1).

(c) For each regular cardinal λ the following λ-directed colimit in $\mathcal{A}$ is not a colimit in **Gra**: let i be an ordinal such that the underlying set of B_i has cardinality greater or equal to λ. Each proper subgraph of B_i clearly lies in $\mathcal{A}$. The λ-directed diagram D of all subgraphs of B_i of cardinality smaller than λ thus lies in $\mathcal{A}$. Since the only compatible cocone of D in $\mathcal{A}$ is the trivial one with codomain T, we have

$$\operatorname{colim} D = T \quad \text{in } \mathcal{A}, \qquad \operatorname{colim} D = B_i \quad \text{in } \mathbf{Gra}.$$

(d) The graph G_1 is not λ-presentable in $\mathcal{A}$ for any regular cardinal λ: consider the diagram D in (c) above, then the morphism $G_1 \to T$ does not have an essentially unique(!) factorization through the colimit maps.

6.13 Example constructed under the assumption of the negation of Vopěnka's principle: A full, finitely accessible subcategory of $\mathbf{Rel}\,\Sigma$, Σ finitary, which is not accessibly embedded.

The category **Gra** can be fully embedded into its full subcategory $\mathbf{Gra}_0$ of connected graphs—see Remark 2.65. Thus, the negation of Vopěnka's principle provides us with a large rigid class of connected graphs, say, D_i ($i \in \mathrm{Ord}$, $i > 0$); put $D_0 = (\emptyset, \emptyset)$. Define a full embedding $F\colon \mathbf{Ord} \to \mathbf{Gra}$ on objects by

$$F_i = \coprod_{j \leq i} D_j .$$

The connecting morphisms $f_{i,j} : F_i \to F_j$ are, necessarily, the coproduct injections for $i \leq j$. Observe that

(i) each $f_{i,j}$ is one-to-one

and

(ii) for each limit ordinal j the cocone $(F_i \xrightarrow{f_{i,j}} F_j)_{i<j}$ is *not* a colimit of the corresponding j-chain (because of the summand D_j).

Let Σ be the one-sorted signature of two binary and two unary relation symbols. For each graph $G = (X, \alpha)$ we define a Σ-structure $\overline{G}$ as follows. If X is infinite, we have a unique ordinal i with $\mathrm{card}\, X = \aleph_i$; if X is finite, put $i = 0$. For the above graph $F_i = (Y_i, \beta_i)$ assume, to simplify the notation, that $Y_i \cap X = \emptyset$. Then define

$$\overline{G} = \big(X \cup Y_i,\ \alpha \cup \beta_i, \neq, X, Y_i\,\big)$$

where $\neq$ denotes the binary relation "non-equal". We claim that the full subcategory $\mathcal{A} = \{\,\overline{G} \mid G \text{ a graph}\,\}$ of $\mathbf{Rel}\,\Sigma$ has the above-mentioned properties.

(a) $\mathcal{A}$ has directed colimits. In fact, let

$$D\colon T \to \mathcal{A}$$

be a directed diagram with

$$D_t = \big(X_t \cup Y_{i(t)},\ \alpha_t \cup \beta_{i(t)}, \neq, X_t, Y_{i(t)}\,\big).$$

We form a colimit of D

$$\big((X_t, \alpha_t) \xrightarrow{d_t} (X, \alpha)\big)_{t \in T}$$

in **Gra**, which is also the colimit in the category of graphs and monomorphisms. (Due to the relation $\neq$, for each $s \leq t$ in T the morphism $D(s \to t)$ is a monomorphism.) Therefore, in $\mathcal{A}$ we have

$$\mathrm{colim}\, D = \big(\,D_t \xrightarrow{d_t \cup f_{i(t),i}} \big(X \cup Y_i,\ \alpha \cup \beta_i, \neq, X, Y_i\,\big)\big)$$

for $\mathrm{card}\, X = \aleph_i$ or $\mathrm{card}\, X < \aleph_0$ and $i = 0$.

(b) Each finite Σ-structure in $\mathcal{A}$ is finitely presentable in $\mathcal{A}$. Since any graph is a directed union of its finite subgraphs, and $\mathcal{A}$-morphisms are one-to-one, $\mathcal{A}$ is finitely accessible.

(c) $\mathcal{A}$ is not accessibly embedded in $\mathbf{Rel}\,\Sigma$. In fact, for each regular cardinal λ, let G be a graph on $\aleph_\lambda$ vertices, and let us express G as a colimit of a λ-chain of subgraphs G_i on $\aleph_i$ vertices ($i < \lambda$). Denote by $g_{i,j}: G_i \to G_j$ the inclusion morphisms. Then the following λ-chain in $\mathcal{A}$

$$g_{i,j} \cup f_{i,j} : \overline{G}_i \to \overline{G}_j \qquad (i \leq j < \lambda)$$

has different colimits in $\mathcal{A}$ and in **Gra**—this follows from (ii) above.

6.14 Characterization Theorem. *Assuming Vopěnka's principle, the following conditions on a category $\mathcal{K}$ are equivalent:*

(i) *$\mathcal{K}$ is locally presentable;*

(ii) *$\mathcal{K}$ is complete and bounded;*

(iii) *$\mathcal{K}$ is cocomplete and bounded.*

Remark. The statement is equivalent to Vopěnka's principle, as Example 6.12 demonstrates.

PROOF. (i) $\Rightarrow$ (ii) is clear.

(ii) $\Rightarrow$ (iii). Let $\mathcal{K}$ be a complete category containing a dense subcategory $\mathcal{A}$ and $E: \mathcal{K} \to \mathbf{Set}^{\mathcal{A}^{\mathrm{op}}}$ be the canonical functor. Since E satisfies the solution-set condition (by Corollary 6.7), $\mathcal{K}$ is equivalent to a reflective subcategory $E(\mathcal{K})$ in $\mathbf{Set}^{\mathcal{A}^{\mathrm{op}}}$. Thus $\mathcal{K}$ is cocomplete.

(iii) $\Rightarrow$ (i) follows from Theorems 1.20 and Corollary 6.11. □

6.15. Recall from Remark 3.23(2) that SP-classes of algebras, i.e., full subcategories of $\mathbf{Alg}\,\Sigma$ closed under subobjects and products, can be described by implications of the form

$$\bigwedge_{i\in I}(\tau_i = \tau_i') \implies (\tau = \tau').$$

We can ask whether a set of implications is sufficient. This is undecidable:

Corollary. *The following statements are equivalent:*

(i) *Every SP-class of finitary algebras is a quasivariety;*

(ii) *Every SP-class of finitary algebras is locally presentable;*

(iii) *Vopěnka's principle holds.*

PROOF. (i) $\Rightarrow$ (ii) follows from Theorem 3.33.

(ii) $\Rightarrow$ (iii). See Example 6.12.

(iii)$\Rightarrow$ (i) follows from Theorem 6.9 and Remark 3.32. □

6.16 Remark. Assuming Vopěnka's principle, every full subcategory of a locally presentable category closed under connected limits is a small cone-orthogonality class.

This follows from the result, proved in [Adámek, Rosický 1994b], that Vopěnka's principle implies that every full subcategory $\mathcal{A}$ of an accessible category $\mathcal{K}$, closed under equalizers, is accessible. In fact, $\mathcal{A}$ is then multireflective in $\mathcal{K}$ by Theorem 4.26 and Corollary 6.7, and the result follows from Theorems 4.29 and 6.9.

Consequently, assuming Vopěnka's principle, the following conditions on a category $\mathcal{K}$ are equivalent:

(1) $\mathcal{K}$ is locally multipresentable;

(2) $\mathcal{K}$ is bounded and has connected limits.

(Use Theorem 4.30 and the above argument applied to the canonical functor $E\colon \mathcal{K} \to \mathbf{Set}^{\mathcal{A}^{\mathrm{op}}}$.)

6.C A Characterization of Accessible and Axiomatizable Categories

In the present section we characterize accessible categories (assuming Vopěnka's principle) as the bounded categories which have λ-directed colimits for some regular cardinal λ. Recall from Corollary 2.36 that the accessibility of accessibly embedded subcategories can be characterized by their closedness under λ-pure subobjects.

6.17 Theorem. *Assuming Vopěnka's principle, the following conditions are equivalent for each full subcategory $\mathcal{A}$ of an accessible category $\mathcal{K}$:*

(i) *$\mathcal{A}$ is accessible;*

(ii) *$\mathcal{A}$ is accessibly embedded into $\mathcal{K}$;*

(iii) *$\mathcal{A}$ is closed in $\mathcal{K}$ under λ-pure subobjects for some regular cardinal λ.*

Remark. The statement (i) $\Leftrightarrow$ (ii) is equivalent to Vopěnka's principle since a rigid class is a full subcategory which is accessibly embedded (by default). The statement (i) $\Leftrightarrow$ (iii) is also equivalent to Vopěnka's principle: In Example 6.12 we exhibited a full subcategory of an accessible category which is closed under all subobjects but is not accessible.

PROOF. (i) $\Rightarrow$ (iii) follows from Theorem 6.9 and Corollary 2.36.

(iii) $\Rightarrow$ (ii). Let $\mathcal{A}$ be closed in $\mathcal{K}$ under λ-pure subobjects, and assume that for any regular cardinal n there is an n-directed diagram $D_n : \mathcal{D}_n \to \mathcal{A}$ such that a colimit $K_n = \operatorname{colim} D_n$ of D_n in $\mathcal{K}$ does not belong to $\mathcal{A}$. We choose a large class C of cardinals such that K_n is m-presentable in $\mathcal{K}$ for $n < m$ in C. It follows from Theorem 2.34 and Lemma 6.3 that there is a λ-pure morphism $h\colon K_n \to K_m$ for some $n < m$, $n, m \in C$. Since K_n is m-presentable, h has a factorization through some colimit morphism $k : D_m d \to K_m$ of D_m, say, $h = k \cdot g$, for some $d \in \mathcal{D}_m^{\mathrm{obj}}$. Since $g\colon K_n \to D_m d$ is λ-pure (by Remark 2.28(2)) and $\mathcal{A}$ is closed in $\mathcal{K}$ under λ-pure subobjects, we conclude that K_n lies in $\mathcal{A}$, which is a contradiction.

(ii) $\Rightarrow$ (i) follows from Theorem 2.53 and Corollary 6.7. □

6.18 Corollary. (Characterization of accessible categories) *Assuming Vopěnka's principle, a category $\mathcal{K}$ is accessible iff it is bounded and has λ-directed colimits for some regular cardinal λ.*

PROOF. Let $\mathcal{K}$ be a bounded category with λ-directed colimits. For a dense subcategory $\mathcal{A}$ of $\mathcal{K}$, the canonical functor $E\colon \mathcal{K} \to \mathbf{Set}^{\mathcal{A}^{\mathrm{op}}}$ (of Proposition 1.26) is accessible (by Corollary 6.11). It suffices to apply Theorem 6.17. □

Remark. The statement is equivalent to Vopěnka's principle, as Example 6.12 demonstrates.

6.19. Recall from Theorem 2.60 that the category

$$\mathbf{Mod}(\mathscr{S}, \mathcal{K})$$

of models of a sketch $\mathscr{S}$ in a category $\mathcal{K}$ is accessible whenever $\mathcal{K}$ is locally presentable. We now turn to the question of the accessibility of $\mathbf{Mod}(\mathscr{S}, \mathcal{K})$ for all accessible categories $\mathcal{K}$:

Corollary. *Assuming Vopěnka's principle, for each sketch $\mathscr{S}$ the category $\mathbf{Mod}(\mathscr{S}, \mathcal{K})$ of models in an accessible category $\mathcal{K}$ is accessible.*

PROOF. This follows from the fact that the categories $\mathcal{C}_D$ of the proof of Theorem 2.60 are accessible, hence, by Theorem 6.17, they are accessibly embedded. With this modification, the proof of Theorem 2.60 does not require more than the accessibility of $\mathcal{K}$. □

Remark. The statement of the above corollary is dependent on set theory, but is not equivalent to Vopěnka's principle. In fact, the situation is analogous to that for co-wellpoweredness (Remark 6.8(2)).

6.20 Corollary. (Characterization of axiomatizable categories) *Assuming Vopěnka's principle, a class of structures (of a given signature) is axiomatizable iff it is closed under λ-elementary submodels for some regular cardinal λ.*

PROOF. Any full, axiomatizable subcategory of $\mathbf{Str}\,\Sigma$ is closed under λ-elementary submodels for some regular cardinal λ, see Proposition 5.39. Conversely, let $\mathcal{K}$ be a full subcategory of $\mathbf{Str}\,\Sigma$ closed under λ-elementary submodels. The category $\mathbf{Elem}_\lambda\,\Sigma$ of Σ-structures and λ-elementary embeddings is accessible (see Theorem 5.42). Let $\overline{\mathcal{K}}$ be the full subcategory of $\mathbf{Elem}_\lambda\,\Sigma$ over all $\mathcal{K}$-objects. Since $\overline{\mathcal{K}}$ is closed in $\mathbf{Elem}_\lambda\,\Sigma$ under subobjects, it follows from Theorem 6.17 that $\overline{\mathcal{K}}$ is accessible and accessibly embedded. Hence $\mathcal{K}$ is axiomatizable (see Theorem 5.44). □

Remark. The statement is equivalent to Vopěnka's principle since in Example 6.12 the class $\mathcal{A}$ of graphs is closed under substructures, but fails to be axiomatizable. In fact, assuming that $\mathcal{A} = \mathbf{Mod}(T)$, the complementary class $\mathcal{C}$ is axiomatizable too: $\mathcal{C} = \mathbf{Mod}(\bigvee_{\varphi\in T} \neg\varphi)$. By Theorem 5.42 there exists a set $\mathcal{D}$ of objects of $\mathcal{C}$ with

$$\bigcup_{D\in\mathcal{D}} \hom(D, C) \neq \emptyset$$

for each object C in $\mathcal{C}$.

For each ordinal i we clearly have $B_i \in \mathcal{C}$, thus, there exists $D_i \in \mathcal{D}$ with $\hom(D_i, B_i) \neq \emptyset$. By the definition of $\mathcal{A}$, since $D_i \notin \mathcal{A}$ there exists an ordinal i' with $\hom(B_{i'}, D_i) \neq \emptyset$. Thus, $\hom(B_{i'}, B_i) \neq \emptyset$, which implies $i = i'$. It follows immediately that if $i \neq j$, then $D_i \neq D_j$ (because $\hom(B_i, B_j) \neq \emptyset$), in contradiction to the smallness of $\mathcal{D}$.

6.D A Characterization of Reflective Subcategories

In this section we will show that, assuming Vopěnka's principle, each locally presentable category $\mathcal{K}$ has the following properties:

(1) every full subcategory of $\mathcal{K}$ closed under limits is reflective in $\mathcal{K}$,

and

(2) every full subcategory of $\mathcal{K}$ closed under colimits is coreflective in $\mathcal{K}$.

There is a basic difference between these two results: whereas (2) is (once again) equivalent to Vopěnka's principle, the situation with (1) is more subtle. Recall that one formulation of Vopěnka's principle is

$$\mathbf{Ord} \text{ cannot be fully embedded into } \mathbf{Gra}$$

(see Lemma 6.3). Now consider the formally similar statement

$$\mathbf{Ord}^{\mathrm{op}} \text{ cannot be fully embedded into } \mathbf{Gra}.$$

It follows from Lemma 6.3(iii) that the latter statement actually follows from Vopěnka's principle. Since that statement is precisely the assumption needed for (1), we give it a name:

6.21 Definition. We say that *weak Vopěnka's principle* holds provided that $\mathbf{Ord}^{\mathrm{op}}$ cannot be fully embedded into any locally presentable category.

Remark. Since, by Lemma 6.3(iii)

$$\text{Vopěnka's principle} \implies \text{weak Vopěnka's principle},$$

it follows that the assumption of the existence of huge cardinals implies consistency of weak Vopěnka's principle (see Theorem A.18). On the other hand,

$$\text{weak Vopěnka's principle} \implies \text{arbitrarily large measurable cardinals exist},$$

see Corollary A.7. Therefore, anybody who refutes measurable cardinals must refute weak Vopěnka's principle.

Unfortunately, it is not known so far whether either of the implications above is an equivalence.

6.22 Theorem. *Assuming weak Vopěnka's principle, every full subcategory of a locally presentable category $\mathcal{K}$ closed in $\mathcal{K}$ under limits is reflective in $\mathcal{K}$.*

Remark. The statement is equivalent to weak Vopěnka's principle, as demonstrated by Example 6.23 below.

PROOF. Let $\mathcal{A}$ be a full subcategory of a locally presentable category $\mathcal{K}$, closed under limits in $\mathcal{K}$. Express $\mathcal{A}$ as a union

$$\mathcal{A} = \bigcup_{i \in \mathrm{Ord}} \mathcal{A}_i$$

of small, full subcategories $\mathcal{A}_0 \subseteq \mathcal{A}_1 \subseteq \cdots$, and for each i let $\overline{\mathcal{A}}_i$ be the smallest full subcategory of $\mathcal{K}$ containing $\mathcal{A}_i$ and closed under limits in $\mathcal{K}$. Since $\mathcal{K}$ is wellpowered (Remark 1.56(1)), it follows that $\overline{\mathcal{A}}_i$ is wellpowered. We have $\mathcal{A} = \bigcup_{i \in \mathrm{Ord}} \overline{\mathcal{A}}_i$. For each i, $\mathcal{A}_i$ is a cogenerator of $\overline{\mathcal{A}}_i$. In fact, let $\mathcal{B}$ be the collection of all objects B in $\mathcal{K}$ such that for any pair of distinct morphisms $p, p' \colon B_0 \to B$ in $\mathcal{K}$ there is a morphism $a \colon B \to A$, $A \in \mathcal{A}_i^{\mathrm{obj}}$, with $a \cdot p \neq a \cdot p'$. Then $\mathcal{B}$ contains $\mathcal{A}_i$ and is closed under limits in $\mathcal{K}$—thus, $\overline{\mathcal{A}}_i \subseteq \mathcal{B}$. By the special adjoint functor theorem applied to the inclusion functor $\overline{\mathcal{A}}_i \hookrightarrow \mathcal{K}$, we see that $\overline{\mathcal{A}}_i$ is a reflective subcategory of $\mathcal{K}$.

Let us prove that every object K of $\mathcal{K}$ has a reflection in $\mathcal{A}$. For each $i \in \mathrm{Ord}$ we have a reflection $r_i \colon K \to A_i$ of K in $\overline{\mathcal{A}}_i$. Since $i \leq j$ implies $\overline{\mathcal{A}}_i \subseteq \overline{\mathcal{A}}_j$, we know that r_i factorizes through r_j. In the case that, conversely, r_j factorizes through r_i we obtain that r_i and r_j are isomorphic (as objects of the comma-category $K \downarrow \mathcal{A}$), thus, r_i is then a reflection in all of $\overline{\mathcal{A}}_j$. Consequently, either some r_i is a reflection of K in $\mathcal{A}$, or there exists a class $\mathcal{C}$ of ordinals such that for $i, j \in \mathcal{C}$, $i < j$, the arrow r_j does not factorize through r_i. To conclude the proof, we will show that the latter contradicts to weak Vopěnka's principle because we can embed $\mathcal{C}^{\mathrm{op}} \cong \mathbf{Ord}^{\mathrm{op}}$ fully into the comma-category $K \downarrow \mathcal{K}$ (which is locally presentable, see Proposition 1.57). In fact, for any morphism from $K \xrightarrow{r_i} A_i$ to $K \xrightarrow{r_j} A_j$ in $K \downarrow \mathcal{K}$ $(i, j \in \mathcal{C})$ we know that, since r_j factorizes through r_i in $\mathcal{K}$, we have $i \geq j$. Then that morphism is the unique factorization of $r_j \colon K \to A_j$ (in $\overline{\mathcal{A}}_i$) through the reflection r_i; hence, we have a full embedding

$$E \colon \mathcal{C}^{\mathrm{op}} \longrightarrow K \downarrow \mathcal{K}, \qquad E(i) = \left(K \xrightarrow{r_i} A_i\right),$$

a contradiction. □

6.23 Example constructed under the assumption of the negation of weak Vopěnka's principle: A non-reflective, full subcategory of **Gra** closed in **Gra** under limits.

Assume that $F \colon \mathbf{Ord}^{\mathrm{op}} \to \mathbf{Gra}$ is a full embedding. Let $\mathcal{A}$ be the full subcategory of **Gra** of objects K having the following property: there exists an ordinal i_K such that

(a) there is no morphism $F(i) \to K$ for $i < i_K$,

(b) for any $i \geq i_K$ there is exactly one morphism $F(i) \to K$.

It is easy to check that $\mathcal{A}$ contains all objects $F(i)$, $i \in \mathrm{Ord}$ and is closed under limits in **Gra**. Further, $\mathcal{A}$ does not have an initial object because there is no morphism $K \to F(i)$ for K in **Gra** and $i > i_K$. Therefore, $\mathcal{A}$ is not reflective in **Gra**.

6.24 Corollary. *Assuming Vopěnka's principle, the following properties of full subcategories $\mathcal{A}$ of a locally presentable category are equivalent:*

(i) *$\mathcal{A}$ is closed under limits;*

(ii) *$\mathcal{A}$ is an orthogonality class;*

(iii) *$\mathcal{A}$ is a small-orthogonality class;*

(iv) *$\mathcal{A}$ is reflective.*

Each of those properties implies that $\mathcal{A}$ is locally presentable.

In fact, (i) $\Rightarrow$ (iv) by Theorem 6.22, and (iv) $\Rightarrow$ (iii) by Theorems 6.9 and 1.39 (which also imply local presentability). The implications (iii) $\Rightarrow$ (ii) $\Rightarrow$ (i) are trivial. □

Remark: The statement that any of the conditions (i)–(iv) above implies local presentability is equivalent to Vopěnka's principle—see Example 6.12.

In contrast, the orthogonal-subcategory problem, i.e., the statement (ii) $\Leftrightarrow$ (iv), is equivalent to weak Vopěnka's principle as we now demonstrate:

6.25 Example constructed under the assumption of the negation of weak Vopěnka's principle: An orthogonality class in a locally presentable category which is not reflective. Consider the following auxiliary category $\mathcal{C}$. Its objects are A_i, B_i ($i \in \mathrm{Ord}$) and C. Its morphisms are freely generated by the morphisms $a_{i,j}: A_i \to A_j$ ($i, j \in \mathrm{Ord}$, $i < j$), $b_{i,k}: A_i \to B_k$ ($i, k \in \mathrm{Ord}$), and $c_i: C \to A_i$ ($i \in \mathrm{Ord}$), subject to the following relations:

$$\begin{aligned} a_{i,j} &= a_{t,j} \cdot a_{i,t} && \text{for all } i < t < j \\ b_{i,k} &= b_{j,k} \cdot a_{i,j} && \text{for all } i < j \text{ and all } k \\ b_{i,k} \cdot c_i &= b_{k,k} \cdot c_k && \text{for all } k < i. \end{aligned}$$

(1) Let us verify that whenever $\mathcal{C}$ is a full subcategory of a category $\mathcal{K}$, then for $\mathcal{M} = \{ a_{0,i} \}_{i \in \mathrm{Ord}}$ the orthogonality class $\mathcal{M}^{\perp}$ is not reflective in $\mathcal{K}$. In fact, suppose that the object A_0 has a reflection

$$r: A_0 \to R$$

in $\mathcal{M}^{\perp}$. We will derive a contradiction by showing that $\hom(C, R)$ is a proper class.

We first observe that each B_k lies in $\mathcal{M}^{\perp}$: for every morphism

$$f\colon A_0 \to B_k$$

we have $f = b_{0,k}$ (since $\mathcal{C}$ is a full subcategory), and $b_{i,k}\colon A_i \to B_k$ is the unique morphism with $f = b_{i,k} \cdot a_{0,i}$. Thus, for each $b_{0,k}\colon A_0 \to B_k$ there exists a unique $b^*_{0,k}\colon R \to B_k$ with $b_{0,k} = b^*_{0,k} \cdot r$.

Next, since R lies in $\mathcal{M}^{\perp}$, for each $i \in \mathrm{Ord}$ we have a unique morphism $r_i\colon A_i \to R$ with $r = r_i \cdot a_{0,i}$. We prove that

$$i < k \qquad \text{implies} \qquad r_i \cdot c_i \neq r_k \cdot c_k\colon C \to R,$$

which will be the desired contradiction. Since

$$b^*_{0,k} \cdot r_i \cdot a_{0,i} = b_{0,k} = b_{i,k} \cdot a_{0,i} \quad (\colon A_0 \to B_k)$$

and R lies in $\mathcal{M}^{\perp}$, we see that $b^*_{0,k} \cdot r_i = b_{i,k}$. Finally, since in $\mathcal{C}$ we have $b_{i,k} \cdot c_i \neq b_{k,k} \cdot c_k$ (for $i < k$), it follows that $b^*_{0,k} \cdot r_i \cdot c_i \neq b^*_{0,k} \cdot r_k \cdot c_k$, thus $r_i \cdot c_i \neq r_k \cdot c_k$.

(2) We will find a full embedding $F\colon \mathcal{C} \to \mathbf{Rel}\,\Sigma$ where Σ consists of two binary and one unary relation symbols.

The negation of weak Vopěnka's principle implies that $\mathbf{Ord}^{\mathrm{op}}$ can be fully embedded into $\mathbf{Gra}$ (see Theorem 2.65); let

$$s_{i,j}\colon S_i \to S_j = (X_j, \sigma_j) \qquad (i \geq j,\ i, j \in \mathrm{Ord})$$

be such a full embedding. Furthermore, the negation of Vopěnka's principle follows too, thus, we have a rigid class

$$T_i = (Y_i, \tau_i) \qquad (i \in \mathrm{Ord})$$

in $\mathbf{Gra}$. We can assume that $X_i \cap Y_i = \emptyset$, and we choose an element $z_i \notin X_i \cup Y_i$. Define relational structures as follows:

$$\overline{A}_i = \big(X_i \cup Y_i \cup \{z_i\},\ \sigma_i \cup \tau_i,\ (X_i \times \{z_i\}) \cup (\{z_i\} \times Y_i),\ \{z_i\}\big).$$

They form a rigid class in $\mathbf{Rel}\,\Sigma$. In fact, each morphism $f\colon \overline{A}_i \to \overline{A}_j$ fulfils $f(z_i) = z_j$ (due to the unary relation), thus $f(X_i) \subseteq X_j$ and $f(Y_i) \subseteq Y_j$ (due to the second binary relation). It follows that $f = f_1 \cup f_2 \cup f_3$ where $f_1\colon S_i \to S_j$ and $f_2\colon T_i \to T_j$ are graph homomorphisms, and $f_3(z_i) = z_j$. Then $i = j$ and $f_2 = id$ (since the T_i's are rigid) and $f_1 = s_{i,i} = id$, hence, $f = id$.

Further, denote by $\overline{B}_i$ the quotient object of $\overline{A}_i$ modulo the equivalence merging Y_i to one point (called y_i):

$$\overline{B}_i = (\, X_i \cup \{y_i, z_i\},\ \sigma_i \cup \{(y_i, y_i)\},\ (X_i \times \{z_i\}) \cup \{(z_i, y_i)\},\ \{z_i\}\,).$$

We have a morphism

$$\overline{b}_{i,j} : \overline{A}_i \to \overline{B}_j \qquad \text{for each } i \geq j$$

mapping all of Y_i to y_j, defined on X_i as $s_{i,j}$, and sending z_i to z_j. These are the only homomorphisms, i.e.,

$$\text{given } f : \overline{A}_i \to \overline{B}_j \text{, then } i \geq j \text{ and } f = \overline{b}_{i,j}.$$

In fact, $f(z_i) = z_j$ (due to the unary relation), thus, $f(y) = y_j$ for each $y \in Y_i$ and $f(X_i) \subseteq X_j$ (due to the second binary relation), and, since the domain-range restriction of f yields a morphism $S_i \to S_j$, we have $i \geq j$ and $f/_{X_i} = s_{i,j}$.

We are now prepared to define a full embedding

$$F : \mathcal{C} \to \mathbf{Rel}\,\Sigma$$

on objects by

$$FA_i = \coprod_{j \leq i} \overline{A}_j$$

$$FB_i = \Big(\coprod_{j < i} \overline{A}_j\Big) + \overline{B}_i$$

$$FC = (\,\{0\}, \emptyset, \emptyset, \{0\}\,).$$

For morphisms we have a unique choice: $Fa_{i,i'}$ is the coproduct injection for $i \leq i'$. (Since $Fa_{i,i'}$ must preserve, due to the second binary relation, the connected components of FA_i, which are $\overline{A}_j$, $j \leq i$, this is the only morphism between FA_i and $FA_{i'}$.) $Fb_{i,k}$ has the following components: coproduct injections of $\overline{A}_j$ to B_k for all $j < k$, and $\overline{b}_{j,k}$ composed with the coproduct injection for all $j \geq k$. (Again, since $Fb_{i,k}$ preserves connected components, this is the only morphism.) Finally, Fc_i sends 0 to z_i. Thus, F is full.

6.26 Theorem. *Assuming Vopěnka's principle, the following conditions on a full subcategory $\mathcal{A}$ of a locally presentable category $\mathcal{K}$ are equivalent:*

(i) *$\mathcal{A}$ is closed under products and split subobjects in $\mathcal{K}$;*

(ii) *$\mathcal{A}$ is an injectivity class of $\mathcal{K}$;*

(iii) *$\mathcal{A}$ is weakly reflective and closed under split subobjects in $\mathcal{K}$.*

PROOF. It is obvious that (iii) $\Rightarrow$ (ii) $\Rightarrow$ (i) (see Remark 4.5(3) and Proposition 4.3), thus it is sufficient to prove (i) $\Rightarrow$ (iii). Express $\mathcal{A}$ as a union

$$\mathcal{A} = \bigcup_{i \in \mathrm{Ord}} \mathcal{A}_i$$

of small, full subcategories $\mathcal{A}_0 \subseteq \mathcal{A}_1 \subseteq \cdots$, and for each i let $\overline{\mathcal{A}}_i$ be the smallest full subcategory of $\mathcal{K}$ containing $\mathcal{A}_i$ and closed under products and split subobjects in $\mathcal{K}$. We have $\mathcal{A} = \bigcup_{i \in \mathrm{Ord}} \overline{\mathcal{A}}_i$. Each $\overline{\mathcal{A}}_i$ is weakly reflective: given an object K in $\mathcal{K}$, the canonical morphism

$$K \to \prod_{A \in \mathcal{A}_i^{\mathrm{obj}}} A^{\mathrm{hom}(K,A)}$$

is a weak reflection of K in $\overline{\mathcal{A}}_i$. We will prove that $\mathcal{A}$ is weakly reflective. For each object K we have weak reflections $r_i \colon K \to K_i$ of K in $\overline{\mathcal{A}}_i$, and it is sufficient to prove that there exists an ordinal i such that all r_j, $j \geq i$, factorize through r_i; then, r_i is a weak reflection in $\mathcal{A}$.

Assuming the contrary, we have ordinals $i_0 < i_1 < \cdots < i_s < \cdots$ ($s \in \mathrm{Ord}$) such that r_{i_t} does not factorize through any r_{i_s} with $s < t$ (in Ord). Observe that since $\mathcal{A}_{i_s}$ is a subcategory of $\mathcal{A}_{i_t}$, we know that r_{i_s} factorizes through r_{i_t}. Thus in the comma-category $K \downarrow \mathcal{A}$ we have objects $A_s = \left(K \xrightarrow{r_{i_s}} K_{i_s}\right)$ such that

$$(*) \qquad \mathrm{hom}(A_s, A_t) \neq \emptyset \quad \text{iff } s \geq t \qquad (s, t \in \mathrm{Ord}).$$

Since $K \downarrow \mathcal{K}$ is locally presentable (see Proposition 1.57), this contradicts Vopěnka's principle: see Lemma 6.3. □

6.27 Remarks

(1) The statement of Theorem 6.26 implies weak Vopěnka's principle. (Whether it implies Vopěnka's principle is an open problem.) In fact, assuming the negation of weak Vopěnka's principle, the category $\mathcal{A}$ of Example 6.23 does not have a *weakly initial object* (where an object A is weakly initial provided that for each object B there exists a morphism from A to B). Thus, $\mathcal{A}$ is not weakly reflective in $\mathcal{K}$ because a weak reflection of an initial object would be weakly initial.

(2) Consider the following statement:

(∗) in any locally presentable category there do not exist objects A_i ($i \in \mathrm{Ord}$) such that $\hom(A_i, A_j) \neq \emptyset$ iff $i \geq j$.

This lies in between the full and weak Vopěnka's principles:

$$\text{Vopěnka's principle} \Rightarrow (*) \Rightarrow \text{weak Vopěnka's principle.}$$

By inspecting the proof of Theorem 6.26, it is obvious that, instead of Vopěnka's principle, (∗) is sufficient. And, conversely, the statement of that theorem implies (∗), which is demonstrated by a straightforward modification of Example 6.23 (omitting (a) and relaxing (b) to the existence of a morphism $F(i) \to K$).

(3) We do not know whether the implication (i) ⇒ (ii) in Theorem 6.26 depends on set theory.

6.28 Theorem. *Assuming Vopěnka's principle, every full subcategory of a locally presentable category $\mathcal{K}$, closed under colimits in $\mathcal{K}$, is coreflective.*

PROOF. Let $\mathcal{A}$ be a full subcategory of $\mathcal{K}$ closed under colimits. Since $\mathcal{K}$ is co-wellpowered (Theorem 1.58), $\mathcal{A}$ is also co-wellpowered. Vopěnka's principle implies that $\mathcal{A}$ is bounded (Theorem 6.6), thus, $\mathcal{A}$ has a generator. By the dual of the special adjoint functor theorem (see 0.7) the inclusion $\mathcal{A} \hookrightarrow \mathcal{K}$ (which preserves colimits) is a left adjoint. □

Remark. The statement is equivalent to Vopěnka's principle since, assuming the negation of Vopěnka's principle, there exists a non-coreflective, full subcategory of **Gra** which is closed under colimits.

In fact, let $\mathcal{D}$ be a large, rigid class of connected graphs in **Gra** (see Theorem 2.65). The closure $\overline{\mathcal{D}}$ of $\mathcal{D}$ under coproducts is already closed under colimits in **Gra**. However, $\overline{\mathcal{D}}$ is not coreflective because it evidently does not have a terminal object.

6.29 Corollary. *Assuming Vopěnka's principle, the following properties of full subcategories of a locally presentable category $\mathcal{A}$ are equivalent:*

(i) *$\mathcal{A}$ is closed under colimits;*

(ii) *$\mathcal{A}$ is a co-orthogonality class (= the dual to orthogonality class);*

(iii) *$\mathcal{A}$ is coreflective.*

Each of these properties implies that $\mathcal{A}$ is locally presentable.

In fact, the equivalence of (i), (ii) and (iii) is an immediate consequence of Theorem 6.28. Every full, coreflective subcategory of a locally presentable category is accessible (by Theorem 6.17), and therefore locally presentable (see Corollary 2.47). □

6.E A Characterization of Accessible Functors

We know that any accessible functor satisfies the solution-set condition (see Corollary 2.45). In the present section we will show that, assuming Vopěnka's principle, any functor between accessible categories satisfying the solution-set condition is accessible.

6.30 Theorem. *Assuming Vopěnka's principle, a functor $F: \mathcal{K} \to \mathcal{L}$ between accessible categories $\mathcal{K}$, $\mathcal{L}$ is accessible iff satisfies the solution-set condition.*

Remarks

(1) The statement is equivalent to Vopěnka's principle, as Example 6.32 demonstrates.

(2) One implication is absolute: every accessible functor satisfies the solution-set condition by Corollary 2.45. The reverse implication will be proved, more generally, for bounded categories $\mathcal{K}$ and $\mathcal{L}$: every functor $F: \mathcal{K} \to \mathcal{L}$ satisfying the solution-set condition preserves λ-directed colimits for some regular cardinal λ.

PROOF. Let $\mathcal{K}, \mathcal{L}$ be bounded categories and let $F: \mathcal{K} \to \mathcal{L}$ be a functor satisfying the solution-set condition. Let $\mathcal{A}$ be a dense subcategory in $\mathcal{L}$. By Corollary 6.11, objects of $\mathcal{A}$ are λ_0-presentable for some regular cardinal λ_0. Thus the canonical functor $E: \mathcal{L} \to \mathbf{Set}^{\mathcal{A}^{\mathrm{op}}}$ preserves λ_0-directed colimits (by Proposition 1.26) and satisfies the solution-set condition (by Corollary 6.7). Therefore, without loss of generality we can assume that $\mathcal{L}$ is finitely accessible. We are going to find a regular cardinal λ such that F preserves λ-directed colimits.

I. We will first find a regular cardinal μ such that, for every μ-directed diagram D in $\mathcal{K}$, the induced morphism

$$h: \operatorname{colim} FD \to F \operatorname{colim} D$$

is an extremal epimorphism.

For each finitely presentable object L of $\mathcal{L}$ choose a solution set $(L \xrightarrow{e_j} FK_j)_{j \in J_L}$. By Corollary 6.11, there exists a regular cardinal μ such that any K_j ($L \in \mathbf{Pres}_\omega \mathcal{L}$, $j \in J_L$) is μ-presentable. Let $D: \mathcal{D} \to \mathcal{K}$ be a μ-directed diagram with a colimit $(Dd \xrightarrow{c_d} \operatorname{colim} D)$ and let $h: \operatorname{colim} FD \to F \operatorname{colim} D$ be the induced morphism. We want to show that h is an extremal epimorphism. To this end, consider a factorization

$$\operatorname{colim} FD \xrightarrow{w} B \xrightarrow{m} F \operatorname{colim} D$$

of h, where m is a monomorphism. In order to prove that m is an isomorphism, it suffices to show that any morphism $f\colon L \to F\operatorname{colim} D$ with L in $\mathbf{Pres}_\omega\,\mathcal{L}$ factorizes through m. The morphism f has a factorization

$$L \xrightarrow{e_j} FK_j \xrightarrow{Fg} F\operatorname{colim} D$$

for some $j \in J_L$ and $g\colon K_j \to \operatorname{colim} D$. Since K_j is μ-presentable, g factorizes as $g = c_d \cdot \overline{g}$ for some $d \in \mathcal{D}^{\mathrm{obj}}$ and some $\overline{g}\colon K_j \to Dd$. If $(\overline{c}_d\colon FDd \to \operatorname{colim} FD)_{d\in\mathcal{D}^{\mathrm{obj}}}$ is a colimit cocone of FD, then we get

$$f = Fg \cdot e_j = Fc_d \cdot F\overline{g} \cdot e_j = h \cdot \overline{c}_d \cdot F\overline{g} \cdot e_j = m \cdot (w \cdot \overline{c}_d \cdot F\overline{g} \cdot e_j).$$

This is the desired factorization of f through m.

II. We are going to find a regular cardinal $\lambda \geq \mu$ such that for each λ-directed diagram D the induced morphism $h\colon \operatorname{colim} FD \to F\operatorname{colim} D$ is a monomorphism; by I, h is then an isomorphism.

The cardinal λ will be constructed by special factorizations of all morphisms in the category $\mathbf{Pres}_\omega\,\mathcal{L}$. Since $\mathbf{Pres}_\omega\,\mathcal{L}$ is small, we can index these morphisms by a set T:

$$(\mathbf{Pres}_\omega\,\mathcal{L})^{\mathrm{mor}} = \{\, P_t \xrightarrow{p_t} Q_t \mid t \in T \,\}$$

For each $t \in T$ denote by $\mathcal{C}_t$ the category whose objects are all pairs (f, u), where $f\colon A \to B$ is a morphism of $\mathcal{K}$ and $u\colon P_t \to FA$ is a morphism of $\mathcal{L}$ such that $Ff \cdot u$ factorizes through p_t. Morphisms of $\mathcal{C}_t$ from (f,u) to (f',u') are all pairs (a,b) of morphisms in $\mathcal{K}$ such that $f' \cdot a = b \cdot f$ holds in $\mathcal{K}$, and $u' = Fa \cdot u$ holds in $\mathcal{L}$:

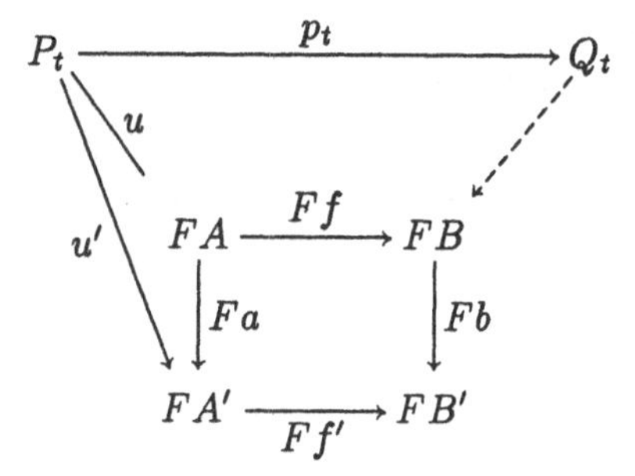

We now make use of the above solution sets $(P_t \xrightarrow{e_{t,j}} FK_{t,j})$, the index set of which will be denoted by J_t instead of J_{P_t}. The subcategory $\mathcal{C}_{t,j}$ of $\mathcal{C}_t$ of all objects (f,u) with $u = e_{t,j}$ and all morphisms $(id_{K_{t,j}}, b)$ is bounded. (In fact, it is isomorphic to a full subcategory of the category $K_{t,j} \downarrow \mathcal{K}$ which is bounded by Theorem 6.6 and Corollary 2.44.) Let $\mathcal{C}^*_{t,j}$ be a dense subcategory of $\mathcal{C}_{t,j}$. Then for each object (f,u) of $\mathcal{C}_t$ there exists a morphism $(a,b)\colon (\hat{f}, e_{t,j}) \to (f,u)$ of $\mathcal{C}_t$ for some $j \in J_t$ and some $\hat{f}$. In fact, u factorizes through a member $e_{t,j}$ of the chosen solution set as $u = Fa \cdot e_{t,j}$ for some $a\colon K_{t,j} \to A$. For the object $(f \cdot a, e_{t,j})$ of $\mathcal{C}_{t,j}$ there exists a morphism $(id_{K_{t,j}}, b)\colon (\hat{f}, e_{t,j}) \to (f \cdot a, e_{t,j})$ with $(\hat{f}, e_{t,j})$ in $\mathcal{C}^*_{t,j}$, since $\mathcal{C}^*_{t,j}$ is dense in $\mathcal{C}_{t,j}$.

Choose a regular cardinal $\lambda \geq \mu$ such that for each $t \in T$, $j \in J_t$, and any object $(f, e_{t,j})$ of $\mathcal{C}^*_{t,j}$, the domain and codomain of f are λ-presentable (see Corollary 6.11). Given a λ-directed diagram $D\colon \mathcal{D} \to \mathcal{K}$, we will prove that the induced morphism $h\colon \operatorname{colim} FD \to F \operatorname{colim} D$ is a monomorphism. To this end, it is sufficient to prove that given morphisms $r, q\colon L \to \operatorname{colim} FD$ with L in $\mathbf{Pres}_\omega\, \mathcal{L}$, then $h \cdot r = h \cdot q$ implies $r = q$.

Denote by

$$(Dd \xrightarrow{c_d} K)_{d \in \mathcal{D}^{\text{obj}}}$$

a colimit of D in $\mathcal{K}$, and by

$$(FDd \xrightarrow{\overline{c}_d} \overline{K})_{d \in \mathcal{D}^{\text{obj}}}$$

a colimit of FD in $\mathcal{L}$. Since the above object L is finitely presentable, there exists $d \in \mathcal{D}^{\text{obj}}$ such that both r and q factorize through $\overline{c}_d\colon FDd \to \operatorname{colim} FD$:

$$r = \overline{c}_d \cdot r', \qquad \text{and} \qquad q = \overline{c}_d \cdot q'.$$

The object FDd (of the finitely accessible category $\mathcal{L}$) is a directed colimit of finitely presentable objects. Consequently, there exists a morphism $u\colon P \to FDd$ with P in $\mathbf{Pres}_\omega\, \mathcal{L}$ through which both r' and q' factorize:

$$r' = u \cdot r'' \qquad \text{and} \qquad q' = u \cdot q''.$$

Analogously, express FK as a directed colimit of finitely presentable objects. Since $h \cdot r = h \cdot q$ implies $Fc_d \cdot r' = Fc_d \cdot q'$ (in fact, $Fc_d \cdot r' = h \cdot \overline{c}_d \cdot r' = h \cdot r$; analogously for $Fc_d \cdot q'$), there exists a morphism $v\colon Q \to FK$ with Q in $\mathbf{Pres}_\omega\, \mathcal{L}$ such that (a) $Fc_d \cdot u$ factorizes through v, say, $Fc_d \cdot u = p \cdot v$ and (b) the morphism p merges r'' with q'':

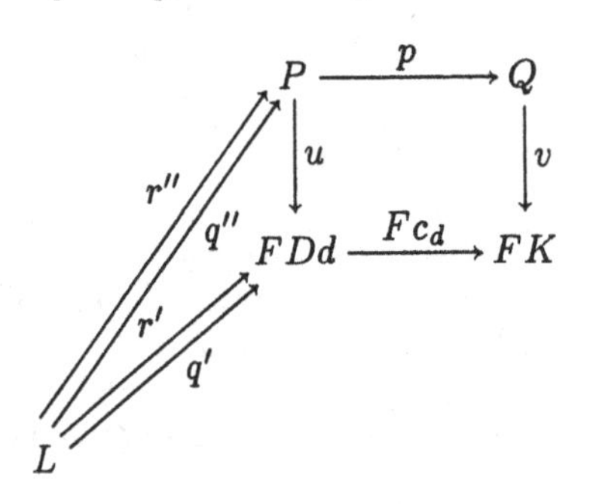

Now p is a morphism of $\mathbf{Pres}_\omega \mathcal{L}$, thus, $p = p_t$ for some $t \in T$. The pair (c_d, u) is an object of $\mathcal{C}_t$. As observed above, we can find a morphism

$$(a, b) : (\hat{f}, e_{t,j}) \to (c_d, u)$$

in $\mathcal{C}_t$ with $(\hat{f}, e_{t,j}) \in \mathcal{C}^*_{t,j}$. Thus, $\hat{f}$ has a λ-presentable domain and codomain, say, $\hat{f} : A \to B$ (see the above choice of λ). By the definition of $\mathcal{C}_t$, there exists $w : Q \to FB$ with $F\hat{f} \cdot e_{t,j} = w \cdot p$:

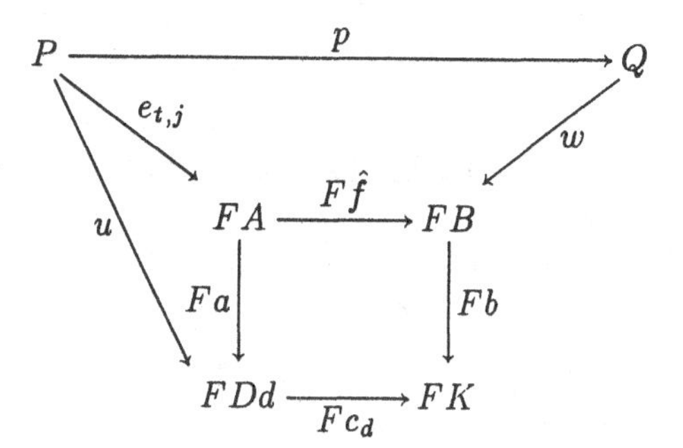

Since B is λ-presentable and $K = \operatorname{colim} D$, there exists $d' \geq d$ such that b factorizes through $c_{d'}$: $b = c_{d'} \cdot b'$. Since A is λ-presentable, there exists $d'' \geq d'$ such that $D(d \to d'') \cdot a = D(d' \to d'') \cdot b' \cdot \hat{f}$. Consequently, we have the following factorization of r:

$$\begin{aligned} r &= (\bar{c}_d \cdot u) \cdot r'' \\ &= \big(\bar{c}_{d''} \cdot FD(d \to d'') \cdot u\big) \cdot r'' \\ &= \big(\bar{c}_{d''} \cdot FD(d \to d'') \cdot Fa \cdot e_{t,j}\big) \cdot r'' \\ &= (\bar{c}_{d''} \cdot FD(d' \to d'') \cdot Fb' \cdot F\hat{f} \cdot e_{t,j}) \cdot r'' \\ &= \bar{c}_{d'} \cdot Fb' \cdot w \cdot p \cdot r''. \end{aligned}$$

Analogously for q. Thus, $p \cdot r'' = p \cdot q''$ implies $r = q$, which concludes the proof. □

6.31 Corollary. *Assuming Vopěnka's principle, every subfunctor of an accessible functor is accessible.*

In fact, let $F : \mathcal{K} \to \mathcal{L}$ be accessible, and let $\sigma : G \to F$ be a mono-transformation. To prove that G satisfies the solution-set condition, apply Theorem 6.6 to the (bounded) comma-category $L \downarrow F$ and observe that $L \downarrow G$ is isomorphic to a full subcategory of $L \downarrow F$. □

Remark. The statement is equivalent to Vopěnka's principle. In fact, assuming the negation of Vopěnka's principle, we have the following non-accessible subfunctor F of $Id_{\mathbf{Gra}}$: choose a large rigid class $\mathcal{A}$ of graphs and for each graph X let FX be the subgraph of X on the set

$$\bigcup_{A \in \mathcal{A}} \bigcup_{f\colon A \to X} \operatorname{im} f.$$

For each cardinal λ the functor F fails to be λ-accessible. In fact, choose a graph $A \in \mathcal{A}$ of cardinality $\geq \lambda$. Then A is a colimit of the λ-directed diagram of all subgraphs of cardinality $< \lambda$, and F does not preserve this colimit: we have $FA = A$ and $FA' = \emptyset$ for any proper subgraph A' of A.

6.32 Example constructed under the assumption of the negation of Vopěnka's principle: A non-accessible functor $F\colon \mathbf{Gra} \to \mathbf{Set}$ satisfying the solution-set condition.

Let $A_i = (X_i, \alpha_i)$, $i \in I$, be a large rigid class of graphs. Define a functor $F\colon \mathbf{Gra} \to \mathbf{Set}$ on objects $K = (X, \alpha)$ by

$$FK = \begin{cases} X & \text{if } \hom(A_i, K) = \emptyset \text{ for all } A_i,\ i \in I \\ \{0\} & \text{otherwise} \end{cases}$$

and on morphisms $f\colon K \to K'$ by

$$Ff = \begin{cases} f & \text{if } \hom(A_i, K') = \emptyset \text{ for all } A_i,\ i \in I \\ \text{const}_{\{0\}} & \text{otherwise.} \end{cases}$$

The solution-set condition is trivial: for each set X the corresponding discrete graph $K = (X, \emptyset)$ satisfies $FK = X$, and $\{ id_X\colon X \to FK \}$ is a solution set.

For every regular cardinal λ choose $i \in I$ with $\operatorname{card} X_i \geq \lambda$. The diagram D of all subgraphs of A_i of cardinality less than λ is λ-directed, and $A_i = \operatorname{colim} D$ in **Gra**. Since for each proper subgraph K of A_i we have $\hom(A_j, K) = \emptyset$ for all A_j, $j \in I$, we see that $\operatorname{colim} FD = \{0\}$, whereas $F \operatorname{colim} D = FA_i = X_i$. Thus F is not λ-accessible.

6.F Colimit-dense Subcategories

We have seen throughout the book the usefulness of the concept of a dense subcategory. However, it is often difficult to verify that a certain category has a dense subcategory (since canonical colimits are often hard to describe). We will now prove that, assuming Vopěnka's principle, canonical colimits can be substituted by arbitrary colimits. The proof is quite complicated since here the canonical functor is not full.

6.33 Definition. A full, small subcategory $\mathcal{A}$ of a category $\mathcal{K}$ is called *colimit-dense* provided that every object of $\mathcal{K}$ is a colimit of a diagram in $\mathcal{A}$.

6.34 Examples

(1) Every dense subcategory is, obviously, colimit-dense.

(2) Although $\mathbb{R}$ does not form a dense singleton subcategory of **Vec** (see Example 1.24(4)), it forms a colimit-dense subcategory.

(3) The group $(\mathbb{Z}, +)$ of integers forms a singleton subcategory of the category of groups which is not colimit-dense. Yet, by forming coproducts we get all free groups, and then coequalizers yield all groups.

6.35 Theorem. *Assuming Vopěnka's principle, each category with a colimit-dense subcategory is bounded.*

Remark. The statement is equivalent with Vopěnka's principle: see Example 6.36.

PROOF. I. We first prove that a category $\mathcal{L}$ which has a colimit-dense subcategory $\mathcal{A}$ cannot have a large, rigid class of objects.

In fact, we will show that the canonical functor $E: \mathcal{L} \to \mathbf{Set}^{\mathcal{A}^{\mathrm{op}}}$ (see Notation 1.25) has the property that, given objects L, L' in $\mathcal{L}$, then

$$\text{(1)} \qquad \hom(L, L') = \emptyset \quad \text{implies} \quad \hom(EL, EL') = \emptyset.$$

Thus, since $\mathbf{Set}^{\mathcal{A}^{\mathrm{op}}}$ does not have a large, rigid class of objects, neither has $\mathcal{L}$. The proof of (1) follows from the fact that L is a colimit of a diagram $D: \mathcal{D} \to \mathcal{A}$. In fact, let $\left(Dd \xrightarrow{l_d} L\right)$ be a colimit cocone. Assume that a morphism $f: EL \to EL'$ exists. Then f is a natural transformation from $\hom(-, L)/_{\mathcal{A}^{\mathrm{op}}}$ to $\hom(-, L')/_{\mathcal{A}^{\mathrm{op}}}$, and the cocone $f_{Dd}(l_d): Dd \to L'$ ($d \in \mathcal{D}^{\mathrm{obj}}$) is obviously compatible with D. This contradicts $\hom(L, L') = \emptyset$.

II. Let $\mathcal{K}$ be a category with a colimit-dense subcategory $\mathcal{A}$. Assuming that $\mathcal{K}$ is not bounded, we will derive a contradiction to the statement I. We can express $\mathcal{K}$ as a union of a transfinite chain of small, full subcategories:

$$\mathcal{K} = \bigcup_{n \in \mathrm{Ord}} \mathcal{K}_n \qquad \text{where} \quad \mathcal{K}_0 = \mathcal{A}.$$

Since $\mathcal{K}_n$ is not dense, there exists an object P_n which is not a canonical colimit of its canonical diagram w.r.t. $\mathcal{K}_n$. That is, the canonical diagram has a compatible cocone.

$$\text{(2)} \qquad A \xrightarrow{f^*} P_n^* \qquad \text{(for all } A \xrightarrow{f} P_n \text{ with } A \text{ in } \mathcal{K}_n\text{)}$$

which does not uniquely factorize through the canonical cocone.

The canonical functor $E\colon \mathcal{K} \to \mathbf{Set}^{\mathcal{A}^{op}}$ with $E(K) = \hom(-,K)/_{\mathcal{A}^{op}}$ is faithful (since $\mathcal{A}$ is a generator of $\mathcal{K}$), but not necessarily full. However

(3) each morphism $EA \to EK$ with A in $\mathcal{A}$ carries a $\mathcal{K}$-morphism

(i.e., it has the form Ef for a unique $\mathcal{K}$-morphism $f\colon A \to K$). Consequently, the density of $E(\mathcal{A})$ in $\mathbf{Set}^{\mathcal{A}^{op}}$ implies that there exists a unique morphism $h_n\colon EP_n \to EP_n^*$ with

(4) $h_n \cdot Ef = Ef^*$ for each $f\colon A \to P_n$ in $\mathcal{K}$ with A in $\mathcal{A}$.

We will prove that, more generally,

(5) $h_n \cdot Ef = Ef^*$ for each $f\colon K \to P_n$ with K in $\mathcal{K}_n$.

In fact, EK is a canonical colimit w.r.t. $E(\mathcal{A})$, and by (3) the colimit morphisms have the form $EA_i \xrightarrow{Ek_i} EK$ $(i \in I)$ for suitable $\mathcal{K}$-morphisms k_i. The compatibility of the cocone (2) implies that

$$(f \cdot k_i)^* = f^* \cdot k_i \qquad \text{for each } i,$$

thus, $(h_n \cdot Ef) \cdot Ek_i = Ef^* \cdot Ek_i$ for each i—this proves (5).

Since $\mathcal{A}$ is colimit-dense, there exists a diagram D_n in $\mathcal{A}$ with a colimit $\left(A_i \xrightarrow{p_{n,i}} P_n\right)_{i\in I_n}$. The cocone of all $h_n \cdot Ep_{n,i}\colon EA_i \to EP_n^*$ carries, by (3), a cocone in $\mathcal{K}$ which (since E is faithful) is compatible with D_n. Thus, there exists a unique morphism $k_n\colon P_n \to P_n^*$ in $\mathcal{K}$ with

(6) $h_n \cdot Ep_{n,i} = E(k_n \cdot p_{n,i})$ for each $i \in I_n$.

The choice of the cocone (2) implies that

(7) $Ek_n \neq h_n$.

We are ready to construct the desired category $\mathcal{L}$ with a colimit-dense subcategory, which, nevertheless, also has a large rigid class of objects. We first extend $\mathcal{K}$ to a category $\widehat{\mathcal{K}}$ by adding a new formal initial object O $(\notin \mathcal{K}^{obj})$. Next we choose an arbitrary faithful functor

$$U_0\colon \mathbf{Set}^{\mathcal{A}^{op}} \to \mathbf{Set}$$

and define a faithful functor

$$U\colon \widehat{\mathcal{K}} \to \mathbf{Set} \qquad \text{by} \qquad UO = \emptyset \quad \text{and} \quad U/_{\mathcal{K}} = U_0E.$$

Next, let Σ be a one-sorted signature of four binary and three unary relation symbols. The category $\mathcal{L}$ has as objects triples

$$(X, K, \alpha),$$

where X is a set, K is an object of $\mathcal{K}$ with $UK \subseteq X$, and α is a Σ-structure on X. The morphisms

$$f\colon (X, K, \alpha) \to (X', K', \alpha')$$

are Σ-homomorphisms $f\colon (X, \alpha) \to (X', \alpha')$ for which there exists a $\mathcal{K}$-morphism $f_0\colon K \to K'$ with f extending Uf_0. Denote by $F\colon \mathbf{Rel}\,\Sigma \to \mathcal{L}$ the embedding with $F(X, \alpha) = (X, O, \alpha)$ and by $G\colon \mathcal{K} \to \mathcal{L}$ the embedding with $GK = (UK, K, \emptyset)$. (We denote, here and below, by $\emptyset$ the discrete Σ-structure on any set, i.e., the structure of seven empty relations.)

(a) $\mathcal{L}$ has a colimit-dense subcategory. To prove this, choose a dense subcategory $\mathcal{B}$ in $\mathbf{Rel}\,\Sigma$ containing the object $B_0 = \big(\{1\}, \emptyset\big)$. We will prove that the set

$$\mathcal{A}^* = \{\, GA \,\}_{A \in \mathcal{A}^{\mathrm{obj}}} \cup \{\, FB \,\}_{B \in \mathcal{B}^{\mathrm{obj}}}$$

is colimit-dense in $\mathcal{L}$. Consider an arbitrary object $L = (X, K, \alpha)$ of $\mathcal{L}$. If $K = O$, then L is a canonical colimit w.r.t. $\mathcal{A}^*$ (since $\mathcal{B}$ is dense in $\mathbf{Rel}\,\Sigma$, and no morphism leads from GA, $A \in \mathcal{A}^{\mathrm{obj}}$, into L). If K is an object of $\mathcal{K}$, we can choose a diagram D' in $\mathcal{A}$ with a colimit $\big(D'_j \xrightarrow{k_j} K\big)_{j \in J}$ in $\mathcal{K}$, and we have the canonical diagram D'' of (X, α) w.r.t $\mathcal{B}$ in $\mathbf{Rel}\,\Sigma$. We now create a diagram D in $\mathcal{L}$ by taking a disjoint union of GD' and FD'' and adding the following morphisms

$$p\colon FD''(B_0 \xrightarrow{f} (X, \alpha)) \longrightarrow GD'_j$$

for all $f\colon B_0 \to (X, \alpha)$, all $j \in J$, and all $p\colon \{1\} \to UD'_j$ with $f(1) = (Uk_j \cdot p)(1)$. Then L is a colimit of D in $\mathcal{L}$ with the colimit maps

$$\tilde{k}_j\colon GD'_j \to L \qquad \text{where} \qquad \tilde{k}_j(x) = Uk_j(x) \quad \text{for all } x \in UD'_j$$

and

$$f\colon FD''(B \xrightarrow{f} (X, \alpha)) \longrightarrow L.$$

In fact, that cocone is clearly compatible. Let $\hat{k}_j\colon GD'_j \to \widehat{L}$ and

$$\hat{f}\colon FD''(B \xrightarrow{f} (X, \alpha)) \longrightarrow \widehat{L}$$

be a compatible cocone. Then there exists a unique Σ-homomorphism $q\colon ((X, \alpha) \to (\widehat{X}, \widehat{\alpha})$, where $\widehat{L} = (\widehat{X}, \widehat{K}, \widehat{\alpha})$, with $\hat{f} = q \cdot f$ for each $f\colon B \to$

(X, α), B in $\mathcal{B}$. It is sufficient to prove that $\hat{k}_j = q \cdot \tilde{k}_j$ for each $j \in J$. Given $x \in UD'_j$ we have a morphism $f: B_0 \to (X, \alpha)$ with $f(1) = \tilde{k}_j(x)$. Then $p: FB_0 \to GD'_j$, defined by $p(1) = x$, is a morphism of the diagram D. Therefore, $\hat{f} = \hat{k}_j \cdot p$, which implies

$$\hat{k}_j(x) = \hat{f}(1) = q \cdot f(1) = q \cdot \tilde{k}_j(x).$$

(b) $\mathcal{L}$ has a large, rigid class of objects. In fact, for each ordinal n define an object

$$L_n = (X_n, P_n, \alpha_n)$$

of $\mathcal{L}$ as follows. Put

$$X_n = (UP_n) \cup (UP_n^*)$$

where we assume, for notational convention, that $(UP_n) \cap (UP_n^*) = \emptyset$, and let $\alpha_n = \{ \alpha_{n,1}, \ldots, \alpha_{n,7} \}$ be the following Σ-structure on X_n. Recall from Lemma 2.64 that X_n carries a rigid binary relation $\alpha_{n,1}$, let $\alpha_{n,2}$ be the relation $\neq$, and put

$$\alpha_{n,3} = \{ (x, (U_0 h_n)(x)) \mid x \in UP_n \};$$
$$\alpha_{n,4} = \{ (x, (Uk_n)(x)) \mid x \in UP_n \}.$$

The unary relations are defined as follows:

$$\alpha_{n,5} = UP_n;$$
$$\alpha_{n,6} = UP_n^*;$$
$$\alpha_{n,7} = \{ x_n \} \quad \text{where } x_n \in UP_n \text{ fulfils } (U_0 h_n)(x_n) \neq (Uk_n)(x_n).$$

(The existence of x_n follows from (7).) We can suppose, without loss of generality, that

$$n < m \quad \text{implies} \quad \operatorname{card} X_n < \operatorname{card} X_m \quad \text{and} \quad P_n \in \mathcal{K}_m^{\text{obj}}.$$

(Otherwise, we pick up from the collection L_n, $n \in \mathrm{Ord}$, a transfinite subsequence with that property.) Then L_n ($n \in \mathrm{Ord}$) is a rigid class in $\mathcal{L}$. In fact, due to $\alpha_{n,2}$ every morphism $L_n \to L_m$ is one-to-one, thus, $n \leq m$. Either $n = m$, and then $\alpha_{n,1}$ implies $\hom(L_n, L_n) = \{ id \}$, or $n < m$, and we must derive a contradiction. The unary relations guarantee that any morphism from L_n to L_m in $\mathcal{L}$ has the form

$$e \cup e^* : L_n \to L_m$$

for some maps $e: UP_n \to UP_m$ and $e^* : UP_n^* \to UP_m^*$ with

$$e(x_n) = x_m. \tag{8}$$

The relation $\alpha_{n,3}$ implies

$$U_0 h_m \cdot e = e^* \cdot U_0 h_n \tag{9}$$

and due to $\alpha_{n,4}$ we have

$$U k_m \cdot e = e^* \cdot U k_n. \tag{10}$$

Let us prove that

$$U_0 h_m \cdot e = U k_m \cdot e \tag{11}$$

Since P_n is a colimit $\left(A_i \xrightarrow{p_{n,i}} P_n\right)_{i \in I_n}$ in $\mathcal{K}$, it is sufficient to prove that $(U_0 h_m \cdot e) \cdot U p_{n,i} = (U k_m \cdot e) \cdot U p_{n,i}$ for each i because k_m is a morphism of $\mathcal{K}$, e carries a morphism $P_n \to P_m$ in $\mathcal{K}$ (by the definition of $\mathcal{L}$-morphisms $e \cup e^*: L_n \to L_m$), and also $U_0 h_m \cdot e$ carries a morphism $P_n \to P_m^*$ in $\mathcal{K}$ (because P_n is in $\mathcal{K}_m$, see (5)). We have

$$\begin{aligned} U_0 h_m \cdot e \cdot U p_{n,i} &= e^* \cdot U_0 h_n \cdot U_0 E p_{n,i} && \text{by (9)} \\ &= e^* \cdot U_0 E(k_n \cdot p_{n,i}) && \text{by (6)} \\ &= e^* \cdot U k_n \cdot U p_{n,i} && \\ &= U k_m \cdot e \cdot U p_{n,i} && \text{by (10).} \end{aligned}$$

Now we obtain a contradiction with the choice of x_m:

$$\begin{aligned} (U_0 h_m)(x_m) &= (U_0 h_m \cdot e)(x_n) && \text{by (8)} \\ &= (U k_m \cdot e)(x_n) && \text{by (11)} \\ &= (U k_m)(x_m). \end{aligned}$$

Consequently, L_n ($n \in \mathrm{Ord}$) is a rigid class. □

6.36 Example constructed under the assumption of the negation of Vopěnka's principle: A cocomplete category $\mathcal{K}$ which is not bounded, although it has a finite, colimit-dense subcategory. Let $\mathcal{A}$ be a large, rigid class of graphs. For each graph $G = (X, \alpha)$ denote by $G^* \subseteq X$ the union of all images of homomorphisms from $\mathcal{A}$-objects into G:

$$G^* = \bigcup_{A \in \mathcal{A}} \bigcup_{f: A \to G} \operatorname{im} f.$$

The objects of the category $\mathcal{K}$ are triples of sets

$$(X, Y, \alpha) \qquad \text{with} \quad Y \subseteq X \quad \text{and} \quad \alpha \subseteq Y \times Y;$$

the morphisms $f: (X, Y, \alpha) \to (X', Y', \alpha')$ are maps $f: X \to X'$ such that

(a) $f(Y) \subseteq Y'$ and the domain–codomain restriction of f is a homomorphism $f_0: (Y, \alpha) \to (Y', \alpha')$ in **Gra**,

and

(b) $f(X - Y) \subseteq (X' - Y') \cup (Y', \alpha')^*$.

(1) $\mathcal{K}$ is a cocomplete category. In fact, let $D: \mathcal{D} \to \mathcal{K}$ be a diagram with objects (X_i, Y_i, α_i), $i \in I$. For the forgetful functor $U: \mathcal{K} \to \mathbf{Set}$ (with $U(X, Y, \alpha) = X$) we form a colimit $\left(X_i \xrightarrow{f_i} X\right)_{i \in I}$ of UD, and we put $Y = \bigcup_{i \in I} f_i(Y_i)$ and $\alpha = \bigcup_{i \in I}(f_i \times f_i)(\alpha_i)$. Then we will prove that $\left((X_i, Y_i, \alpha_i) \xrightarrow{f_i} (X, Y, \alpha)\right)_{i \in I}$ is a colimit of D in $\mathcal{K}$.

It is easy to verify that the above construction describes coproducts well; thus, we just verify its correctness in the case where f is a coequalizer of a pair $p, q: (X_1, Y_1, \alpha_1) \to (X_2, Y_2, \alpha_2)$ of morphisms in $\mathcal{K}$. Thus, $f: X_2 \to X$ is a coequalizer of p, q in **Set**, and $Y = f(Y_2)$, $\alpha = (f \times f)(\alpha_2)$. It is obvious that $f: (X_2, Y_2, \alpha_2) \to (X, Y, \alpha)$ fulfils the condition (a) above. To verify (b), let $x \in X_2 - Y_2$ be an element with $f(x) \in Y$. Then there exists $y \in Y_2$ with $f(x) = f(y)$. By the well-known description of coequalizers in **Set**, this implies that there exist points $x = x_0$, x_1, $\ldots$, $x_n = y$ in X_2 with $\{x_i, x_{i+1}\}$ having the form $\{p(z_i), q(z_i)\}$ for a suitable $z_i \in X_1$ ($i = 0, \ldots, n-1$). Let i be the largest index with $x_i \in X_2 - Y_2$. Assume for example that $x_i = p(z_i)$ and $x_{i+1} = q(z_i)$. The former implies $z_i \in X_1 - Y_1$, and since $x_{i+1} \in Y_2$, the latter implies $x_{i+1} \in (Y_2, \alpha_2)^*$. The domain–codomain restriction of f yields a graph homomorphism from (Y_2, α_2) to (Y, α), thus, $f(x_{i+1}) \in (Y, \alpha)^*$. We have $f(x) = f(x_{i+1})$.

The universal property of f is easy to verify.

(2) $\mathcal{K}$ has a finite colimit-dense subcategory. In fact, let

$$G_1 = (\{0\}, \emptyset) \quad \text{and} \quad G_2 = (\{0,1\}, \{(0,1)\})$$

be graphs forming a dense subcategory of **Gra**, and let E: **Gra** $\to \mathcal{K}$ be the embedding $(Y, \alpha) \mapsto (Y, Y, \alpha)$. Then EG_1, EG_2, and $B = (\{0\}, \emptyset, \emptyset)$ form a colimit-dense subcategory of $\mathcal{K}$. For every object (X, Y, α) of $\mathcal{K}$ we thus have a canonical diagram D of (Y, α) w.r.t. $\{G_1, G_2\}$ in **Gra**. We form a diagram in $\mathcal{K}$ by adding to ED a discrete category of card$(X - Y)$ copies of the object B. A colimit of this diagram is, as follows from the description of colimits in (1), the object $\operatorname{colim} ED + \coprod_{X-Y} B \cong (X, Y, \alpha)$.

(3) $\mathcal{K}$ does not have a dense subcategory. Assume that, on the contrary, $\mathcal{B}$ is dense in $\mathcal{K}$. We will prove that for each object A of $\mathcal{A}$ there exists an injective morphism from EA into some $\mathcal{B}$-object. Since the cardinalities of

the underlying sets of graphs in $\mathcal{A}$ are unbounded, we get a contradiction to the smallness of $\mathcal{B}$.

Since $B = (\{0\}, \emptyset, \emptyset)$ is a colimit of a diagram in $\mathcal{B}$, it is clear that $\mathcal{B}$ contains an object of the form $B_0 = (Z_0, \emptyset, \emptyset)$ for some $Z_0 \neq \emptyset$. Choose $z_0 \in Z_0$.

Given an object A in $\mathcal{A}$, let D denote the canonical diagram of EA w.r.t. $\mathcal{B}$. We can clearly choose a morphism $h: B_0 \to EA$. Let us verify that h has a factorization $h = h' \cdot g$ in $\mathcal{K}$ for some $g: B_0 \to B_1 = (X, Y, \alpha)$ with B_1 from B and $g(z_0) \in (Y, \alpha)^*$. In fact, if we suppose the contrary, then every morphism $g: B_0 \to B_1 = (X, Y, \alpha)$ of D fulfils $g(z_0) \in X - Y$. By the above description of colimits, it follows that the colimit map $h: B_0 \to \operatorname{colim} D = (\overline{X}, \overline{Y}, \overline{\alpha})$ must fulfil $h(z_0) \in \overline{X} - \overline{Y}$. However, since $\operatorname{colim} D = EA$, we have $\overline{X} = \overline{Y}$. Thus, a factorization $h = h' \cdot g$ as above exists. Since $g(z_0) \in (Y, \alpha)^*$, there exists a graph homomorphism $k: A_0 \to (Y, \alpha)$ with A_0 in $\mathcal{A}$. The domain–codomain restriction of h' is a graph homomorphism $h'_0: (Y, \alpha) \to A$. Thus, $h'_0 \cdot k: A_0 \to A$ is a graph homomorphism, which implies (since $\mathcal{A}$ is rigid) that $A_0 = A$ and $h'_0 \cdot k = id$. Therefore, $k: EA \to E(Y, \alpha)$ is an injective morphism. Composing it with the inclusion morphism of $E(Y, \alpha)$ into $B_1 = (X, Y, \alpha)$, we obtain the desired injective morphism from EA into $B_1 \in \mathcal{B}^{\text{obj}}$.

6.37 Corollary. *Assuming Vopěnka's principle, a category is locally presentable iff it is cocomplete, and has a colimit-dense subcategory.*

This follows from Theorems 6.14 and 6.35.

6.38 Example. In Theorem 6.35, colimit-denseness cannot be weakened to the following concept: A full subcategory $\mathcal{A}$ of a category $\mathcal{K}$ is called *weakly colimit-dense* provided that the smallest full subcategory of $\mathcal{K}$ containing $\mathcal{A}$ and closed under colimits in all of $\mathcal{K}$. In fact, independently of set theory, there exists a cocomplete category with a small, weakly colimit-dense subcategory which is not locally presentable. To show this, we will construct a category $\mathcal{K}$ which has a weakly colimit-dense object, and which is cocomplete, but has no terminal object.

Consider the following "large essentially algebraic" theory of unary algebras: an algebra is a set X together with partial unary operations α_i ($i \in \text{Ord}$) such that α_0 is everywhere defined, and for $i > 0$,

$$\alpha_i(x) \quad \text{is defined iff } \alpha_j(x) = x \text{ for all } j < i.$$

The category of these algebras and homomorphisms $f: (X, \alpha_i) \to (X', \alpha'_i)$ (i.e., maps $f: X \to X'$ with $f(\alpha_i(x)) = \alpha_i(f(x))$ whenever $\alpha_i(x)$ is defined) is denoted by $\mathcal{L}$. It is not difficult to verify that $\mathcal{L}$ is cocomplete.

Let A be the algebra of natural numbers with α_0 the successor operation and $\alpha_i = \emptyset$ for $i > 0$. Denote by $\mathcal{K}$ the closure of $\{A\}$ under colimits in $\mathcal{L}$, i.e., the smallest full subcategory of $\mathcal{L}$ containing A and closed under colimits. Then $\{A\}$ is weakly colimit-dense in $\mathcal{K}$. We will prove that $\mathcal{K}$ does not have a terminal object. In fact, for each algebra (X, α_i) in $\mathcal{K}$ there exists an ordinal i_0 with $\alpha_{i_0} = \emptyset$. (Proof: A has this property, and the full subcategory of $\mathcal{L}$ of all algebras with this property is obviously closed under colimits in $\mathcal{L}$.) Suppose that $(\overline{X}, \overline{\alpha}_i)$ is a terminal object of $\mathcal{K}$, and choose an ordinal i_0 with $\overline{\alpha}_{i_0} = \emptyset$. Then each object (X, α_i) of $\mathcal{K}$ fulfils $\alpha_{i_0} = \emptyset$ for this chosen ordinal i_0. This, however, is not the case since for each ordinal $j > 0$ the algebra A_j of natural numbers with the following operations $\alpha_{j,i}$ ($i \in \mathrm{Ord}$):

$$\begin{aligned}
\alpha_{j,0}(0) &= 0 \quad \text{and} \quad \alpha_{j,0}(n) = n+1 \qquad (n > 0),\\
\alpha_{j,i}(0) &= 0 \quad \text{and} \quad \alpha_{j,i}(n) \text{ undefined} \quad (n > 0), \text{ for all } 0 < i < j,\\
\alpha_{j,j}(0) &= 1 \quad \text{and} \quad \alpha_{j,j}(n) \text{ undefined} \quad (n > 0),\\
\alpha_{j,i} &= \emptyset \quad \text{for all } i > j
\end{aligned}$$

is an object of $\mathcal{K}$. We prove this by transfinite induction on j.

First step: A coequalizer of the homomorphisms

$$p, q \colon A \to A, \quad p(n) = n, \quad \text{and} \quad q(n) = n+1 \qquad (\text{for all } n)$$

in $\mathcal{L}$ is the homomorphism

$$c \colon A \to A_0, \quad c(n) = 0 \qquad \text{for all } n.$$

Thus, since $\mathcal{K}$ contains A, it contains A_0.

Isolated step: A coequalizer of the homomorphisms

$$p, q \colon A \to A_j, \quad p(n) = n, \quad \text{and} \quad q(n) = n+1 \qquad (\text{for all } n)$$

in $\mathcal{L}$ is the homomorphism

$$c_{j+1} \colon A_j \to A_{j+1}, \quad c_{j+1}(n) = 0 \qquad (\text{for all } n).$$

Thus, since $\mathcal{K}$ contains A and A_j, it contains A_{j+1}.

Limit step: If j is a limit ordinal, then the chain

$$A_0 \to A_1 \to \cdots A_i \to \cdots \qquad (i < j)$$

formed by the constant maps with value 0 has A_j as a colimit object.

Exercises

6.a Almost Rigid Classes

(1) A class of topological spaces is *almost rigid* if it admits no morphisms except the identity morphisms and constant morphisms. Prove that, assuming Vopěnka's principle, there does not exist a proper almost rigid class of compact Hausdorff spaces. (Hint: by Gelfand duality, the dual of the category of compact Hausdorff spaces is locally presentable. Use Example 2.3(6).)

(2) A class of abelian groups is almost rigid if it admits no morphisms except the identity morphisms and the constant morphisms to 0. Prove that, assuming Vopěnka's principle, there does not exist a proper almost rigid class of abelian groups.

6.b λ-free Abelian Groups. For an uncountable cardinal λ, an abelian group A is called *λ-free* if every subgroup of less than λ elements is free. Prove that, under Vopěnka's principle, there is a regular cardinal λ such that every λ-free abelian group is free. (Hint: apply Theorem 6.17 to the full subcategory of **Ab** consisting of free abelian groups. This subcategory is closed under all subobjects.)

6.c Torsion Classes. A class $\mathcal{A}$ of abelian groups is a *torsion class* provided that $\mathcal{A}$ is closed under extensions, coproducts, and quotients (the first property means that given abelian groups $A \subseteq B$, then $A, B/A \in \mathcal{A}$ implies $B \in \mathcal{A}$). Prove that, assuming Vopěnka's principle, every torsion class $\mathcal{A}$ of abelian groups is singly generated, i.e., it contains a group such that $\mathcal{A}$ is the smallest torsion class in which that group lies. (Hint: $\mathcal{A}$ is coreflective. Apply Corollary 6.24, Remark 1.39, and Exercise 1.j(5) to prove that the class $\mathcal{B}$ of all abelian groups whose coreflection is 0 has the form $\mathcal{B} = \{h\}^{\perp}$ for a reflection map $h: C \to C/A$ of some abelian group C. Then A is the desired member of $\mathcal{A}$.)

6.d Mono-injectivity Classes of Abelian Groups. A *mono-injectivity class* is an injectivity class $\mathcal{M}$-**Inj** with $\mathcal{M} \subseteq$ monomorphisms. Prove that, under Vopěnka's principle, mono-injectivity classes in **Ab** are precisely classes of abelian groups closed under products and split subobjects which contain all divisible groups. (Hint: apply Theorems 6.17 and 4.8.)

6.e Reflective Subcategories

(1) Prove that, assuming Vopěnka's principle, every full subcategory of the category of compact Hausdorff spaces closed under limits is reflective.

(2) Prove that weak Vopěnka's principle is equivalent to the following statement:

> Every intersection of full, reflective subcategories of a locally presentable category $\mathcal{K}$ is reflective in $\mathcal{K}$.

6.f Weakly Locally Presentable Categories. Prove that the statement $(*)$ from Remark 6.27(2) is equivalent to the following assertion:

> A category $\mathcal{K}$ is weakly locally presentable iff it is equivalent to an accessibly embedded subcategory of a locally presentable category closed under products.

(Hint: the modification of Example 6.23 indicated in Remark 6.27(2) is a subcategory of **Gra** closed under products, split subobjects, and directed colimits.)

6.g Preradicals. A *preradical* T is a subfunctor of the identity functor $\mathbf{Ab} \to \mathbf{Ab}$. For a cardinal λ let

$$T^{[\lambda]}(A) = \coprod \{ T(X) \mid X \text{ is a subgroup of } A \text{ generated by} < \lambda \text{ elements} \}.$$

Show that, assuming Vopěnka's principle, for every preradical T there exists a cardinal λ such that $T = T^{[\lambda]}$.

6.h Cone-injectivity Classes. Show that assuming Vopěnka's principle the following statements are equivalent for each class $\mathcal{A}$ of objects of a locally presentable category $\mathcal{K}$:

(i) $\mathcal{A}$ is a cone-injectivity class;

(ii) $\mathcal{A}$ is closed under split subobjects.

6.i Weakly Initial Sets. Prove that Vopěnka's principle is equivalent to the following statement:

> Every full subcategory of a locally presentable category contains a weakly initial set of objects.

(Hint: inspect the proof of Theorem 6.30.)

Historical Remarks

The story of Vopěnka's principle (as related to the authors by Petr Vopěnka) is that of a practical joke which misfired: In the 1960's P. Vopěnka was repelled by the multitude of large cardinals which emerged in set theory. When he constructed, in collaboration with Z. Hedrlín and A. Pultr, a rigid graph on every set (see Lemma 2.64), he came to the conclusion that, with some more effort, a large rigid class of graphs must surely be also constructible. He then decided to tease set-theorists: he introduced a new principle (known today as Vopěnka's principle), and proved some consequences concerning large cardinals. He hoped that some set-theorists would continue this line of research (which they did) until somebody showed that the principle is nonsense. However, the latter never materialized—after a number of unsuccessful attempts at constructing a large rigid class of graphs, Vopěnka's principle received its name from Vopěnka's disciples. One of them, T. J. Jech, made Vopěnka's principle widely known. Later the consistency of this principle was derived from the existence of huge cardinals: see [Powell 1972]; our account (in the Appendix) is taken from [Jech 1978]. Thus, today this principle has a firm position in the theory of large cardinals. Petr Vopěnka himself never published anything related to that principle.

The question of rigid classes in various categories arose in Prague in the 1960's and 1970's within the research initiated by John Isbell, see [Isbell 1964], concerning the possibility of embedding categories into "everyday" categories (such as those of graphs, semigroups, topological spaces, etc.). These results are summarized in the monograph [Pultr, Trnková 1980]. One of the crucial results is that whenever measurable cardinals are not arbitrarily large, then every concrete category has a full embedding into the category of graphs. (Compare this with Theorem 2.65.)

Counterexamples based (essentially) on the negation of Vopěnka's principle were constructed by [Gabriel, Ulmer 1971], the source of our Example 6.12, and [Isbell 1971], the source of Example 6.23. We first used large-cardinal principles in [Adámek, Rosický 1988]: based on embedding theorems of the "Prague School" we proved that an intersection of full, reflective subcategories of **Top** (the category of topological spaces) need not be reflective; if **Top** is substituted by **Gra** then we observed that the answer depends on the existence of measurable cardinals. The discovery that the answer is equivalent to weak Vopěnka's principle was made in [Adámek, Rosický, Trnková 1988]. In the course of this research we discovered a short announcement by E. R. Fisher [Fisher 1977] of results equivalent to Vopěnka's principle. He never published the proof, but upon our request he has kindly written up the results, see [Fisher 1987]. That preprint contains full model-theoretic proofs of our Theorem 6.14 and Corollaries 6.15, 6.18, and 6.24. The presentation of Chapter 6 is based on the following publications in which, inter alia, new (categorical) proofs of Fisher's results are presented:

[Adámek, Rosický 1988]: Example 6.25
[Adámek, Rosický, Trnková 1988]: Theorem 6.22
[Adámek, Herrlich, Reiterman 1989]: Example 6.36
[Rosický, Trnková, Adámek 1990]: Theorems 6.6, 6.9, 6.14, 6.28, and 6.35
[Adámek, Rosický 1993]: Theorem 6.26
[Rosický, Tholen 1994]: Theorem 6.30

We have not included some closely related (but rather technical) results in Chapter 6. For example, assuming the negation of weak Vopěnka's principle, it is possible to find two full, reflective subcategories of **Gra** whose intersection is non-reflective, see [Trnková, Adámek, Rosický 1990]. (Moreover, the analogous result for topological spaces instead of graphs is proved to hold without any set-theoretical assumptions.) The fact that weak Vopěnka's principle follows from the equivalence of the conditions (i) and (ii) in Corollary 6.24 is proved in [Adámek, Rosický 1992].

Some further subtle points are touched on in the exercises. Whereas almost rigid proper classes of topological spaces do exist, see [Koubek 1975], the existence of almost rigid proper classes of abelian groups is set-theoretical, see [Dugas, Herden 1983b] and [Göbel, Shelah 1985]. Exercise 6.c stems from [Dugas, Herden 1983a] and Exercise 6.g from [Eda 1989]. Almost free abelian groups are studied in [Eklof, Mekler 1990]. The characterization of injectivity classes of abelian groups in Exercise 6.d solves an older problem included in [Fuchs 1970]; this solution was published in [Adámek, Rosický 1993].

Vopěnka's principle has also been used in model theory by [Makowsky 1985a,b].

Appendix: Large Cardinals

The aim of this Appendix is to explain the role of Vopěnka's principle in set theory: we will prove that

(1) Vopěnka's principle implies that measurable cardinals exist (Theorem A.6);

(2) If huge cardinals exist, then Vopěnka's principle is consistent (Theorem A.18).

Whereas (1) is simple, (2) requires some more advanced set-theoretical techniques (which we explain in detail). Using the same technique we finally present a category $\mathcal{K}$ which is accessible but not co-wellpowered, and we conclude that the category of models of some finite sketch in $\mathcal{K}$ is not accessible.

Vopěnka's Principle Implies Measurable Cardinals

A.1 Definition. By a λ-*additive* (two-valued) *measure* on a set X, where λ is an infinite cardinal, is meant a mapping σ assigning to each subset of X a value 0 or 1 in such a way that

(i) $\sigma(X) = 1$,

and

(ii) $\sigma\left(\bigcup_{i\in I} A_i\right) = \sum_{i\in I} \sigma(A_i)$ for each collection A_i $(i \in I)$ of pairwise disjoint subsets of X with $\operatorname{card} I < \lambda$.

A.2 Remarks

(1) Condition (ii) is equivalent to the following two statements: σ is monotone (i.e., if $A \subseteq B$ and $\sigma(A) = 1$ then $\sigma(B) = 1$) and whenever $\sigma(\bigcup_{i \in I} A_i) = 1$, with the A_i pairwise disjoint and $\operatorname{card} I < \lambda$, then $\sigma(A_i) = 1$ holds for a unique $i \in I$.

(2) The set $\mathcal{U}$ of all subsets $A \subseteq X$ with $\sigma(A) = 1$ is an ultrafilter (see Remark 5.17) which is *λ-complete*, i.e., closed under intersections of less than λ members.

In fact, since σ is monotone, $\mathcal{U}$ is closed under supersets. For each subset $A \subseteq X$ either $A \in \mathcal{U}$ or $X - A \in \mathcal{U}$ because

$$1 = \sigma(X) = \sigma\big(A \cup (X - A)\big) = \sigma(A) + \sigma(X - A).$$

Finally, let $\alpha < \lambda$ and let A_i $(i < \alpha)$ be members of $\mathcal{U}$. The sets $B_i = (X - A_i) - \bigcup_{j<i}(X - A_j)$ are pairwise disjoint, and fulfil $\sigma(B_i) = 0$ and $\bigcup_{i<\alpha} B_i = X - \bigcap_{i<\alpha} A_i$. Therefore, $\bigcap_{i<\alpha} A_i$ is a member of $\mathcal{U}$:

$$\sigma\Big(\bigcap_{i<\alpha} A_i\Big) = 1 - \sigma\Big(X - \bigcap_{i<\alpha} A_i\Big) = 1 - \sigma\Big(\bigcup_{i<\alpha} B_i\Big) = 1 - \sum_{i<\alpha} \sigma(B_i) = 1.$$

(3) Conversely, every λ-complete ultrafilter $\mathcal{U}$ on a set X defines a λ-additive measure σ by

$$\sigma(A) = 1 \qquad \text{iff} \qquad A \in \mathcal{U}.$$

Example. For each point $x \in X$ we can define a λ-additive measure σ by $\sigma(A) = 1$ iff $x \in A$. Such measures are called *trivial*, and all other measures are called *non-trivial*.

A.3 Definition. An uncountable cardinal λ is called *measurable* if each set of cardinality λ has a non-trivial λ-additive measure.

Remark. We will see below that it is consistent with set theory to assume that no measurable cardinal exists. We will need the following weaker assumption:

A.4 Definition. We denote by (M) the following statement:

(M) There do not exist arbitrarily large measurable cardinals.

(That is, we can find a cardinal λ such that no cardinal larger or equal to λ is measurable.)

A.5 Theorem. *The category* $\mathbf{Set}^{\mathrm{op}}$ *is bounded iff* (M) *holds.*

PROOF. I. It is clear that $\mathbf{Set}^{\mathrm{op}}$ is bounded iff there exists a cardinal λ such that the (essentially small) collection $\mathbf{Set}_\lambda$ of all sets of cardinality $< \lambda$ is dense in $\mathbf{Set}^{\mathrm{op}}$. Let us reformulate this more concretely.

For each set X the canonical diagram in $\mathbf{Set}^{\mathrm{op}}$ can, obviously, be substituted by the poset $Q_\lambda(X)$ of all quotients $q\colon X \to X_q$ of X in $\mathbf{Set}$ with $\operatorname{card} X_q < \lambda$. The canonical colimit of $Q_\lambda(X)$ (i.e., the appropriate limit in $\mathbf{Set}$) is the set $\lim Q_\lambda(X)$ of all tuples $(t_q)_{q\in Q_\lambda(X)}$ of elements $t_q \in X_q$ such that for any map $f\colon X_q \to X_{q'}$ with $f \cdot q = q'$ we have $f(t_q) = t_{q'}$. (The limit cone is formed by the projections.) Now, the canonical map $h\colon X \to \lim Q_\lambda(X)$ is given by $h(x) = (q(x))_{q\in Q_\lambda(X)}$. If $\lambda > 1$, then h is one-to-one. Thus, $\mathbf{Set}_\lambda$ is dense iff h is always onto, i.e., iff the following holds for each set X:

(*) for every $(t_q) \in \lim Q_\lambda(X)$ there exists $a \in X$ with $(t_q) = \big(q(a)\big)$.

II. If $\mathbf{Set}^{\mathrm{op}}$ is bounded, we prove (M). Let λ be a cardinal satisfying (*), then we show that every λ-additive measure is trivial (thus, no measurable cardinal is larger or equal to λ). In fact, let σ be a λ-additive measure on a set X. For each $q\colon X \to X_q$ in $Q_\lambda(X)$ we have $X = \bigcup_{t\in X_q} q^{-1}(t)$ and $\operatorname{card} X_q < \lambda$; thus, from the λ-additivity of σ we conclude that there exists a unique $t_q \in X_q$ with $\sigma\big(q^{-1}(t_q)\big) = 1$ for each q. Let a be an element as in (*), and let $q \in Q_\lambda(X)$ be a quotient with $q^{-1}\big(q(a)\big) = \{a\}$. Then

$$\sigma\big(\{a\}\big) = \sigma\big(q^{-1}(t_q)\big) = 1,$$

thus, σ is trivial.

III. (M) implies that (*) holds, whenever no cardinal $\geq \lambda$ is measurable. To prove this, we first verify that every λ-additive measure is trivial. Let σ be a λ-additive measure on a set X. Since $1 = \sigma\big(\bigcup_{x\in X}\{x\}\big) \neq \sum_{x\in X} \sigma\big(\{x\}\big) = 0$, we know that $\operatorname{card} X \geq \lambda$. Therefore, it is sufficient to prove that for $\mu = \operatorname{card} X$ the measure σ is μ-additive (thus, trivial, since μ is not measurable). Suppose the contrary, and suppose we choose the smallest possible μ with a λ-additive measure which is not μ-additive. We will derive a contradiction. Let A_i $(i \in I)$ be pairwise disjoint sets with $\operatorname{card} I < \mu$, $\sigma(A_i) = 0$ for all $i \in I$, and $\sigma(\bigcup_{i\in I} A_i) = 1$. Define a measure $\overline{\sigma}$ on the set I by $\overline{\sigma}(J) = \sigma(\bigcup_{i\in J} A_i)$ for all $J \subseteq I$. It is clear that $\overline{\sigma}$ is λ-additive and, since $\operatorname{card} I < \mu$, it is also $(\operatorname{card} I)$-additive. This contradicts $1 = \overline{\sigma}\big(\bigcup_{i\in I}\{i\}\big) \neq \sum_{i\in I} \overline{\sigma}\big(\{i\}\big) = 0$.

We are ready to prove (*). Define a measure σ on X by

$$\sigma(A) = 1 \quad \text{iff} \quad t_q \in q(A) \quad \text{for each} \quad q\colon X \to X_q \text{ in } Q_\lambda(X).$$

This is a λ-additive measure: $\sigma(X) = 1$, the monotonicity is obvious, and given pairwise disjoint sets $A_j \subseteq X$ for $j \in J$ with $\operatorname{card} J < \lambda$, from $\sigma(\bigcup_{j\in J} A_j) = 1$ we conclude that $\sigma(A_j) = 1$ for a unique j as follows:

(i) Existence. Suppose that, on the contrary, $\sigma(A_j) = 0$ for each $j \in J$. Then there exists $q_j \in Q_\lambda(X)$ with $t_{q_j} \notin q_j(A_j)$. Since the number of all these quotients q_j is $\operatorname{card} J < \lambda$, it is clear that they have a common refinement in $Q_\lambda(X)$. That is, there exists $q \in Q_\lambda(X)$ and maps $f_j\colon X_q \to X_{q_j}$ with $q_j = f_j \cdot q$ for all $j \in J$. This is a contradiction of $\sigma(\bigcup_{j\in J} A_j) = 1$: since $t_q \in q(\bigcup_{j\in J} A_j)$, there exists $j \in J$ with $t_q \in q(A_j)$, and then $t_{q_j} = f_j(t_q) \in (f_j \cdot q)(A_j)$.

(ii) Uniqueness. Suppose that, on the contrary, $\sigma(A_j) = 1 = \sigma(A_{j'})$ for $j \neq j'$. There clearly exists a quotient $q\colon X \to X_q$ in $Q_\lambda(X)$ with $q(A_j) \cap q(A_{j'}) = \emptyset$, which contradicts $t_q \in q(A_j) \cap q(A_{j'})$.

Thus, σ is trivial, which proves (*). □

A.6 Theorem. *Vopěnka's principle implies* $\neg$(M), *i.e., the existence of arbitrarily large measurable cardinals.*

PROOF. Assuming (M), we derive a contradiction with Vopěnka's principle. By Theorem A.5 the category $\mathbf{Set}^{\mathrm{op}}$ has a dense subcategory with objects X_i $(i \in I)$. It follows that the category $\mathbf{Gra}^{\mathrm{op}}$ has a dense subcategory consisting of the graphs $(X_i, X_i \times X_i)$ $(i \in I)$ together with the graph

$$G^* = (Y, \sigma), \qquad Y = \{0,1,2\} \quad \text{and} \quad \sigma = Y \times Y - \{(0,1)\}.$$

(To verify the denseness, it is sufficient to use the following property of G^*: given graphs K, K' and a map $f\colon |K| \to |K'|$, then $f\colon K \to K'$ is a homomorphism iff $g \cdot f\colon K \to G^*$ is a homomorphism for each homomorphism $g\colon K' \to G^*$.) Consequently, there exists a full and faithful functor $E\colon \mathbf{Gra}^{\mathrm{op}} \to \mathbf{Gra}$, see Proposition 1.26 and Theorem 2.65. For each cardinal λ let $A_\lambda = (X_\lambda, \neq)$ be a graph with $\operatorname{card} X_\lambda = \lambda$, here $\neq$ means the complement of the diagonal relation. Then

$$\lambda < \lambda' \quad \text{implies} \quad \hom(EA_\lambda, EA_{\lambda'}) = \emptyset$$

(because $\hom(A_{\lambda'}, A_\lambda) = \emptyset$). Put $EA_\lambda = (Y_\lambda, \beta_\lambda)$ and choose a large class C of cardinals such that $\lambda < \lambda'$ implies $\operatorname{card} Y_\lambda < \operatorname{card} Y_{\lambda'}$ (for all $\lambda, \lambda' \in C$). For the one-sorted signature Σ of three binary relations the Σ-structures

$$A^*_\lambda = (Y_\lambda, \beta_\lambda, \neq, \gamma_\lambda), \qquad \lambda \in C,$$

where γ_λ is a rigid binary relation on the set Y_λ (see Lemma 2.64), are rigid in $\mathbf{Rel}\,\Sigma$. In fact, for each homomorphism $f\colon A^*_\lambda \to A^*_{\lambda'}$ we have

$\lambda \geq \lambda'$ because $f \in \hom(EA_\lambda, EA_{\lambda'})$, and $\lambda \leq \lambda'$ because of the relation $\neq$. Therefore, $\lambda = \lambda'$ and $f = id$ due to the last relation. □

A.7 Corollary. *Weak Vopěnka's principle implies* $\neg$(M).

In fact, assuming (M), we have a full embedding of **Ord** to **Gra** (see Theorem A.6 and Lemma 6.3), thus, a full embedding of $\mathbf{Ord}^{\mathrm{op}}$ to $\mathbf{Gra}^{\mathrm{op}}$. By composing the latter with the functor E in the proof of Theorem A.6, we obtain a full embedding of $\mathbf{Ord}^{\mathrm{op}}$ to **Gra**, in contradiction to weak Vopěnka's principle. □

Vopěnka's Principle Cannot be Proved

We are now going to show that in the set theory BGC (i.e., Bernays–Gödel theory with the axiom of choice for classes) in which we work, Vopěnka's principle cannot be proved from the axioms.

It is sufficient to show that the existence of measurable cardinals is not provable from BGC, see Theorem A.6. This follows from the fact that each measurable cardinal λ is (strongly) inaccessible, thus, the class V_λ of all sets of rank smaller than λ is a model of BGC. If λ is the smallest inaccessible cardinal, then V_λ is a model of BGC without measurable cardinals—thus, without (even weak) Vopěnka's principle.

Let us formalize this procedure.

A.8 Remark. Define sets V_α ($\alpha \in$ Ord) by the following transfinite induction:

$$V_0 = \emptyset;$$

$$V_{\alpha+1} = \{\, A \mid A \subseteq V_\alpha \,\};$$

$$V_\alpha = \bigcup_{\beta<\alpha} V_\beta \text{ for limit ordinals } \alpha.$$

By the axiom of regularity, $V = \bigcup_{\alpha\in\mathrm{Ord}} V_\alpha$ is the class of all sets.

By the *rank* of a set A is meant the smallest ordinal α such that $A \subseteq V_\alpha$ (i.e., $A \in V_{\alpha+1}$).

Observe that

$$A \in B \quad \text{implies} \quad \operatorname{rank} A < \operatorname{rank} B,$$

and for each ordinal α (considered as the set of all smaller ordinals) we have

$$\operatorname{rank} \alpha = \alpha.$$

A.9 Definition. A cardinal λ is called (strongly) *inaccessible* provided that λ is uncountable, regular, and

$$\alpha < \lambda \quad \text{implies} \quad 2^\alpha < \lambda \qquad \text{for each cardinal } \alpha.$$

Remark. For each inaccessible cardinal λ the set V_λ is a model of BGC with $\operatorname{card} V_\lambda = \lambda$.

A.10 Theorem. *Every measurable cardinal is inaccessible.*

PROOF. Let λ be a measurable cardinal, and let σ be a non-trivial λ-additive measure on a set X of cardinality λ.

I. λ is regular, i.e., given a cardinal $\alpha < \lambda$, then whenever $X = \bigcup_{i<\alpha} A_i$, there exists $i < \alpha$ with $\operatorname{card} A_i = \lambda$. Suppose the contrary is true, then $\operatorname{card} A_i < \lambda$ implies $\sigma(A_i) = \sigma\big(\bigcup_{a \in A_i}\{a\}\big) = \sum_{a \in A_i} \sigma(\{a\}) = 0$. Thus, the pairwise disjoint sets $A'_i = A_i - \bigcup_{j<i} A_j$ fulfil $\sigma(A'_i) = 0$. This contradicts $X = \bigcup_{i<\alpha} A'_i$:

$$1 = \sigma(X) = \sigma\Big(\bigcup_{i<\alpha} A'_i\Big) = \sum_{i<\alpha} \sigma(A'_i).$$

II. If $\alpha < \lambda$ then $2^\alpha < \lambda$. In fact, suppose $2^\alpha \geq \lambda$. That is, there exists a set Y of cardinality $\alpha < \lambda$ which has pairwise distinct subsets $B_x \subseteq Y$ $(x \in X)$. For each $y \in Y$ put

$$P_y = \{\, x \in X \mid y \in B_x \,\},$$

and define

$$Y_0 = \big\{\, y \in Y \mid \sigma(P_y) = 1 \,\big\}.$$

Then the set

$$A = \bigcap_{y \in Y_0} P_y \cap \bigcap_{y \in Y - Y_0} (X - P_y)$$

is either empty (if $Y_0 \neq B_x$ for each $x \in X$) or $\{x\}$ (if $Y_0 = B_x$). However, A is an intersection of less than λ sets of measure 1; consequently, $\sigma(A) = 1$, which is a contradiction with being non-trivial. □

A.11 Proposition. *The non-existence of inaccessible cardinals is consistent with the set theory BGC.*

PROOF. If our model V of BGC has no inaccessible cardinals, there is nothing to prove. If it has some, let λ be the smallest inaccessible cardinal. Then V_λ is a model of BGC. For any ordinal $\alpha < \lambda$ we have that α is an ordinal (i.e., the set of all smaller ordinals) in the model V_λ, thus

$$V_\lambda \models (\alpha \text{ is an ordinal}) \quad \text{iff} \quad V \models (\alpha \text{ is an ordinal}).$$

A cardinal is an ordinal not isomorphic to any smaller ordinal, and we thus see that

$$V_\lambda \models (\alpha \text{ is a cardinal}) \quad \text{iff} \quad V \models (\alpha \text{ is a cardinal}).$$

Analogous proofs hold for the properties "α is a regular cardinal", "α is uncountable", and, finally, "α is inaccessible".

Consequently, V_λ is a model of BGC without inaccessible cardinals. □

A.12 Corollary. *The negation of Vopěnka's principle is consistent with the set theory BGC.*

See Theorems A.6 and A.10, and Proposition A.11. □

Remark. A much stronger result can be proved: the negation of Vopěnka's principle is consistent with BGC + ¬(M), i.e., the theory BGC plus the existence of arbitrarily large measurable cardinals. This follows from the fact that Vopěnka's principle implies the existence of supercompact cardinals: if λ is supercompact then V_λ is a model of BGC + ¬(M), see [Jech 1978] 33.10, 33.14(a) and 33.15.

Huge Cardinals Imply Vopěnka's Principle

In the rest of the Appendix we are going to show that the existence of cardinals called huge implies that Vopěnka's principle is consistent with BGC: if λ is huge, then V_λ is a model of the set theory BGC satisfying Vopěnka's principle. Our treatment, which now becomes more involved with the intricacies of set theory, follows closely the corresponding part of [Jech 1978].

A.13 Remarks

(1) Recall that ZFC (Zermelo–Fraenkel set theory with the axiom of choice) is a (one-sorted) theory of one binary relation $\in$ that satisfies axioms listed e.g. in [Jech 1978] p. 1. It is a weaker set theory than BGC, i.e., any model of BGC is a model of ZFC. (BGC is a conservative extension of ZFC, which means that every ZFC-sentence provable in ZFC is also provable in BGC, see [Jech 1978] p. 77.)

(2) A model of ZFC is a *class* M together with a binary relation $E \subseteq M \times M$ (which is a class again) satisfying all the axioms of ZFC. By a *transitive model* is meant a model M which is transitive in the sense that

$$(*) \qquad a \in b \in M \quad \text{implies} \quad a \in M,$$

and which is such that the binary relation E is the restriction of the relation $\in$ to $M \times M$.

Example: V is a transitive model of ZFC.

(3) If M_1 and M_2 are models of ZFC, then a map $h\colon M_1 \to M_2$ is called an *elementary embedding* provided that for each L_ω-formula $\varphi(x_1, \ldots, x_n)$ of ZFC and each assignment $x_i \mapsto a_i$, $i = 1, \ldots, n$, of the free variables of φ in M_1 we have

$$M_1 \models \varphi[a_1, \ldots, a_n] \quad \text{iff} \quad M_2 \models \varphi[ha_1, \ldots, ha_n].$$

(See Definition 5.37.)

A.14 Remark. Let M be a transitive model of ZFC. Each elementary embedding $h\colon V \to M$ has the following properties.

(1) $X \subseteq Y$ implies $h(X) \subseteq h(Y)$.

In fact, consider the following formula $\varphi(x, y)$:

$$(\forall z)\big(z \in x \Rightarrow z \in y\big).$$

Since $X \subseteq Y$ implies $V \models \varphi[X, Y]$, we have $M \models \varphi\big[h(X), h(Y)\big]$, i.e., $h(X) \subseteq h(Y)$.

(2) h preserves union: $h(X \cup Y) = h(X) \cup h(Y)$.

In fact, consider the formula

$$(\forall z)\big((z \in x \cup y) \Leftrightarrow ((z \in x) \vee (z \in y))\big).$$

(3) h preserves complement: $h(X - Y) = h(X) - h(Y)$ (this is analogous to (2)), thus, h also preserves intersection: $h(X \cap Y) = h(X) \cap h(Y)$.

(4) h preserves cartesian products: $h(X \times Y) = h(X) \times h(Y)$.

In fact, recall that an ordered pair (x, y) is defined to be the set $\{\{x\}, \{x, y\}\}$. For each $x \in X$ and $y \in Y$ we have $h(\{x, y\}) = \{h(x), h(y)\}$, due to the formula

$$(\forall z)\big((z \in \{x, y\}) \Leftrightarrow (z = x) \vee (z = y\}\big).$$

It follows that for ordered pairs (x, y) we have

$$h\big((x, y)\big) = \big(h(x), h(y)\big),$$

thus, $h(X \times Y) = h(X) \times h(Y)$.

(5) h preserves mappings: if $f\colon X \to Y$ is a mapping, then

$$h(f)\colon h(X) \to h(Y)$$

is a mapping.

In fact, a mapping is a subset f of $X \times Y$ satisfying the formula

$$(\forall x)(\exists! y)\big((x, y) \in f\big).$$

Moreover, if f is one-to-one, then $h(f)$ is also one-to-one.

(6) If α is an ordinal, then $h(\alpha)$ is an ordinal. An ordinal can be defined as a transitive set (i.e., a set α satisfying $(*)$ of Remark A.13) which is linearly ordered by the relation $\in$. Since transitivity means satisfaction of the formula

$$(\forall x)\big((x \in \alpha) \Rightarrow (x \subseteq \alpha)\big),$$

$h(\alpha)$ is transitive. Linearity means satisfaction of the sentence

$$(\forall\, x, y)\big[\big((x \in \alpha) \wedge (y \in \alpha)\big) \Rightarrow \big((x \in y) \vee (y \in x)\big)\big],$$

thus, $h(\alpha)$ is an ordinal.

(7) For each ordinal $\alpha \le \omega$ we have $h(\alpha) = \alpha$.

In fact, for $\alpha = 0$ this follows from the fact that $0\ (= \emptyset)$ is defined by $(\forall x)(x \notin 0)$. If $h(n) = n$, then

$$h(n+1) = h\big(n \cup \{n\}\big) = h(n) \cup \big\{h(n)\big\} = h(n) + 1,$$

due to (2) above. Thus, $h(n) = n$ for all $n < \omega$. To prove $h(\omega) = \omega$, it is sufficient to show that $\alpha \le \omega$ implies $h(\alpha) \le \omega$, and this follows from the fact that $\alpha \le \omega$ holds iff each ordinal β with $0 < \beta < \alpha$ has a predecessor.

(8) h preserves rank:

$$h(\operatorname{rank} X) = \operatorname{rank} h(X).$$

In fact, since every set is an element of a transitive set Y, it is sufficient to prove that

$$Y \text{ transitive} \implies h(\operatorname{rank} X) = \operatorname{rank} h(X) \quad \text{for all } X \in Y.$$

Put $\alpha = \operatorname{rank} Y$. The function $r\colon Y \to \alpha$ defined by $r(X) = \operatorname{rank} X$ is uniquely determined by the satisfaction of the following formulas:

$$\begin{aligned}
&r(0) = 0;\\
&(\forall\, x, y)\big((x \in y) \Rightarrow (r(x) < r(y))\big);\\
&(\forall z)\big([(z \text{ is an ordinal}) \wedge (\forall\, x, y)((x \in y) \Rightarrow (r(x) < z))]\\
&\qquad \Longrightarrow (r(y) \leq z)\big).
\end{aligned}$$

Consequently, $h(r)\colon h(Y) \to h(\alpha)$ is a function satisfying the above three formulas, and since $h(Y)$ is transitive, this proves $h(\operatorname{rank} X) = \operatorname{rank} h(X)$ for all $X \in Y$.

A.15 Proposition. *Let M be a transitive model. Given an elementary embedding $h\colon V \to M$ such that*

$$h(\alpha) = \alpha \qquad \text{for each ordinal } \alpha,$$

then $M = V$ and $h = id_V$.

PROOF. Since $M \subseteq V$, it is sufficient to prove that $h(X) = X$ for each set X. We prove this by transfinite induction on the rank of X. For $\operatorname{rank} X = 0$, see (7) in Remark A.14. Suppose that $\operatorname{rank} X = \alpha$ and for each set Y of a smaller rank we have $h(Y) = Y$. Then $Y \in X$ implies $Y = h(Y) \in h(X)$. Consequently,

$$X \subseteq h(X).$$

Conversely, we use the fact that $h(\alpha) = \alpha$ and $\operatorname{rank} h(X) = h(\operatorname{rank} X)$, see (8) in Remark A.14; thus,

$$\operatorname{rank} h(X) = \alpha.$$

Given $Z \in h(X)$, then $\operatorname{rank} Z < \operatorname{rank} h(X) = \alpha$ implies $h(Z) = Z$. Consequently, $h(Z) \in h(X)$ which, since h is elementary, implies $Z \in X$. Thus $X = h(X)$. □

A.16 Proposition. *Let M be a transitive model, and let $h\colon V \to M$ be an elementary embedding, $h \neq id_V$. Then the first ordinal λ with $h(\lambda) \neq \lambda$ is a measurable cardinal.*

PROOF. By Proposition A.15, λ exists. It is uncountable by Remark A.14(7). We will prove that the measure σ defined on the set λ by

$$\sigma(A) = 1 \qquad \text{iff} \qquad \lambda \in h(A)$$

is non-trivial and λ-additive.

I. Since $h(\alpha) = \alpha$ for every $\alpha \in \lambda$, we have $\lambda \subseteq h(\lambda)$. Since $h(\lambda)$ is an ordinal (see Remark A.14(6)), we conclude that $\lambda \in h(\lambda)$, i.e., $\sigma(\lambda) = 1$.

II. σ is obviously monotone.

III. Let A_i $(i \in \alpha)$ be pairwise disjoint subsets of λ for some cardinal $\alpha < \lambda$. Since the $h(A_i)$ are pairwise disjoint (by Remarks A.14(3) and (7)), we have $\lambda \in h(A_i)$ for at most one i. Therefore, it is sufficient to show that $\sigma(\bigcup_{i\in\alpha} A_i) = 1$ implies $\sigma(A_i) = 1$ for some i. Since $h(\alpha) = \alpha$, the following function:

$$f\colon \alpha \to \exp\lambda, \qquad f(i) = A_i$$

is mapped by h to a function

$$h(f)\colon \alpha \to h(\exp\lambda),$$

see Remark A.14(5).

Since $f(i) = A_i$ means $(i, A_i) \in f$, we have $\big(h(i), h(A_i)\big) \in h(f)$ (by Remark A.14(4)) and, since $h(i) = i$ for each $i \in \alpha$, we conclude that $\big(i, h(A_i)\big) \in h(f)$. Therefore, $h(f)$ is the function

$$h(f)\colon \alpha \to h(\exp\lambda)$$

given by $h(f)(i) = h(A_i)$ for every $i \in \alpha$.

From the formula

$$(\forall x)\Big[x \in \bigcup_{i\in I} A_i \iff (\exists\, i)\big((i \in \alpha) \wedge \big(x \in f(i)\big)\big)\Big]$$

we conclude that

$$(\forall x)\Big[x \in h\Big(\bigcup_{i\in I} A_i\Big) \iff (\exists\, i)\big((i \in \alpha) \wedge \big(x \in h(f)(i)\big)\big)\Big].$$

Consequently, $\lambda \in h(\bigcup_{i\in\alpha} A_i)$ implies that $\lambda \in h(A_i)$, i.e., $\sigma(A_i) = 1$, for some $i \in \alpha$.

IV. σ is non-trivial because for each $\alpha \in \lambda$ we have

$$h\big(\{\alpha\}\big) = \big\{h(\alpha)\big\} = \{\alpha\}. \qquad \square$$

A.17 Definition. A cardinal λ is called *huge* if there exists a transitive model M and an elementary embedding $h\colon V \to M$ such that

(1) λ is the smallest ordinal with $h(\lambda) \neq \lambda$;

(2) $M^{h(\lambda)} \subseteq M$, i.e., M contains each function from $h(\lambda)$ to M.

Remark. Every huge cardinal is measurable (by Proposition A.16). Thus, the non-existence of huge cardinals is consistent with BGC.

A.18 Theorem. *The consistency of Vopěnka's principle follows from the existence of huge cardinals: if λ is a huge cardinal, then V_λ is a model of the set theory BGC in which Vopěnka's principle holds.*

PROOF. We are going to prove that for each set C of graphs with $\operatorname{rank} C = \lambda$ there exists a homomorphism $G \to G'$ for two distinct members G, G' of C. This proves Vopěnka's principle in V_λ (which is a model of BGC by Remarks A.9 and A.17 and Theorem A.10).

Let $h: V \to M$ be as in Definition A.17. Recall that a graph (X, α) is an ordered pair of sets with $\alpha \subseteq X \times X$. Since every element of C is a graph, the following formula

$$(\forall x)\big[x \in C \implies (\exists\, y, z)\big((x = (y, z)) \wedge (z \subseteq y \times y)\big)\big]$$

is satisfied, which implies that every element of $h(C)$ is a graph.

Let us choose a graph

$$G_0 \in h(C) - C, \qquad G_0 = (X_0, \alpha_0).$$

This is possible, since $\lambda < h(\lambda)$, thus, by Remark A.14(8), the rank of $h(C)$ is larger than λ, consequently, $h(C) \nsubseteq C$. Since for each $x \in X_0$ we have $h(x) \in h(X_0)$, we can define a mapping $h_0: X_0 \to h(X_0)$ as a restriction of h. Furthermore, $(x, y) \in \alpha_0$ implies $\big(h(x), h(y)\big) \in h(\alpha_0)$, see Remark A.14(4), thus, we get a graph homomorphism

$$h_0: G_0 \to h(G_0) = \big(h(X_0), h(\alpha_0)\big).$$

We observe that $G_0 \neq h(G_0)$: otherwise $h(G_0) \in h(C)$ would imply $G_0 \in C$ (since h is elementary). We will prove that $h_0 \in M$.

Let us show that for every graph homomorphism $f: H \to H'$ we have

$$(*) \qquad \text{if } H \in h(C) \text{ and } H' \in M, \text{ then } f \in M.$$

For this purpose, consider an arbitrary graph (X, α) in C. We know that $\operatorname{rank} X < \lambda$, thus, $\operatorname{card} X \leq \lambda$: in fact, λ is inaccessible by Proposition A.16 and Theorem A.10, thus, $\operatorname{card} V_\lambda = \lambda$ by Remark A.9 and we have $X \subseteq V_\lambda$. Consequently, there is a one-to-one mapping $X \to \lambda$. Since h is elementary, for every graph $H = (X, \alpha) \in h(C)$ there is a one-to-one mapping $m: X \to h(\lambda)$ in M. Now, $(*)$ follows from (2) in Definition A.17 (since f factorizes through m and $f = f' \cdot m$ implies $f' \in M$).

Since $h_0 \colon G_0 \to h(G_0)$ belongs to M, we have that

$$M \models \text{there exists a homomorphism } A \to h(G_0) \\ \text{of graphs satisfying } h(G_0) \neq A \in h(C)$$

and, since h is elementary, this implies that

$$V \models \text{there exists a homomorphism } g \colon A \to G_0 \\ \text{of graphs satisfying } G_0 \neq A \in C.$$

In the last statement we have $\operatorname{rank} A < \lambda$. Thus, it follows from the proof of Proposition A.15 that $h(A) = A$. Therefore $g \colon h(A) \to G_0$ is a homomorphism of graphs and $h(A)$ and G_0 are distinct members of $h(C)$. Since $g \in M$ (by $(*)$) and h is elementary, there is a homomorphism $G \to G'$ for two distinct members of C. □

Accessible Categories Need Not Be Co-Wellpowered

We conclude this appendix by exhibiting, under the assumption (M), a category $\mathcal{K}$ such that

(i) $\mathcal{K}$ is accessible,

(ii) $\mathcal{K}$ is not co-wellpowered,

and

(iii) $\mathbf{Mod}(\mathscr{S}, \mathcal{K})$ is not accessible for some sketch $\mathscr{S}$.

(Here $\mathbf{Mod}(\mathscr{S}, \mathcal{K})$ is the category of all $\mathcal{K}$-valued models of $\mathscr{S}$, see 2.60.) Recall from Corollary 6.19 that no such category $\mathcal{K}$ can exist if Vopěnka's principle holds. This, then, is another proof that Vopěnka's principle implies ¬(M); see Theorem A.6.

A.19 Example constructed under the assumption (M). Let λ be a cardinal such that no cardinal larger or equal to λ is measurable. We denote by ZC the theory obtained from the axioms of ZFC by deleting the axiom-scheme of replacement. Let $\mathcal{K}$ be the following category:

its objects are all transitive models of ZC and a formal initial object I;

its morphisms are all λ-elementary embeddings of models of ZC and formal morphisms with the domain I.

(1) $\mathcal{K}$ is accessible. By the Mostowski collapsing theorem ([Jech 1978], Theorem 28) every model of ZC is isomorphic to a transitive model. Since $\mathbf{Elem}_\lambda(\mathrm{ZC})$ is accessible by Theorem 5.42, it follows that $\mathcal{K}$ is accessible.

(2) $\mathcal{K}$ is not co-wellpowered. In fact, since the axiom-scheme of replacement has been deleted, for each limit cardinal γ the set V_γ is a transitive model of ZC. We will prove that the formal morphism $I \to V_\gamma$ is an epimorphism for each limit cardinal γ.

In other words, for each transitive model M of ZC we are to show that there exists at most one λ-elementary embedding $h: V_\gamma \to M$. By Proposition A.15, it is sufficient to prove that $h(\alpha) = \alpha$ for each ordinal $\alpha \in V_\gamma$. By Proposition A.16, the first ordinal moved by h is a measurable cardinal—thus, we just have to show that $h(\alpha) = \alpha$ for each ordinal $\alpha < \lambda$. This follows from the fact that h is λ-elementary and α is characterized by the following formula in L_λ:

$$(\exists x_i)_{i<\alpha}\Big((x_i \in \alpha) \wedge \bigwedge_{i<j<\alpha} (x_i \in x_j) \wedge (\forall y)\Big(y \in \alpha \Rightarrow \bigvee_{i<\alpha} (y = x_i)\Big)\Big).$$

(3) $\mathbf{Mod}(\mathscr{S}, \mathcal{K})$ is not accessible for the sketch $\mathscr{S}$ consisting of a single pushout square. More precisely, for the sketch $\mathscr{S} = (\mathcal{A}, \emptyset, \mathbf{C}, \sigma)$, where $\mathcal{A}$ is the Boolean algebra $\{0, a, \overline{a}, 1\}$, $\mathbf{C}$ has the span $a \leftarrow 0 \rightarrow \overline{a}$ as the unique member, and σ assigns to the span the cocone with codomain 1. Thus, $\mathbf{Mod}(\mathscr{S}, \mathcal{K})$ is the category of all pushout squares in $\mathcal{K}$. If this category were accessible, then $\mathcal{K}$ would be co-wellpowered—this follows from an inspection of the proof of Theorem 2.49.

Historical Remarks

For Vopěnka's principle see Chapter 6. Theorem A.5 is due to [Isbell 1960] and Example A.19 is from [Makkai, Paré 1989] (although (2) appeared in [Mekler 1978]) where the non-existence of measurable cardinals is assumed instead of (M).

Open Problems

1. Is each small-orthogonality class in the dual of a locally presentable category reflective?

Comments. The answer is affirmative under the assumption of Vopěnka's principle: if $\mathcal{K}$ is locally presentable, then every small-orthogonality class in $\mathcal{K}^{op}$ is closed under colimits in $\mathcal{K}$ (see Observation 1.34), thus, it is coreflective in $\mathcal{K}$ by Theorem 6.28.

The question of which categories $\mathcal{K}$ have the property that each small-orthogonality class in $\mathcal{K}$ is reflective has been studied in [Freyd, Kelly 1972], [Kelly 1980], and [Adámek, Rosický 1993]. It has been proved, e.g., that this is true for the category $\mathcal{K}$ of topological spaces.

2. Is each orthogonality class in the dual of a locally presentable category (a) reflective? (b) an intersection of two full reflective subcategories?

Comments. (a) The affirmative answer lies between Vopěnka's principle and $\neg$(M). In fact, assuming (M) every concrete category can be fully embedded into **Gra**, see [Pultr, Trnková 1980]. Thus, the dual of the category $\mathcal{C}$ in Example 6.25 can be fully embedded into **Gra**, which implies that $\mathbf{Gra}^{op}$ has a non-reflective orthogonality class.

(b) Rather mild conditions on a category $\mathcal{K}$ such that each orthogonality class in $\mathcal{K}$ is equal to an intersection of two full, reflective subcategories are exhibited in [Trnková, Adámek, Rosický 1990]. All locally presentable categories, but also the category of topological spaces, satisfy those conditions.

3. Can the definition of a locally generated category be stated without co-wellpoweredness? More precisely, is every cocomplete category with a strong generator formed by generated objects locally presentable?

Comments. See Remark 1.72. The answer is affirmative under the assumption of Vopěnka's principle: this follows from Corollary 6.8, since boundedness can be proved analogously to Proposition 1.22.

The problem is equivalent to the following: is every full, reflective subcategory of a locally finitely presentable category $\mathcal{K}$, closed under λ-directed unions in $\mathcal{K}$, locally presentable? (In fact, let $\mathcal{K}$ be cocomplete and let it have a set $\mathcal{A}_0$ of λ-generated objects forming a strong generator. The closure $\mathcal{A}$ of $\mathcal{A}_0$ under λ-small colimits is dense (and consists of λ-generated objects), this is analogous to Theorem 1.11 and Proposition 1.16. The canonical functor $E\colon \mathcal{K} \to \mathbf{Set}^{\mathcal{A}^{op}}$ is a full, faithful, right adjoint preserving λ-directed colimits of monomorphisms (see Propositions 1.26 and 1.27).)

4. Is every full subcategory of a locally presentable category which is closed under limits and directed unions reflective?

Comments. Compare with the reflection theorem, 2.48. The answer is affirmative under the assumption of weak Vopěnka's principle, see Theorem 6.22.

5. Is every full subcategory of the category **Ab** of abelian groups which is closed under limits reflective?

Is every full subcategory of **Ab** which is closed under products and split subobjects weakly reflective?

Comments. The first problem was stated in [Isbell 1966]. By Theorem 6.22, the answer is affirmative under the assumption of weak Vopěnka's principle. Analogously, the answer to the second problem is affirmative under the assumption (*) from Remark 6.27(2).

6. Is every full subcategory of a locally λ-presentable category which is closed under λ-directed colimits cone-reflective?

Comments. Compare with the (absolute!) reflection theorem 2.48. The answer is affirmative under the assumption of Vopěnka's principle, see Corollary 6.7. Conversely, the answer is negative if weak Vopěnka's principle does not hold, see Exercise 6.f (i.e., even if (*) in Remark 6.27(2) does not hold).

7. Is every full subcategory of a locally presentable category which is closed under products and split subobjects an injectivity class?

Comments. The answer is affirmative under the assumption of the statement (*) from Remark 6.27(2).

8. Is every full subcategory of the category **Top** of topological spaces which is closed under products and directed colimits an injectivity class?

Is every full subcategory of **Top** which is closed under limits an orthogonality class?

9. Is every full subcategory of the category **Ab** of abelian groups, closed under limits, reflective?

Is every full subcategory of **Ab**, closed under products and split subobjects, weakly reflective?

Coments. The first problem was formulated in [Isbell 1966]. By Theorem 6.22, the answer is affirmative under weak Vopěnka's principle. Analogously, the answer to the second problem is affirmative under the assumption (*) from Remark 6.27(2).

10. Is every injectivity class of a locally presentable category which is closed under equalizers an orthogonality class?

Comments. The answer is affirmative under the assumption of weak Vopěnka's principle, see Theorem 6.22.

11. Is each accessible category co-wellpowered w.r.t. strong epimorphisms?

What is the precise status of the statement "every accessible category is co-wellpowered"?

Comments. The answer to the first question is affirmative under the assumption of Vopěnka's principle, see Corollary 6.8. A discussion of the latter question is presented in Remark 6.8.

12. Is every λ-pure morphism in a λ-accessible category a regular monomorphism?

Comments. By Proposition 2.31, the answer is affirmative in each locally λ-presentable category.

13. Is every complete, axiomatizable category locally presentable?

Comments. The answer is affirmative under the assumption of Vopěnka's principle, since axiomatizable categories are bounded. The affirmative answer under the (weaker) assumption that there exist arbitrarily large compact cardinals was proved in [Rosický 1994b].

14. Find a large, accessible category $\mathcal{T}$ which has a faithful, accessible functor into an arbitrary accessible category.

Comments. The category of linearly ordered sets has a faithful, finitely accessible functor into any large finitely accessible category, as observed in [Makkai, Paré 1989], 3.4.1.

15. Is every finitely accessible category sketchable by a finite sketch?

16. Is Vopěnka's principle equivalent to weak Vopěnka's principle?

Or is the existence of arbitrarily large measurable cardinals equivalent to weak Vopěnka's principle?

Bibliography

Adámek, J. (1983): *Theory of Mathematical structures.* Reidel, Dordrecht.

Adámek, J. (1986): Classification of concrete categories. *Houston J. Math.* 12, 305–326.

Adámek, J. (1990): How many variables does a quasivariety need? *Alg. Univ.* 27, 44–48.

Adámek, J. (1994): The existence and non-existence of a regular generator. *Can. Math. Bull.*, to appear.

Adámek, J., H. Herrlich and J. Reiterman (1989): Cocompleteness almost implies completeness. *Proc. Conf. "Categorical Topology, Prague 1988"*, World Sci. Publ., Singapore, 246–256.

Adámek, J., H. Herrlich and J. Rosický (1988): Essentially equational categories. *Cahiers Top. Géom. Diff. Cat.* 29, 175–197.

Adámek, J., H. Herrlich and G. Strecker (1990): *Abstract and Concrete Categories.* John Wiley, New York.

Adámek, J., H. Herrlich and W. Tholen (1989): Monadic decompositions. *J. Pure Appl. Alg.* 59, 111–123.

Adámek, J. and J. Rosický (1988): Intersections of reflective subcategories. *Proc. Amer. Math. Soc.* 103, 710–712.

Adámek, J. and J. Rosický (1989): Reflections in locally presentable categories. *Arch. Math. (Brno)* 25, 89–94.

Adámek, J. and J. Rosický (1991): What is a locally generated category? In *Proc. Conf. "Category Theory, Como 1990"*, Lect. Notes in Math. 1488, Springer-Verlag, Berlin, 14–19.

Adámek, J. and J. Rosický (1992): On orthogonal subcategories of locally presentable categories. *Discrete Math.* 108, 133–137.

Adámek, J. and J. Rosický (1993): On injectivity in locally presentable categories. *Trans. Amer. Math. Soc*, to appear.

Adámek, J. and J. Rosický (1994a): On weakly locally presentable categories. *Cahiers Top. Géom. Diff. Cat.*, to appear.

Adámek, J. and J. Rosický (1994b): A remark on locally multipresentable categories. To appear.

Adámek, J., J. Rosický and V. Trnková (1988): Are all limit-closed subcategories of locally presentable categories reflective? In *Proc. Conf. "Category Theory, Louvain-la-Neuve 1987"*, Lect. Notes in Math. 1348, Berlin, 1–18.

Adámek, J. and W. Tholen (1990): Total categories with generators. *J. Alg. 133*, 63–78.

Adámek, J. and H. Volger (1994): On locally reflective categories of structures. Submitted.

Ageron, P. (1992): The logic of structures. *J. Pure Appl. Alg.* 79, 15–34.

Andréka, H. and I. Németi (1979a): Formulas and ultraproducts in categories. *Beitr. Alg. Geom.* 8, 133–151.

Andréka, H. and I. Németi (1979b): Injectivity in categories to represent all first order formulas, I. *Demonstratio Math.* 12, 717–732.

Andréka, H. and I. Németi (1982): A general axiomatizability theorem formulated in terms of cone-injective subcategories. In *Proc. Conf. "Universal Algebra Esztergom 1977"*, Coll. Math. Soc. J. Bolyai 29, North-Holland, Amsterdam, 13–15.

Artin, M., A. Grothendieck and J. L. Verdier (1972): *Théorie des topos et cohomologie étale des schémas.* Lect. Notes in Math. 269, Springer-Verlag, Berlin.

Bacsich, P. D. (1972): Injectivity in model theory. *Coll. Math.* 25, 165–176.

Banaschewski, B. and G. Bruns (1967): Categorical characterization of MacNeille completion. *Arch. Math.* 18, 369–377.

Banaschewski, B. and G. Bruns (1968): Injective hulls in the category of distributive lattices. *J. Reine Angew. Math.* 232, 102–109.

Banaschewski, B. and H. Herrlich (1976): Subcategories defined by implications. *Houston J. Math.* 2, 149–171.

Barr, M. (1986): Models of sketches. *Cahiers Top. Géom. Diff. Cat.* 27, 93–108.

Barr, M. (1989): Models of Horn theories. *Contemporary Math.* 92, 1–7.

Barr, M. (1990): Accessible categories and models of linear logic. *J. Pure Appl. Alg.* 69, 219–232.

Barr, M. and C. Wells (1985): *Toposes, Triples and Theories.* Springer-Verlag, Berlin.

Bastiani, A. and C. Ehresmann (1972): Categories of sketched structures. *Cahiers Top. Géom. Diff. Cat.* 13, 104–215.

Benecke, K. and H. Reichel (1983): Equational partiality. *Alg. Univ.* 16, 219–232.

Bergman, G. and R. M. Solovay (1987): Generalized Horn sentences and compact cardinals. *Abstracts Amer. Math. Soc.* 49, 832–04–13.

Bird, G. J. (1984): Limits in 2-categories of locally presentable categories. Ph.D. Thesis, University of Sydney.

Birkhoff, G. (1935): On the structure of abstract algebras. *Proc. Cambridge Phil. Soc.* 31, 433–454.

Birkhoff, G. and J. D. Lipson (1970): Heterogenous algebras, *J. Combinatorial Theory* 8, 115–133.

Borceux, F. and G. M. Kelly (1987): On locales of localizations. *J. Pure Appl. Alg.* 46, 1–34.

Burroni, A. (1970): Esquisse des catégories a limites et des quasi-topologies. *Esquisses Math.* 5.

Chang, C. C. and J. Keisler (1973): *Model Theory.* North-Holland, Amsterdam.

Cohn, P. M. (1965): *Universal Algebra.* Harper and Row, New York.

Coquand, T. (1989): Categories of embeddings. *Theor. Comp. Sci.* 68, 221–237.

Coste, M. (1979): Localisation, spectra and sheaf representation. *Appl. of Sheaves*, Lect. Notes in Math. 753, Springer-Verlag, Berlin, 212–238.

Dickmann, M. A. (1975): *Large Infinitary Languages.* North-Holland, Amsterdam.

Diers, Y. (1979): Familles universelles des morphismes. *Ann. Soc. Sci. Bruxelles* 93, 175–195.

Diers, Y. (1980a): Catégories localement multiprésentables. *Arch. Math.* 34, 344–356.

Diers, Y. (1980b): Quelques constructions de catégories localement multiprésentables. *Ann. Sc. Math. Québec* 4, 79–101.

Diers, Y. (1984): Une construction universelle des spectres, topologies spectrales et faisceau structuraux. *Commun. Alg.* 12, 2141–2183.

Dugas, M. and G. Herden (1983a): Arbitrary torsion classes of abelian groups. *Commun. Alg.* 11, 1455–1472.

Dugas, M. and G. Herden (1983b): Arbitrary torsion classes of almost free abelian groups. *Israel J. Math.* 44, 332–334.

Eda, K. (1989): Cardinality restrictions on preradicals. *"Abelian Group Theory", Contemp. Math.* 87, 277–283.

Eda, K. and Y. Abe (1987): Compact cardinals and abelian groups. *Tsukuba J. Math.* 11, 353–360.

Ehresmann, C. (1966): Introduction to the theory of structured categories. Technical Report 10, University of Kansas, Lawrence.

Ehresmann, C. (1968): Esquisses et types des structures algébraiques. *Bull. Inst. Pol. Iasi* 14, 1–14.

Eilenberg, S. and J. C. Moore (1965): Adjoint functors and triples. *Illinois J. Math.* 9, 381–398.

Eklof, P. C. (1975): Categories of local functors. *Model Theory and Algebra*, Lect. Notes in Math. 498, Springer-Verlag, Berlin.

Eklof, P. C. and A. Mekler (1990): *Almost Free Modules*. North-Holland, Amsterdam.

Fakir, S. (1975): Objects algébraiquement clos et injectifs dans les catégories localement présentables. In *Bull. Soc. Math. France*, Mém. 42.

Feferman, S. (1972): Infinitary properties, local functors and systems of ordinal functions. *Conf. Math. Logic*, Lect. Notes in Math. 255, Springer-Verlag, Berlin.

Felscher, W. (1968): Kennzeichung von primitiven und quasiprimitiven Kategorien von Algebren. *Arch. Math.* 19, 390–397.

Fisher, E. R. (1977): Vopěnka's principle, category theory and universal algebra. *Notices Amer. Math. Soc.* 24, A–44.

Fisher, E. R. (1987): Vopěnka's principle, universal algebra and categories. Personal communication.

Foltz, F. (1969): Sur la catégorie des foncteurs dominés. *Cahiers Top. Geom. Diff. Cat.* 11, 101–130.

Freyd, P. J. (1964): *Abelian Categories*. Harper and Row, New York.

Freyd, P. J. (1965): The theories of functors and models. In *Proc. Conf. Symp. Theory of Models*, North-Holland, Amsterdam, 107–120.

Freyd, P. J. (1972): Aspects of topoi. *Bull. Austral. Math. Soc.* 7, 1–76 and 467–480.

Freyd, P. J. and G. M. Kelly (1972): Categories of continuous functors I. *J. Pure Appl. Alg.* 2, 169–191.

Fuchs, L. (1970): *Infinite Abelian Groups* I. Academic Press, New York.

Gabriel, P. and F. Ulmer (1971): *Lokal Präsentierbare Kategorien*. Lect. Notes in Math. 221, Springer-Verlag, Berlin.

Girard, J.-Y. (1988): Normal functors, power series and lambda calculus. *Ann. Pure Appl. Logic* 37, 129–177.

Göbel, R. and S. Shelah (1985): Semi-rigid classes of cotorsion-free abelian groups. *J. Alg.* 93, 136–150.

Grätzer, G. (1968): *Universal Algebra*. Van Nostrand, Princeton.

Guitart, R. and C. Lair (1980): Calcul syntaxique des modèles et calcule des formules intérnes. *Diagrammes* 4, 1–106.

Guitart, R. and C. Lair (1982): Limites et colimites pour répresenter les formules. *Diagrammes* 7, 1–24.

Hales, A. W. (1964): On the non-existence of free complete Boolean algebras. *Fund. Math.* 56, 45–66.

Hatcher, W. S. (1970): Quasiprimitive subcategories. *Math. Ann.* 190, 93–96.

Hébert, M. (1991): Syntactic characterization of closure under connected limits. *Arch. Math. Logic* 31, 133–143.

Hébert, M. (1994): Syntactic characterization of closure under pullbacks and of locally polypresentable categories. To appear.

Hébert, M. and W. S. Hatcher (1994): *Preservation Theory.* To appear.

Hébert, M., R. N. McKenzie and G. E. Weaver (1989): Two definability results in the equational context. *Proc. Amer. Math. Soc.* 107, 47–53.

Herrlich, H. (1994): Almost reflective subcategories of Top. *Topol. and Appl.*, to appear.

Herrlich, H. and C. M. Ringel (1972): Identities in categories. *Can. Math. Bull.* 15, 297–299.

Hodges, W. (1981): In singular cardinality, locally free algebras are free. *Alg. Univ.* 12, 205–220.

Hodges, W. (1984): Models built on linear orderings. *Ann. Discr. Math.* 23, 207–234.

Hu, H. (1992a): Dualities for accessible categories. In *Category Theory* 1991, CMS Conf. Proc. 13, Amer. Math. Soc., Providence, 211–242.

Hu, H. (1992b): Dualities for accessible categories. Ph.D. thesis, McGill University, Montréal.

Hu, H. (1994): A duality theorem on accessible exact categories. To appear.

Hu, H. and M. Makkai (1994): Accessible embeddings and the solution set condition. To appear.

Isbell, J. R. (1960): Adequate subcategories. *Illinois J. Math.* 4, 541–552.

Isbell, J. R. (1964): Subobjects, adequacy, completeness and categories of algebras. *Rozprawy Matem.* 36, Warszawa.

Isbell, J. R. (1966): Structure of categories. *Bull. Amer. Math. Soc.* 72, 619–655.

Isbell, J. R. (1968): Small adequate subcategories. *J. London Math. Soc.* 43, 242–246.

Isbell, J. R. (1971): *Math. Rev.* 41 #5444, p. 1000.

Isbell, J. R. (1972): General functorial semantics I. *Amer. J. Math.* 94, 535–596.

Iwamura, T. (1944): A lemma on directed sets (in Japanese),*Zenkoku Shijo Sugaku Danwakai* 262, 107–111.

Jarzembski, G. (1989): Finitary fibrations. *Cahiers Top. Géom. Diff. Cat.* 30, 111–126.

Jech, T. (1978): *Set Theory.* Academic Press, New York.

John, R. (1977): A note on implicational subcategories. In *Proc. Colloq. "Szeged, 1975"*, Coll. Math. J. Bolyai 17, North-Holland, Amsterdam, 213–222.

Johnstone, P. T. (1979): A syntactic approach to Diers' localizable categories. *Applications of Sheaves*, Lect. Notes in Math. 753, Springer-Verlag, Berlin, 466–478.

Johnstone, P. T. and A. Joyal (1982): Continuous categories and exponentiable toposes. *J. Pure Appl. Alg.* 25, 255–296.

Johnstone, P. T. and S. Vickers (1991): Preframe presentations present. In *Proc. Conf. "Category Theory, Como 1990"*, Lect. Notes in Math. 1488, Springer-Verlag, Berlin, 193–212.

Kanamori, A. (1978): On Vopěnka's and related principles. In *Proc. "Logic Coll. Wroclaw* 1977" North-Holland, Amsterdam, 145–153.

Kanamori, A. and M. Magidor (1978): The evolution of large cardinal axioms in set theory. In *Higher Set Theory*, Lect. Notes in Math. 669, Springer-Verlag, Berlin, 99–275.

Keane, O. (1975): Abstract Horn theories. In *Model Theory and Topoi*, Lect. Notes in Math. 445, Springer-Verlag, Berlin, 15–50.

Kelly, G. M. (1980): A unified treatment of transfinite constructions for free algebras, free monoids, colimits, associated sheaves, and so on. *Bull. Austral. Math. Soc.* 22, 1–84.

Kelly, G. M. (1982a): *Basic Concepts of Enriched Category Theory.* Cambridge Univ. Press, Cambridge.

Kelly, G. M. (1982b): On the essentially-algebraic theory generated by a sketch. *Bull. Austral. Math. Soc.* 26, 45–56.

Kelly, G. M. (1987): On the ordered set of reflective subcategories. *Bull. Austral. Math. Soc.* 36, 137–152.

Kennison, J. F. (1968): On limit preserving functors. *Illinois J. Math.* 12, 616–619.

Kiss, E. W., L. Márki, P. Pröhle and W. Tholen (1983): Categorical algebraic properties. A compendium on amalgamation, congruence extension, epimorphisms, residual smallness, and injectivity. *Studia Sci. Math. Hung.* 18, 79–140.

Koubek, V. (1975): Every concrete category has a representation by T_2 paracompact topological spaces. *Comment. Math. Univ. Carolinae* 15, 655–664.

Kučera, L. and A.Pultr (1973): Non-algebraic concrete categories. *J. Pure Appl. Alg.* 3, 95–102.

Kueker, D. W. (1977): Countable approximations and Löwenheim–Skolem theorems. *Ann. Math. Logic* 11, 57–103.

Lair, C. (1971): Foncteurs d'omission de structures algébraiques. *Cahiers Top. Géom. Diff. Cat.* 12, 147–185.

Lair, C. (1981): Catégories modélables et catégories esquissables. *Diagrammes* 6, 1–20.

Lair, C. (1987): Catégories qualifiables et catégories esquissables. *Diagrammes* 17, 1–153.

Lamarche, F. (1988): Modelling polymorphism with categories. Thesis, McGill University, Montréal.

Lawvere, F. W. (1963): Functorial semantics of algebraic theories. Dissertation, Columbia University.

Linton, F. E. J. (1966): Some aspects of equational categories. In "*Proc. Conf. Categ. Alg. La Jolla* 1965", Springer-Verlag, Berlin, 84–94.

MacDonald, J. L. and A. Stone (1982): The tower and regular decomposition. *Cahiers Top. Géom. Diff.* 23, 197–213.

MacLane, S. (1971): *Categories for the Working Mathematician.* Springer-Verlag, Berlin.

Magidor, M. (1971): On the role of supercompact and extendible cardinals in logic. *Israel J. Math.* 10, 147–157.

Makkai, M. (1982): Full continuous embeddings of toposes. *Trans. Amer. Math. Soc.* 269, 167–196.

Makkai, M. (1988): Strong conceptual completeness for first-order logic. *Ann. Pure Appl. Logic* 40, 167–215.

Makkai, M. (1990): A theorem on Barr-exact categories, with an infinitary generalization. *Ann. Pure Appl. Logic* 47, 225–268.

Makkai, M. (1994): The syntax of categorical logic. To appear.

Makkai, M. and R.Paré (1989): *Accessible categories: the foundations of categorical model theory.* Contemporary Math. 104, Amer. Math. Soc., Providence.

Makkai, M. and A. M. Pitts (1987): Some results on locally finitely presentable categories. *Trans. Amer. Math. Soc.* 299, 473–496.

Makkai, M. and G. Reyes (1977): *First-order Categorical Logic.* Lect. Notes in Math. 611, Springer-Verlag, Berlin.

Makowsky, J. A. (1985a): Vopěnka's principle and compact logics. *J. Symb. Logic* 50, 42–48.

Makowsky, J. A. (1985b): Compactness, embeddings and definability. In *Model-theoretic Logics*, Springer-Verlag, Berlin, 645–716.

Makowsky, J. A. (1985c): Abstract embedding relations. In *Model-theoretic Logics*, Springer-Verlag, Berlin, 718–746.

Mal'cev, A. I. (1956): Quasiprimitive classes of abstract algebras (in Russian). *Dokl. Akad. Nauk SSSR* 108, 187–189.

Mal'cev, A. I. (1958): Structural characterization of some classes of algebras (in Russian). *Dokl. Akad. Nauk SSSR* 120, 29–32.

Manes, E. G. (1976): *Algebraic Theories.* Springer-Verlag, Berlin.

Maranda, J. M. (1964): Injective structures. *Trans. Amer. Math. Soc.* 110, 98–135.

Mekler, A. (1978): The size of epimorphic extensions. *Alg. Univ.* 8, 228–232.

Mouen, F. (1984): Caractérisation sémantique des catégories de structures. *Diagrammes* 11

Németi, I. and I. Sain (1982): Cone-implicational subcategories and some Birkhoff-type theorems. *Proc. Conf. "Universal Algebra, Esztergom 1977"*, Coll. Math. Soc. J. Bolyai 29, Budapest, 535–578.

Paré, R. (1989): Some applications of categorical model theory. *Contemporary Math.* 92, 325–339.

Powell, D. (1972): Almost huge cardinals and Vopěnka's principle. *Notices Amer. Math. Soc.* 19, A-616.

Pultr, A. and V. Trnková (1980): *Combinatorial, Algebraic and Topological Representations of Groups, Semigroups and Categories.* North-Holland, Amsterdam.

Pumplün, D. and H. Röhrl (1984): Banach spaces and totally convex spaces I. *Commun. in Alg.* 12, 953–1019.

Reichel, H. (1984): *Structural Induction on Partial Algebras.* Akademie-Verlag, Berlin.

Reyes, G. (1974): *From Sheaves to Logic.* M.A.A. Studies in Math. 9, New York, 143–204.

Richter, M. (1971): Limites in Kategorien von Relationalsystemen. *Z. Math. Log. Grundl. Math.* 17, 75–90.

Rosický, J. (1975): Codensity and binding categories. *Comment. Math. Univ. Carolinae* 16, 515–529.

Rosický, J. (1981a): Concrete categories and infinitary languages. *J. Pure Appl. Alg.* 22, 309–339.

Rosický, J. (1981b): Categories of models of languages $L_{\kappa,\lambda}$. In *Abstr. 9th Winter School on Abstract Analysis*, Math. Inst. Prague, 153–157.

Rosický, J. (1983): Representace konkrétních kategorií. Doctoral thesis, University of Brno.

Rosický, J. (1989): Elementary categories. *Arch. Math.* 52, 284–288.

Rosický, J. (1994a): Models of Horn theories revisited. Submitted to *J. Pure Appl. Alg.*

Rosický, J. (1994b): More on directed colimits of models. Submitted to *Appl. Cat. Structures.*

Rosický, J. and W. Tholen (1988): Orthogonal and prereflective subcategories. *Cahiers Top. Géom. Diff. Cat.* 29, 203–215.

Rosický, J. and W. Tholen (1994): Accessibility and solution set condition. Submitted to *J. Pure Appl. Alg.*

Rosický, J. and V. Trnková (1989): Representability of concrete categories by non-constant morphisms. *Arch. Math. (Brno)* 25, 115–118.

Rosický, J., V. Trnková and J. Adámek (1990): Unexpected properties of locally presentable categories. *Alg. Univ.* 27, 153–170.

Shafaat, A. (1969): On implicationally defined classes of algebras. *J. London Math. Soc.* 44, 137–140.

Shelah, S. and M. Makkai (1990): Categoricity of theories in $L_{\kappa,\omega}$, with κ a compact cardinal. *Ann. Pure Appl. Logic* 47, 41–97.

Skornyakov, L. A. (1964): *Complemented Modular Lattices and Regular Rings*, Oliver and Boyd, Edinburgh.

Slominski, J. (1959): The theory of abstract algebras with infinitary operations. *Rozprawy Matem.* 18.

Smyth, M. B. (1978): Powerdomains. *J. Comp. Syst. Sci.* 16, 23–26.

Smyth, M. B. and G. D. Plotkin (1982): The category-theoretic solution of recursive domain equations. *SIAM J. Computing* 11, 761–783.

Taylor, P. (1990): Locally finitely poly-presentable categories. Preprint, Imperial College, London.

Taylor, P. (1992): The trace factorisations of stable functors. Preprint, Imperial College, London.

Tholen, W. (1987): Reflective subcategories. *Topol. Appl.* 27, 201–212.

Trnková, W., J. Adámek and J. Rosický (1990): Topological reflections revisited. *Proc. Amer. Math. Soc.* 108, 605–612.

Ulmer, F. (1975): On categories of cocontinuous functors. Unpublished manuscript.

Volger, H. (1979): Preservation theorems for limits of structures and global sections of sheaves of structures. *Math. Zeitschr.* 166, 27–53.

Whitman, P. M. (1941): Free lattices. *Ann. Math.* 42, 325–330.

List of Symbols

Index